The Mobile Radio Propagation Channel

The Mobile Radio Propagation Channel

Second Edition

J. D. Parsons, *DSc (Eng), FREng, FIEE*
Emeritus Professor of Electrical Engineering
University of Liverpool, UK

JOHN WILEY & SONS LTD
Chichester · New York · Weinheim · Brisbane · Singapore · Toronto

First Edition published in 1992 by Pentech Press

Copyright © 2000 by John Wiley & Sons, Ltd
          Baffins Lane, Chichester,
          West Sussex PO19 1UD, England

          National   01243 779777
          International (+44) 1243 779777
e-mail (for orders and customer service enquiries): cs-books@wiley.co.uk

Visit our Home Page on: http://www.wiley.co.uk or http://www.wiley.com

Other Wiley Editorial Offices

John Wiley & Sons, Inc., 605 Third Avenue,
New York, NY 10158-0012, USA

WILEY-VCH Verlag GmbH,
Pappelallee 3, D-69469 Weinheim, Germany

Jacaranda Wiley Ltd, 33 Park Road, Milton,
Queensland 4064, Australia

John Wiley & Sons (Canada) Ltd, 22 Worcester Road,
Rexdale, Ontario M9W 1L1, Canada

John Wiley & Sons (Asia) Pte Ltd, 2 Clementi Loop #02-01,
Jin Xing Distripark, Singapore 129809

Library of Congress Cataloging-in-Publication Data

Parsons, J.D. (John David)
   The mobile radio propagation channel/J.D. Parsons – 2nd ed.
      p. cm.
   Includes bibliographical references and index.
   ISBN 0-471-98857-X (alk. paper)
   1. Mobile radio stations. 2. Radio – Transmitters and transmission. 3. Radio wave propagation. I. Title

TK6570.M6 P38 2000
621.3845 – dc21   00-032482

British Library Cataloguing in Publication Data

A catalogue record for this book is available from the British Library

ISBN 0 471 98857 X

Typeset in 10/12pt Times by Dobbie Typesetting Limited, Devon
Printed and bound in Great Britain by Bookcraft (Bath) Ltd.
This book is printed on acid-free paper responsibly manufactured from sustainable forestry, in which at least two trees are planted for each one used for paper production.

To my wife, Mary
and in memory of my parents
Doris and Oswald

Contents

Preface

Some time ago it became apparent that the first edition of this book was rapidly approaching its sell-by date, since many aspects needed revision. There were two obvious courses of action: to forget the whole thing and concentrate my energies on other pursuits such as golf or fishing, or to embark on a new edition. For several reasons I was persuaded that a new edition was a worthwhile endeavour; many people had made complimentary remarks or written complimentary letters about the first edition and I understood that it had become a recommended text for several postgraduate courses. The independent reviewers who had been contacted by the publisher were also very kind; they were unanimous in their opinion that the structure of the book should remain unchanged and that its appeal might be jeopardised by attempting to make it much more system oriented. This is not to say, of course, that there was no need for updating and the inclusion of some new material. This was very much in line with my own thinking, so I was happy to accept that advice. And so, work began.

Soon after I started, it became obvious that major scrutiny, rather than minor attention, was needed. Of course there were some sections covering basic and well-established theory which only needed small amendments, but I have revisited every paragraph of the original book to correct errors, to improve the explanation and to provide further clarification. In most of the chapters I have updated parts that were in need of such action and I have added several new sections, particularly in Chapter 4. A section on ray tracing has been added to Chapter 7, and in Chapter 8 I have extensively revised the sections describing practical channel sounders. The emphasis in Chapter 9 has changed; I have considerably shortened the sections on man-made noise measurements and I have extended the sections on how to predict system performance in the presence of man-made noise. At one stage I was tempted to be far more ruthless in cutting this chapter, since noise is not the limiting factor in cellular radio systems, but in the end I decided to stick to my original theme and not be constrained by considerations of one type of system, important though it may be.

The really major change, however, is the addition of two new chapters at the end of the book. I realised that throughout the first edition I had emphasised the way multipath influences system performance yet I had not described how to mitigate its deleterious effects. Chapter 10 is an attempt to correct that omission, without straying too far from the main theme and getting very system-specific. Again, having said that cellular systems are not the whole of mobile radio, it is difficult to overemphasise their importance in the modern world, and this led to Chapter 11 on system planning. Much of Chapter 11 is applicable beyond cellular radio-telephone systems, so I hope it will prove useful. I have been considerably constrained by my own thoughts about the total length of the book, so again I have gone into detail only on those aspects of planning connected with the main theme.

Once again there are several people who have contributed to the book, directly or indirectly. To those mentioned in the preface to the first edition I must add several new names. Andrew Arowojolu, formerly at the University of Liverpool, but now at Freshfield Communications, has been most helpful. He, and my friend Adel Turkmani, devoted many hours to a discussion of radio planning tools and provided both information and sound advice. The same must be said for Robin Potter and Claes Malmberg of MSI plc; they willingly gave of their time and expertise, particularly in connection with 3D propagation modelling and UMTS planning. George Tsoulos has been extremely helpful with regard to smart antennas, and on DS/SS signalling and associated concepts there is nobody better to contact than Frank Amoroso. My former students Chi Nche and Gary Davies designed and built channel sounders for their own research work and I am pleased to acknowledge their contribution. Paul Leather wrote an award-winning doctoral thesis on diversity for hand-portable equipment and was pleased to allow a brief extract from that work to be included. The assistance of all these people has considerably enhanced the book in many ways. Jane Bainbridge and Alison Reid were enormously helpful with the illustrations and typing new sections of the manuscript; I am in their debt.

My greatest thanks, however, must go to my wife, Mary, whose love, encouragement and understanding have been constant and unwavering. She has, once again, put up with my 'eat and run' existence for several months when, in all honesty, following my retirement from the university, she might have hoped for, and certainly deserved, something better. Never mind – there's always tomorrow!

David Parsons

Preface to the First Edition

Although the demand for mobile radio services has continued to increase for many years, research into mobile radio, as distinct from the development of systems to meet specific operational and economic requirements, was a minority activity on an international scale until the mid-1960s. However, about that time it became apparent that the contribution that civil land mobile radio could make to national economies was very large. Furthermore it was obvious that existing systems had reached the limits of development that could be supported by the relatively unsophisticated technology of the day. These factors, amongst others, made it obvious that a major strategic research effort was justified and the results of that research are now apparent in all the developed countries of the world. Policemen, taxi-drivers and security guards all have individual pocket radios, the general public have access to world-wide telephone services via hand-held and vehicle-borne cellular radio transceivers and pan-European digital systems using wideband TDMA techniques are but a year or two away.

Of all the research activities that have taken place over the years, those involving characterisation and modelling of the radio propagation channel are amongst the most important and fundamental. The propagation channel is the principal contributor to many of the problems and limitations that beset mobile radio systems. One obvious example is multipath propagation which is a major characteristic of mobile radio channels. It is caused by diffraction and scattering from terrain features and buildings; it leads to distortion in analogue communication systems and severely affects the performance of digital systems by reducing the carrier-to-noise and carrier-to-interference ratios. A physical understanding and consequent mathematical modelling of the channel is very important because it facilitates a more accurate prediction of system performance and provides the mechanism to test and evaluate methods for mitigating deleterious effects caused by the radio channel.

This book is an attempt to bring together basic information about the mobile radio channel, some of which has hitherto only been available in published technical papers. The initial concept was that of a fairly slim volume but as the work progressed, so it grew. Even so, the eventual decisions that had to be made were more concerned with what to leave out rather than what to include!

The first two chapters are introductory in nature and attempt to establish the context in which the subject is to be treated. We then move on to propagation over irregular terrain in Chapter 3 and introduce some of the well-known path-loss prediction models. Chapter 4 deals with the problem of urban areas and introduces a number of prediction models that have been specifically developed with the urban area in mind.

Multipath has already been mentioned as a principal feature of mobile radio channels and the characterisation of multipath phenomena is a central topic. Chapter 5 deals with so-called narrowband channels although it must be emphasised that it is the signal that is narrowband, not the channel! In truth, Chapter 5 provides a characterisation which is adequate when frequency-selective fading is not a problem, whilst frequency-selective (wideband) channels are the subject of Chapter 6. Mathematically speaking, these are the 'heaviest' chapters although I have attempted to emphasise physical understanding rather than mathematic rigour.

The conventional elevated base station communicating with vehicle-borne radio transceivers is, nowadays, not the only radio scenario of importance. Hand-portable equipment is commonplace and can be taken into buildings and Chapter 7 is where problems such as this are addressed.

Chapter 8 is exactly what it's title says. At one stage in the writing process it seemed to be in danger of becoming a 'rag-bag' of unrelated topics, but I hope that ultimately it will prove to be useful. Several more practical issues are discussed, without which the treatment would be incomplete. Finally, Chapter 9 covers noise and interference, the latter being a very important topic in the context of current and future cellular systems where frequency re-use is a premier consideration.

There are several people who have contributed directly or indirectly to the writing of this book. My own interest in mobile communications was first aroused by Professor Ramsay Shearman, over 20 years ago and he has been a source of constant encouragement, particularly so in the early part of my academic career. My own research students have taught me much ('teachers teach and students learn' is a half-truth) and in the context of this book it is appropriate for me to mention Anwar Bajwa, Mohamed Ibrahim, Andy Demery and Tumu Kafaru who have allowed me to draw freely on their work. The same can be said for my colleague and friend Adel Turkmani who has, in addition, given generously of his time and expertise. He has advised with regard to content, order of presentation and depth of treatment; I am truly in his debt. Likewise I owe much to my secretary Mrs Brenda Lussey who has typed the manuscript (several times) and has incorporated modifications, second thoughts and other alterations in her usual cheerful manner without so much as a frown of frustration or disapproval.

Finally, this work would never have been completed without the encouragement, understanding and love of my wife, Mary. She has put up, uncomplainingly, with my seemingly unending periods in my study both during the evenings and at weekends. She has been an unpaid coffee-maker and proof-reader who relieved me of many of the less interesting jobs that befall the aspiring author. Will Sundays ever be the same again?

David Parsons

Chapter 1

Introduction

1.1 BACKGROUND

The history of mobile radio goes back almost to the origins of radio communication itself. The very early work of Hertz in the 1880s showed that electromagnetic wave propagation was possible in free space and hence demonstrated the practicality of radio communications. In 1892, less than five years later, a paper written by the British scientist Sir William Crookes [1] predicted telegraphic communication over long distances using tuned receiving and transmitting apparatus. Although the first radio message appears to have been transmitted by Oliver Lodge in 1894 [2], it was the entrepreneur Marconi [3] who initially demonstrated the potential of radio as a powerful means of long-distance communication. In 1895, using two elevated antennas, he established a radio link over a distance of a few miles, and technological progress thereafter was such that only two years later he succeeded in communicating from The Needles, Isle of Wight, to a tugboat over a distance of some 18 miles (29 km). Although it seems highly unlikely that Marconi thought of this experiment in terms of mobile radio, mobile radio it certainly was.

Nowadays the term 'mobile radio' is deemed to embrace almost any situation where the transmitter or receiver is capable of being moved, whether it actually moves or not. It therefore encompasses satellite mobile, aeromobile and maritime mobile, as well as cordless telephones, radio paging, traditional private mobile radio (PMR) and cellular systems. Any book which attempted to cover all these areas would have to be very bulky and the present volume will therefore be concerned principally with the latter categories of use, which are covered by the generic term 'land mobile radio'. This, however, is not a book that deals with the systems and techniques that are used in land mobile communications; it is restricted primarily to a discussion of the radio channel – the transmission medium – a vital and central feature which places fundamental limitations on the performance of radio systems. The majority of the book is concerned with the way in which the radio channel affects the signal that propagates through it, but there are other chapters treating related topics. These have been included to make the book more comprehensive without straying too far from the main theme.

It is not profitable at this point to discuss details; they can be left until later. Suffice it to say that in the vast majority of cases, because of complexity and variability, a deterministic approach to establishing the parameters of the propagation channel is not feasible. Almost invariably it is necessary to resort to measurements and to the powerful tools of statistical communication theory. One point worth clarifying at this stage, however, is that signal transmission over a mobile radio path is reciprocal in the sense that the locations of the transmitter and receiver can be interchanged without changing the received signal characteristics. The discussion can therefore proceed on the basis of transmission in either direction without loss of generality. However, a word of caution is needed. The levels of ambient noise and interference at the two ends of the link may not be the same, so reciprocity with respect to the signal characteristics does not imply reciprocity with respect to the signal-to-noise or signal-to-interference ratios.

Some years ago the primary concern of a book such as this would undoubtedly have been the propagation aspects related to traditional mobile radio services which are based on the concept of an elevated base station on a good site, communicating with a number of mobiles in the surrounding area. Such systems, known as PMR systems, developed rapidly following World War II, especially once the transistor made it possible to design and build compact, lightweight equipment that could easily be installed in a vehicle and powered directly from the vehicle battery. These are often termed dispatch systems because of their popularity with police forces, taxi companies and service organisations who operate fleets of vehicles. The frequency bands used for dispatch systems lie in the range 70–470 MHz and have been chosen because the propagation characteristics are suitable, the antennas have a convenient size and adequate radio frequency (RF) power can be generated easily and efficiently.

The operational strategy is to divide the available spectrum into convenient channels with each user, or user group, having access to one or more of these channels in order to transmit a message, usually speech, by amplitude modulation or frequency modulation. The technique of providing a service to a number of users in this way is known as *frequency division multiple access* (FDMA), and because each channel carries only one message the term *single channel per carrier* (SCPC) is also used. In the early post-war days, channels were spaced by 100 kHz, but advances in technology, coupled with an ever increasing demand for licences, has led to several reductions to the point where currently in the UK, channels in the VHF band (30–300 MHz) are 12.5 kHz apart, whereas 25 kHz separation is still used for some channels in the UHF band (300–3000 MHz).

For these PMR systems, indeed for any mobile radio system with a similar operating scenario, the major propagation-related factors that have to be taken into consideration are the effect of irregular terrain and the influence on the signal of trees, buildings and other natural and man-made obstacles. In recent years, however, expanded services have become available, for example radio pagers, which are now in common use. Hand-portable, rather than vehicle-borne equipment is also being used by security guards, police officers and by subscribers to cellular radio-telephone systems. Hand-portable equipment can easily be taken into buildings, so a book concerned with propagation must also consider the properties of the signal inside buildings and in the surrounding areas. For cordless telephones and the like, there is also a need to study propagation totally within buildings. Neither can we restrict attention to frequencies below 470 MHz; first- and second-generation analogue and digital cellular radio telephone systems, e.g. AMPS, TACS, GSM and DCS1800, use frequencies up to 1900 MHz, and third-generation wideband systems will probably use even higher frequencies to solve the problems of spectrum congestion and required bandwidth.

What then are the matters of primary concern? For transmissions of the traditional type, in which the signals are restricted to fairly narrow radio channels, two major factors have to be quantified:

- Median signal strength
- Signal variability

The ability to predict the minimum power a transmitter must radiate to provide an acceptable quality of coverage over a predetermined service area and the ability to estimate the likely effect of such transmissions on services in adjacent areas, are both critical for improving frequency reuse techniques, for implementing band-sharing schemes between different services and for the success of radio-telephone systems. This is not easy and there is a vital and continuing need for a better understanding of the influence of the different urban and terrain factors on the mobile radio signal.

As far as signal variability is concerned, it is often convenient to separate the effects into those which occur over a short distance and those which are apparent only over much longer distances. Short-distance effects include the rapid fading caused by multipath propagation in urban areas; longer-distance effects include the much slower variations in average signal strength as the receiver moves from one area to another.

For digital systems it is neither efficient nor desirable to use FDMA/SCPC as a multiple-access technique, and spectrum utilisation is substantially improved by allowing each user access to a wider-bandwidth radio channel, but only for a small percentage of the time. This *time division multiple access* (TDMA) strategy is used in the GSM and DCS1800 systems. Third-generation systems will be based around wideband *code division multiple access* (CDMA) and these spread-spectrum systems will offer even greater capacity and security together with access to multimedia communications. First developed for military purposes, CDMA has virtually no noise or crosstalk and is well suited to high-quality multimedia services. The characterisation of wideband channels will be discussed in Chapter 6; for now it will suffice to note that if digital (pulse) signals propagate in a multipath environment then interference can occur between a given pulse and a delayed version of an earlier pulse (an echo) that has travelled via a longer path. This is known as *intersymbol interference* (ISI) and can cause errors. The extent of the problem can be quantified by propagation studies which measure parameters such as the average delay and the spread of delays.

Finally, in this introductory section, it is worth making two further points. Firstly, the geographical service area of many mobile radio systems is too large to be economically covered using a single base station, and various methods exist to provide 'area coverage' using a number of transmitters. We will return to this topic in Section 1.3.2. Secondly, in order to maximise the use of the available spectrum, channels that are allocated to one user in a certain geographical area are reallocated to a different user in another area some distance away. The most common example of this is cellular radio, which relies on frequency reuse to achieve high spectrum efficiency. However, whenever frequencies are reallocated, there is always the possibility that interference will be caused and it should therefore be understood that adequate reception conditions require not only an acceptable signal-to-noise ratio but also, simultaneously, an acceptable signal-to-interference ratio. This subject will be treated in Chapter 9. Throughout the book the term 'base station' will be used when referring to the fixed terminal and the term 'mobile' to describe the moving terminal, whether it be hand-portable or installed in a vehicle.

1.2   FREQUENCY BANDS

Having set the scene, we can now discuss some of the topics in a little more detail. It is very important to understand how RF energy propagates and in preparation for a brief general discussion let us define more clearly what is meant by the term 'radio wave' and how waves of different frequencies are classified. The part of the electromagnetic spectrum that includes radio frequencies extends from about 30 kHz to 300 GHz, although radio wave propagation is actually possible down to a few kilohertz. By international agreement the radio frequency spectrum is divided into *bands*, and each band is given a designation as in Table 1.1.

Electromagnetic energy in the form of radio waves propagates outwards from a transmitting antenna and there are several ways in which these waves travel, largely depending on the transmission frequency. Waves propagating via the layers of the ionosphere are known as *ionospheric waves or sky waves*; those that propagate over other paths in the lower atmosphere (the troposphere) are termed *tropospheric waves*, and those that propagate very close to the Earth's surface are known as *ground waves*. Ground waves can be conveniently divided into *space waves* and *surface waves*, and space waves can be further subdivided into direct waves which propagate via the direct path between transmitting and receiving antennas and ground-reflected waves that reach the receiving antenna after reflection from the ground. Figure 1.1 gives a simple picture. The surface waves are guided along the Earth's surface and because the Earth is not a perfect conductor, energy is extracted from the wave, as it propagates, to supply losses in the ground itself. The attenuation of this wave (sometimes known as the Norton surface wave) is therefore directly affected by the ground constants

Table 1.1   Designation of frequency bands

Frequency band	Frequency range
Extremely low frequency (ELF)	<3 kHz
Very low frequency (VLF)	3–30 kHz
Low frequency (LF)	30–300 kHz
Medium frequency (MF)	300 kHz–3 MHz
High frequency (HF)	3–30 MHz
Very high frequency (VHF)	30–300 MHz
Ultra high frequency (UHF)	300 MHz–3 GHz
Super high frequency (SHF)	3–30 GHz
Extra high frequency (EHF)	30–300 GHz

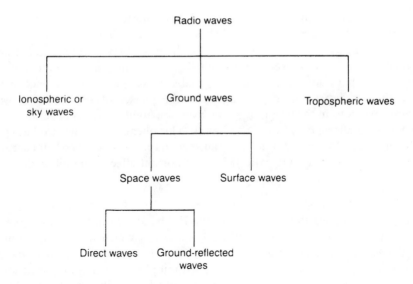

Figure 1.1   Modes of radio wave propagation.

(conductivity and dielectric constant) along the transmission path. The importance of each of these waves in any particular case depends upon the length of the propagation path and the frequency of transmission. We can now discuss each frequency band in turn.

1.2.1   VLF

In the VLF range the wavelength is very long, typically 10^5 m, and antennas are therefore very large. They have to be very close to the Earth and are often buried in the ground. The radio waves are reflected from the ionosphere and a form of Earth–ionosphere waveguide exists that guides the wave as it propagates. Because of diurnal variations in the height of the ionospheric D-layer, the effective height of the terrestrial waveguide also varies around the surface of the Earth. The uses of VLF include long-distance worldwide telegraphy and navigation systems. Frequencies in the VLF range are also useful for communication with submerged submarines, as higher frequencies are very rapidly attenuated by conducting sea water. Digital transmissions are always used but the available bandwidth in this frequency range is very small and the data rate is therefore extremely low.

1.2.2   LF and MF

At frequencies in the range between a few kilohertz and a few megahertz (the LF and MF bands) ground wave propagation is the dominant mode and the radiation characteristics are strongly influenced by the presence of the Earth. At LF, the surface wave component of the ground wave is successfully utilised for long-distance communication and navigation. Physically, antennas are still quite large and high-power transmitters are used. The increased bandwidth available in the MF band allows it to be used for commercial AM broadcasting, and although the attenuation of the surface wave is higher than in the LF band, broadcasting over distances of several hundred kilometres is still possible, particularly during the daytime. At night, sky wave propagation via the D-layer is possible in the MF band and this leads to the possibility of interference between signals arriving at a given point, one via a ground wave path and the other via a sky wave path. Interference can be constructive or destructive depending upon the phases of the incoming waves; temporal variations in the height of the D-layer, apparent over tens of seconds, cause the signal to be alternatively strong and weak. This phenomenon, termed fading, can also be produced by several other mechanisms and always occurs when energy can propagate via more than one path. It is an important effect in mobile radio.

1.2.3   HF

Ground wave propagation also exists in the HF band, but here the ionospheric or sky wave is often the dominant feature. For reasons which will become apparent later, the HF band is not used for civilian land mobile radio and it is therefore inappropriate to go into details of the propagation phenomena. Suffice it to say that the layers of ionised gases within the ionosphere (the so-called D, E and F layers) exist at heights up to several hundred kilometres above the Earth's surface, and single and multiple hops via the various ionospheric layers permit almost worldwide communications. The height of the different layers varies with the time of day, the season of the year and the geographical location [4]; this causes severe problems which have attracted the attention of researchers over many years and are still of great interest.

1.2.4   VHF and UHF

Frequencies in the VHF and UHF bands are usually too high for ionospheric propagation to occur, and communication takes place via the direct and ground-reflected components of the space wave. In these bands, antennas are relatively small in physical size and can be mounted on masts several wavelengths above the ground. Under these conditions the space wave is predominant. Although the space wave is often a negligible factor in communication at lower frequencies, it is the dominant feature of ground wave communication at VHF and UHF. The bandwidth available is such that high-quality FM radio and television channels can be accommodated, but propagation is normally restricted to points within the radio horizon and coverage is therefore essentially local. The analysis of space wave propagation at VHF and UHF needs to take into account the problems of reflections both from the ground and from natural and man-made obstacles. Diffraction over hilltops and buildings, and refraction in the lower atmosphere are also important.

1.2.5   SHF

Frequencies in the SHF band are commonly termed *microwaves*, and this term may also be used to describe that part of the UHF band above about 1.5 GHz. Propagation paths must have line-of-sight between the transmitting and receiving antennas, otherwise losses are extremely high. At these frequencies, however, it is possible to design compact high-gain antennas, normally of the reflector type, which concentrate the radiation in the required direction. Microwave frequencies are used for satellite communication (since they penetrate the ionosphere with little or no effect), point-to-point terrestrial links, radars and short-range communication systems.

1.2.6 EHF

The term 'millimetre wave' is often used to describe frequencies in the EHF band between 30 and 300 GHz. In comparison with lower frequencies, enormous bandwidths are available in this part of the spectrum. Line-of-sight propagation is now predominant and although interference from ground-reflected waves is possible, it is often insignificant, because the roughness of the ground is now much greater in comparison with the wavelength involved. It is only when the ground is very smooth, or a water surface is present, that the ground-reflected waves play a significant role. This topic will be treated in Chapter 2. In the millimetre waveband the most important effects that have to be taken into account are scattering by precipitation (rain and snow) and, at certain frequencies, absorption by fog, water vapour and other atmospheric gases.

A detailed treatment of millimetre wave propagation is well beyond the scope of this book and, in any case, is not directly relevant to current mobile radio systems. However, Figure 1.2 shows the attenuation by oxygen and uncondensed water vapour [5] as a function of frequency. At some frequencies there are strong absorption lines, e.g. the water vapour absorption at 22 GHz and the oxygen absorption at 60 GHz. However, between these lines there are windows where the attenuation is much less. Specialised applications such as very short range secure communication systems and satellite-to-satellite links are where millimetre waves

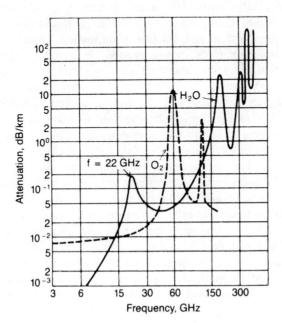

Figure 1.2 Attenuation by oxygen and water vapour at sea level, $T = 20°C$; water content $= 7.5 \, \text{g/m}^3$.

find application, although in the 1980s there was some interest in the absorption bands as they appeared to have some potential for future microcellular systems. At present there is no volume market in this frequency range, so component and system costs are very high.

1.3 MOBILE RADIO FREQUENCIES

There are several factors that have to be taken into account in deciding what frequency band should be used for a particular type of radio communication service. For the specific application of interest, two-way mobile radio operations, communication is required over ranges that do not normally exceed a few tens of kilometres, often much less. Clearly, unnecessary interference would be caused to other users if the signals propagated too far. It is also evident that if mobiles are to communicate freely with their base, or with each other, throughout a given area (which may or may not be the total service area of the system) the transmitters involved must be able to provide an adequate signal strength over the entire area concerned.

Operating frequencies must be chosen in a region of the RF spectrum where it is possible to design efficient antennas of a size suitable for mounting on base station masts, on vehicles and on hand-portable equipment. Since the mobiles can move around freely within the area covered by the radio system, their exact location is unknown and the antennas must therefore radiate energy uniformly in all directions in the horizontal (azimuth) plane; technically this is known as *omnidirectional* radiation.* It is also vital that the frequencies chosen are such that the transmitters used at base stations and mobiles can generate the necessary RF power while remaining fairly small in physical size.

For two-way mobile radio, particularly in urban areas, it is seldom that the mobile antenna has a direct line-of-sight path to the base station. Radio waves will penetrate into buildings to a limited extent and, because of diffraction, appear to bend slightly over minor undulations or folds in the ground. Fortunately, due to multiple scattering and reflection, the waves also propagate into built-up areas, and although the signal strength is substantially reduced by all these effects, sensitive receivers are able to detect the signals even in heavily built-up areas and within buildings. The choice of frequency is therefore limited by the need to minimise the losses due to buildings while continuing to satisfy the other constraints mentioned above.

For these reasons, traditional two-way mobile radio originally developed almost exclusively around the VHF and latterly UHF bands. In a city, for example, there are many mobile radio users such as emergency services and taxi companies. In the case of a police force, the central control room receives reports of incidents in the city area, often by emergency telephone calls. The control room radio operator puts out a call to a police officer believed to be in the appropriate area; who may be on foot with a personal radio or in a vehicle equipped with mobile radio. On receipt of the call, the officer acknowledges it, investigates the incident and reports back by radio. Because of the FDMA/SCPC method of operation, police forces have radio channels allocated for their exclusive use and there is no mutual interference between them

---

Omnidirectional is not to be confused with *isotropic* which means 'in all directions'.

and other users on different channels in the same frequency band. However, all police officers who carry a receiver tuned to the appropriate frequency will hear the calls as they are broadcast.

The range over which signals propagate is also a fundamental consideration since in order to use the available spectrum efficiently, it is necessary to reallocate radio channels to other users operating some distance away. If, in the above example, the message from the control room had been radiated on HF, then it is possible that the signals could have been detected at distances of several hundred kilometres, which is unnecessary, undesirable and would cause interference to other users. The VHF and UHF bands therefore represent an optimum choice for mobile radio because of their relatively short-range propagation characteristics and because radio equipment designed for these bands is reasonably compact and inexpensive.

Vertical polarisation is always used for mobile communications; at frequencies in the VHF band it is preferable to horizontal polarisation because it produces a higher field strength near the ground [6]. Furthermore, mobile and hand-portable antennas for vertical polarisation are more robust and more convenient to use. In an overall plan for frequency reuse, no worthwhile improvement can be achieved by employing both polarisations (as in television broadcasting) because scattering in urban areas tends to cause a cross-polar component to appear. Although this may have some advantages, for example it is often inconvenient to hold the antenna of a hand-portable radio-telephone in a truly vertical position, it is apparent that no general benefit would result from the transmission of horizontally polarised signals.

There are many other services, however, which also operate in the VHF and UHF bands, for example, television, domestic radio, Citizens' Band radio, marine radio, aeromobile radio (including instrument landing systems) and military radio. Several of these services have a 'safety of life' element and it is vital that their use is tightly regulated to ensure maximum efficiency and freedom from interference. The exact frequencies within the VHF and UHF bands that are allocated for various radio systems are agreed at meetings of the International Telecommunications Union (ITU) and are legally binding on the member states. Every twenty years the ITU organises a world administrative radio conference (WARC) at which regulations are revised and updated and changes in allocations and usage are agreed.

In each country, use of the radio frequency spectrum is controlled by a regulatory authority; in the UK this is the Department of Trade and Industry (DTI) and in the USA it is the Federal Communications Commission (FCC). The regulatory authority is responsible for allocating specific portions of the available spectrum for particular purposes and for licensing the use of individual channels or groups of channels by legitimate users. Because of the attractive propagation characteristics of VHF and UHF, it is possible to allocate the same channel to different users in areas separated by distances of 50–100 km with a substantial degree of confidence that, except under anomalous propagation conditions, they will not interfere with each other.

1.3.1 Radio links

For obvious reasons, VHF or UHF radio transmitters intended to provide coverage over a fairly large area are located at strategic points (usually high, uncluttered sites) within the intended area. However, the control room may be at some completely different location, so a method has to be found to get the intended message information (which may be voice or data) to the transmitter sites. This can be achieved by using telephone lines or by a further radio link. The technical specifications for telephone lines and the policy for their use often rule out this possibility, and the necessary quality and reliability of service can only be achieved by using a radio link between the control room and each of the VHF/UHF transmitter sites.

The kind of radio link used for this purpose has requirements quite different from those of the two-way VHF/UHF systems used to communicate with mobiles. In this case we are only communicating between one fixed point (the control room) and another fixed point (the site concerned), and for this reason such links are commonly termed point-to-point links. Omnidirectional radio coverage is not required, in fact it is undesirable, so it is possible to use directional antennas which concentrate the radio frequency energy in the required direction only. In addition, there is a substantial degree of freedom to locate the link transmitters and receivers at favourable locations where a line-of-sight path exists and the radio path does not need to rely on the propagation mechanisms, discussed earlier, which make the VHF and UHF bands so attractive for communications to and from mobiles.

These features have been exploited extensively in link planning, particularly with regard to allocation of frequencies. Because of congestion in the frequency bands best suited to communications with mobiles, link activity has been moved into higher frequency bands and modern links operate typically at frequencies above 2 GHz. This presents no problems since compact high-gain directional antennas are readily available at these frequencies. Two frequencies are necessary for 'go' and 'return' paths, since if a link serves more than one base transceiver then one may be transmitting while others are receiving; this means that full-duplex operation is needed, i.e. messages can pass both ways along the link simultaneously.

When several channels are operated from the same transmitter site, a choice has to be made between using several link frequencies, one for each transmitter, or using a multiplexed link in which the messages for the different transmitters at the remote site are assembled into an FDM baseband signal which is then modulated onto the radio bearer. The multiplex approach can be more efficient than the SCPC alternative in requiring only one transceiver at each end of the link, and this technique is widely implemented. Naturally, the bandwidth occupied by a multiplexed link transmission is proportionally greater than an SCPC signal, but a 10-channel multiplexed link connection occupies no more spectrum than 10 separate links spaced out in frequency.

Certain conditions have to be satisfied for radio links to operate satisfactorily. Firstly it is vital that the direct path between the two antennas (the line-of-sight path) is clear of obstructions. However, this in itself is not enough; it is highly desirable that there are no obstructions close to the line-of-sight path since they could cause reflections and spoil reception. Figure 1.3 shows a simple link path of the kind we are considering; the dotted line defines a region known as the first Fresnel zone. The theory in Chapter 2 enables us to calculate the dimensions of this zone, and shows that for satisfactory radio link operation it should be almost free of obstructions.

1.3.2   Area coverage

A traditional mobile radio system comprises several transceivers which communicate with a single, fixed base station. In most cases the base station is centrally located

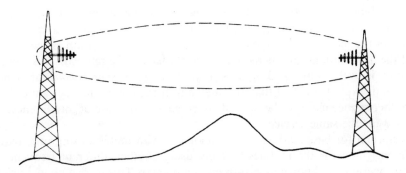

Figure 1.3   Simple point-to-point radio link.

within the area to be served and is connected to the control room via a telephone line or radio link. A straightforward approach to the problem of providing coverage over very large areas would therefore be to erect a very high tower somewhere near the centre of the required coverage area and install a powerful transmitter. This technique is used by the broadcasting authorities; their transmitting masts may be over 300 m high and they radiate signals of many kilowatts. But the broadcasting problem has little in common with the communications problem. Broadcasting aims to deliver a strong signal to many receivers all tuned to the same broadcast and with no capability to transmit back. In communications a relatively small number of users are involved in any one radio network and mobiles normally need to transmit back to the central base station. The requirement for many user groups to use the radio spectrum for independent and unrelated services is the dominant issue here, since there are far too many user groups and far too little spectrum available to allocate a unique segment of it to each group. The same frequencies therefore have to be reused many times in different parts of the country.

The question is therefore, how far away from a transmitter it is necessary to go before its frequency can be reused without risk of mutual interference in either direction. This will be discussed in Chapter 9 but the distance is in fact quite large, at least five times the radius of the coverage area, depending on how comprehensively the service area is provided with strong receivable signals. If a single high mast were situated in the middle of, say, the London area with sufficient transmitter power to cover all of Greater London, then that frequency would not be reusable anywhere in the south of England, nor in a large part of Wales.

What are the alternatives? An obvious one is to have a large number of low-power transmitters radiating from short masts, each covering a small territory but permitting reuse of the frequencies assigned to them many times in a defined geographical area. This is the basis of the 'cellular radio' approach to area coverage and is extremely effective. However, implementing this technique requires firstly a large number of available channels, and secondly a complicated and costly infrastructure [7,8]. Although this is acceptable for a high-quality nationwide radio-telephone network, it is not attractive for a more localised PMR dispatch system.

For traditional mobile radio services, if the area is too large to be economically covered by one base station or if geographical conditions produce difficulties, a more suitable solution is to transmit from several locations at once. In this case the transmitters are all operated at nominally the same frequency so that whatever the location of the mobile within the overall coverage area, it is within range of at least one of the base stations and its receiver does not have to be retuned. This method of operation is well established and is known as *quasi-synchronous* or *simulcast* operation. It exploits the fact that although a transmission frequency cannot be used for another service close to the desired coverage area because of interference, it can be reused for the same service.

It is normal for base stations to communicate with mobiles using one frequency (or channel) and for the mobiles to reply using another, different frequency; this mode of operation is known as *two-frequency simplex*. To ensure a good service it is essential to provide an adequate signal to mobiles located anywhere within the intended coverage area of a given base station. It is impossible to cover 100% of locations, for reasons that will become apparent later on, so in practice this requirement amounts to covering a large percentage of locations (more than 95%).

Early attempts to operate quasi-synchronous systems using AM or FM revealed problems in areas where transmissions from more than one site could be received (the so-called overlap areas). For satisfactory operation it is necessary to operate the various transmitters at frequencies a few hertz apart (hence the term quasi-synchronous). In addition, the modulation needs to be synchronised between the various transmitters so that in the overlap areas, where equally strong signals arrive at the mobile from two or more transmitters, the information part of the signal is coherent regardless of the source from which it originates. Since the message being transmitted originates from a central control point, the synchronisation requirement means that the time delay involved in sending the message from control to the various transmitters in the system must be the same. In other words, all the radio links in the system must be delay-equalised. The situation seems to be even more critical in digital systems using the TETRA standard, which are now reaching the implementation stage. Difficulties are likely to arise as a result of timing and synchronisation problems and to minimise such problems it is necessary for designers to aim at truly synchronous operation of the various transmitters that make up the area coverage system [9].

Linking all the base station transmitters and receivers to the control room may be achieved by either ring or star connections as shown in Figure 1.4. In this type of system there is an additional requirement associated with the way in which mobiles communicate with 'control'. In the earlier discussion of this type of operation, the principal consideration was to ensure that the transmitter network provided an adequate signal at a high percentage of locations. But in considering the receive problem, it is clear that a mobile wishing to access control, transmits on a vacant channel and a signal is received at each of the various base station sites. Usually, one base receiver will receive a stronger signal than the others because the mobile is nearest to the site in question.

The radio system needs to decide which site is nearest to the mobile and to establish communications via that site. This means the system must compare the radio signals from the mobile at all the base station receivers and then choose the strongest. This is known as receiver *voting*. In the absence of other factors, comparison of received signal strengths around a ring connection might be efficient

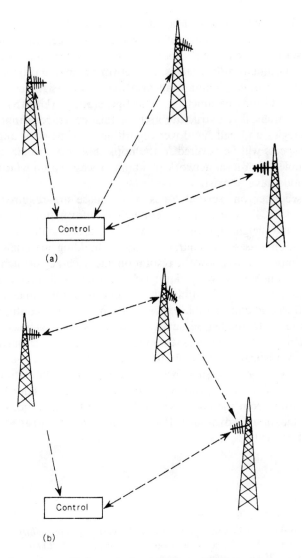

Figure 1.4 Configuration of link networks: (a) star connection, (b) ring connection.

in terms of link deployment, but this type of connection involves accumulated delay in reaching a decision on which is the 'best' receiver and this delay is unacceptable in emergency service applications. Star connections are therefore preferable.

1.4 POSTSCRIPT

In the context of mobile radio systems in general, and channel characterisation in particular, propagation models are required to deal with a number of situations as outlined in Section 1.1. These models are necessary for accurate coverage planning, the characterisation of multipath effects and for interference calculations. Moreover, they are required for a wide variety of environments from rural areas to in-building situations, and for special cases such as in tunnels and along railways. The overall scenario encompasses the full range of macrocells, microcells and picocells; some have the base station antenna well above the local clutter and others do not. In second-generation cellular radio systems, the network planning process (Chapter 11) includes not only coverage planning but also frequency assignment strategies and aspects of base station parameters. Third-generation (UMTS) systems will incorporate a hierarchical cell structure and for this, coverage planning, frequency assignment strategies and call handover algorithms will be very important. Only some of these aspects will be covered in the book, but they are mentioned here to identify the complex high-level network planning process within which propagation prediction methods have to exist.

As we will see later on, several types of database are required to underpin propagation models. They include terrain height information, land usage data, building shape and height information, and vegetation data. For determining building penetration losses, the characteristics of building materials may well be important. It is important to know the resolution and accuracy of such databases, as well as the relationship between database accuracy and prediction accuracy. Although a clear relationship is intuitively present, it is not immediately apparent that the time, effort and considerable expense of acquiring and continually updating such databases leads to predictions with an accuracy that justifies the outlay. A fundamental rethink of approaches to modelling – perhaps a move away from empirical and statistical modelling to a deterministic or semideterministic approach – may well be necessary before accurate, multidimensional databases can be used to full effect. It is probably also necessary to consider how to extract relevant information from the databases and how best to incorporate it into such models to gain maximum advantage. It may well be helpful to read the following chapters in that context.

REFERENCES

1. Sir William Crookes (1892) Some possibilities of electricity. *Fortnightly Review*, 173–81.
2. Austin B.A. (1994) Oliver Lodge – the forgotten man of radio? *Radioscientist*, **5**(1), 12–16.
3. Marconi Co. Ltd. (1981) *Gugliemo Marconi*.
4. Betts J.A. (1967) *High Frequency Communications*. English Universities Press, London.
5. Collin R.E. (1985) *Antennas and Radiowave Propagation*. McGraw-Hill, New York.
6. Knight P. (1969) *Field strength near the ground at VHF and UHF: theoretical dependence on polarisation*. BBC Research Report 1969/3.
7. Appleby M.S. and Garrett J. (1985) Cellnet cellular radio network. *Br. Telecommun. Engng*, **4**, 62–9.
8. Department of Trade and Industry (1985) *A Guide to the Total Access Communication System*. DTI, London.
9. Dernikas D. (1999) Performance evaluation of the TETRA radio interface employing diversity reception in adverse conditions. PhD thesis, University of Bradford.

Chapter 2

Fundamentals of VHF and UHF Propagation

2.1 INTRODUCTION

Having established the suitability of the VHF and UHF bands for mobile communications and the need to characterise the radio channel, we can now develop some fundamental relationships between the transmitted and received power, distance (range) and carrier frequency. We begin with a few relevant definitions.

At frequencies below 1 GHz, antennas normally consist of a wire or wires of a suitable length coupled to the transmitter via a transmission line. At these frequencies it is relatively easy to design an assembly of wire radiators which form an array, in order to beam the radiation in a particular direction. For distances large in comparison with the wavelength and the dimensions of the array, the field strength in free space decreases with an increase in distance, and a plot of the field strength as a function of spatial angle is known as the radiation pattern of the antenna.

Antennas can be designed to have radiation patterns which are not omnidirectional, and it is convenient to have a figure of merit to quantify the ability of the antenna to concentrate the radiated energy in a particular direction. The directivity D of an antenna is defined as

$$D = \frac{\text{power density at a distance } d \text{ in the direction of maximum radiation}}{\text{mean power density at a distance } d}$$

This is a measure of the extent to which the power density in the direction of maximum radiation exceeds the average power density at the same distance. The directivity involves knowing the power actually transmitted by the antenna and this differs from the power supplied at the terminals by the losses in the antenna itself. From the system designer's viewpoint, it is more convenient to work in terms of terminal power, and a power gain G can be defined as

$$G = \frac{\text{power density at a distance } d \text{ in the direction of maximum radiation}}{P_\mathrm{T}/4\pi d^2}$$

where P_T is the power supplied to the antenna.

So, given P_T and G it is possible to calculate the power density at any point in the far field that lies in the direction of maximum radiation. A knowledge of the radiation pattern is necessary to determine the power density at other points.

The power gain is unity for an isotropic antenna, i.e. one which radiates uniformly in all directions, and an alternative definition of power gain is therefore the ratio of power density, from the specified antenna, at a given distance in the direction of maximum radiation, to the power density at the same point, from an isotropic antenna which radiates the same power. As an example, the power gain of a half-wave dipole is 1.64 (2.15 dB) in a direction normal to the dipole and is the same whether the antenna is used for transmission or reception.

There is a concept known as *effective area* which is useful when dealing with antennas in the receiving mode. If an antenna is irradiated by an electromagnetic wave, the received power available at its terminals is the power per unit area carried by the wave × the effective area, i.e. $P = WA$. It can be shown [1, Ch. 11] that the effective area of an antenna and its power gain are related by

$$A = \frac{\lambda^2 G}{4\pi} \tag{2.1}$$

2.2 PROPAGATION IN FREE SPACE

Radio propagation is a subject where deterministic analysis can only be applied in a few rather simple cases. The extent to which these cases represent practical conditions is a matter for individual interpretation, but they do give an insight into the basic propagation mechanisms and establish bounds.

If a transmitting antenna is located in free space, i.e. remote from the Earth or any obstructions, then if it has a gain G_T in the direction to a receiving antenna, the power density (i.e. power per unit area) at a distance (range) d in the chosen direction is

$$W = \frac{P_T G_T}{4\pi d^2} \tag{2.2}$$

The available power at the receiving antenna, which has an effective area A is therefore

$$P_R = \frac{P_T G_T}{4\pi d^2} A$$

$$= \frac{P_T G_T}{4\pi d^2} \left(\frac{\lambda^2 G_R}{4\pi} \right)$$

where G_R is the gain of the receiving antenna.

Thus, we obtain

$$\frac{P_R}{P_T} = G_T G_R \left(\frac{\lambda}{4\pi d} \right)^2 \tag{2.3}$$

which is a fundamental relationship known as the *free space or Friis equation* [2]. The well-known relationship between wavelength λ, frequency f and velocity of propagation c ($c = f\lambda$) can be used to write this equation in the alternative form

$$\frac{P_R}{P_T} = G_T G_R \left(\frac{c}{4\pi f d}\right)^2 \tag{2.4}$$

The propagation loss (or path loss) is conveniently expressed as a positive quantity and from eqn. (2.4) we can write

$$L_F \text{ (dB)} = 10 \log_{10}(P_T/P_R)$$

$$= -10 \log_{10} G_T - 10 \log_{10} G_R + 20 \log_{10} f + 20 \log_{10} d + k \tag{2.5}$$

$$\text{where} \quad k = 20 \log_{10}\left(\frac{4\pi}{3 \times 10^8}\right) = -147.56$$

It is often useful to compare path loss with the basic path loss L_B between isotropic antennas, which is

$$L_B \text{ (dB)} = 32.44 + 20 \log_{10} f_{MHz} + 20 \log_{10} d_{km} \tag{2.6}$$

If the receiving antenna is connected to a matched receiver, then the available signal power at the receiver input is P_R. It is well known that the available noise power is kTB, so the input signal-to-noise ratio is

$$\text{SNR}_i = \frac{P_R}{kTB} = \frac{P_T G_T G_R}{kTB}\left(\frac{c}{4\pi f d}\right)^2$$

If the noise figure of the matched receiver is F, then the output signal-to-noise ratio is given by

$$\text{SNR}_o = \text{SNR}_i / F$$

or, more usefully,

$$(\text{SNR}_o)_{dB} = (\text{SNR}_i)_{dB} - F_{dB}$$

Equation (2.4) shows that free space propagation obeys an inverse square law with range d, so the received power falls by 6 dB when the range is doubled (or reduces by 20 dB per decade). Similarly, the path loss increases with the square of the transmission frequency, so losses also increase by 6 dB if the frequency is doubled. High-gain antennas can be used to make up for this loss, and fortunately they are relatively easily designed at frequencies in and above the VHF band. This provides a solution for fixed (point-to-point) links, but not for VHF and UHF mobile links where omnidirectional coverage is required.

Sometimes it is convenient to write an expression for the electric field strength at a known distance from a transmitting antenna rather than the power density. This can be done by noting that the relationship between field strength and power density is

$$W = \frac{E^2}{\eta}$$

where η is the characteristic wave impedance of free space. Its value is 120π ($\sim 377\,\Omega$) and so eqn. (2.2) can be written

$$\frac{E^2}{120\pi} = \frac{P_T G_T}{4\pi d^2}$$

giving

$$E = \frac{\sqrt{30 P_T G_T}}{d} \tag{2.7}$$

Finally, we note that the maximum useful power that can be delivered to the terminals of a matched receiver is

$$P = \frac{E^2 A}{\eta} = \left(\frac{E^2}{120\pi} \right) \frac{\lambda^2 G_R}{4\pi} = \left(\frac{E\lambda}{2\pi} \right)^2 \frac{G_R}{120} \tag{2.8}$$

2.3 PROPAGATION OVER A REFLECTING SURFACE

The free space propagation equation applies only under very restricted conditions; in practical situations there are almost always obstructions in or near the propagation path or surfaces from which the radio waves can be reflected. A very simple case, but one of practical interest, is the propagation between two elevated antennas within line-of-sight of each other, above the surface of the Earth. We will consider two cases, firstly propagation over a spherical reflecting surface and secondly when the distance between the antennas is small enough for us to neglect curvature and assume the reflecting surface to be flat. In these cases, illustrated in Figures 2.1 and 2.4 the received signal is a combination of direct and ground-reflected waves. To determine the resultant, we need to know the reflection coefficient.

2.3.1 The reflection coefficient of the Earth

The amplitude and phase of the ground-reflected wave depends on the reflection coefficient of the Earth at the point of reflection and differs for horizontal and vertical polarisation. In practice the Earth is neither a perfect conductor nor a perfect dielectric, so the reflection coefficient depends on the ground constants, in particular the dielectric constant ε and the conductivity σ.

For a horizontally polarised wave incident on the surface of the Earth (assumed to be perfectly smooth), the reflection coefficient is given by [1, Ch. 16]:

$$\rho_h = \frac{\sin\psi - \sqrt{(\varepsilon/\varepsilon_0 - \mathrm{j}\sigma/\omega\varepsilon_0) - \cos^2\psi}}{\sin\psi + \sqrt{(\varepsilon/\varepsilon_0 - \mathrm{j}\sigma/\omega\varepsilon_0) - \cos^2\psi}}$$

where ω is the angular frequency of the transmission and ε_0 is the dielectric constant of free space. Writing ε_r as the relative dielectric constant of the Earth yields

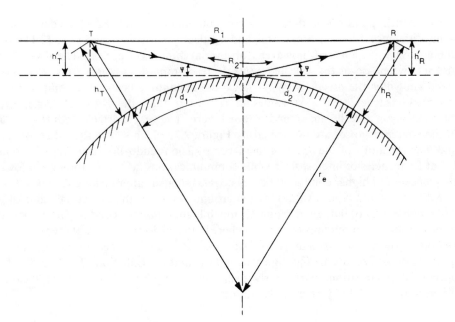

Figure 2.1 Two mutually visible antennas located above a smooth, spherical Earth of effective radius r_e.

$$\rho_h = \frac{\sin\psi - \sqrt{(\varepsilon_r - jx) - \cos^2\psi}}{\sin\psi + \sqrt{(\varepsilon_r - jx) - \cos^2\psi}} \qquad (2.9)$$

where

$$x = \frac{\sigma}{\omega\varepsilon_0} = \frac{18 \times 10^9 \sigma}{f}$$

For vertical polarisation the corresponding expression is

$$\rho_v = \frac{(\varepsilon_r - jx)\sin\psi - \sqrt{(\varepsilon_r - jx) - \cos^2\psi}}{(\varepsilon_r - jx)\sin\psi + \sqrt{(\varepsilon_r - jx) - \cos^2\psi}} \qquad (2.10)$$

The reflection coefficients ρ_h and ρ_v are complex, so the reflected wave will differ from the incident wave in both magnitude and phase. Examination of eqns (2.9) and (2.10) reveals some quite interesting differences. For horizontal polarisation the relative phase of the incident and reflected waves is nearly 180° for all angles of incidence. For very small values of ψ (near-grazing incidence), eqn. (2.9) shows that the reflected wave is equal in magnitude and 180° out of phase with the incident wave for all frequencies and all ground conductivities. In other words, for grazing incidence

$$\rho_h = |\rho_h| e^{j\theta} = 1 e^{j\pi} = -1 \qquad (2.11)$$

As the angle of incidence is increased then $|\rho_h|$ and θ change, but only by relatively small amounts. The change is greatest at higher frequencies and when the ground conductivity is poor.

For vertical polarisation the results are quite different. At grazing incidence there is no difference between horizontal and vertical polarisation and eqn. (2.11) still applies. As ψ is increased, however, substantial differences appear. The magnitude and relative phase of the reflected wave decrease rapidly as ψ increases, and at an angle known as the *pseudo-Brewster angle* the magnitude becomes a minimum and the phase reaches $-90°$. At values of ψ greater than the Brewster angle, $|\rho_v|$ increases again and the phase tends towards zero. The very sharp changes that occur in these circumstances are illustrated by Figure 2.2, which shows the values of $|\rho_v|$ and θ as functions of the angle of incidence ψ. The pseudo-Brewster angle is about $15°$ at frequencies of interest for mobile communications ($x \ll \varepsilon_r$), although at lower frequencies and higher conductivities it becomes smaller, approaching zero if $x \gg \varepsilon_r$.

Table 2.1 shows typical values for the ground constants that affect the value of ρ. The conductivity of flat, good ground is much higher than the conductivity of poorer ground found in mountainous areas, whereas the dielectric constant, typically 15, can be as low as 4 or as high as 30. Over lakes or seas the reflection properties are quite different because of the high values of σ and ε_r. Equation (2.11) applies for horizontal polarisation, particularly over sea water, but ρ may be significantly different from -1 for vertical polarisation.

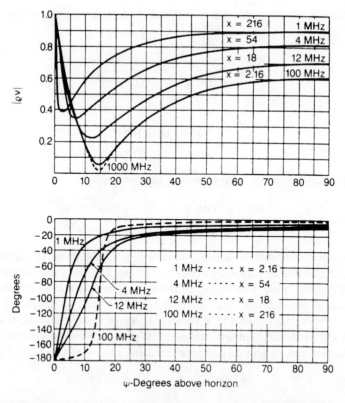

Figure 2.2 Magnitude and phase of the plane wave reflection coefficient for vertical polarisation. Curves drawn for $\sigma = 12 \times 10^{-3}$, $\varepsilon_r = 15$. Approximate results for other frequencies and conductivities can be obtained by calculating the value of x as $18 \times 10^3 \sigma / f_{MHz}$.

Table 2.1 Typical values of ground constants

Surface	Conductivity σ (S)	Dielectric constant ε_r
Poor ground (dry)	1×10^{-3}	4–7
Average ground	5×10^{-3}	15
Good ground (wet)	2×10^{-2}	25–30
Sea water	5×10^{0}	81
Fresh water	1×10^{-2}	81

2.3.2 Propagation over a curved reflecting surface

The situation of two mutually visible antennas sited on a smooth Earth of effective radius r_e is shown in Figure 2.1. The heights of the antennas above the Earth's surface are h_T and h_R, and above the tangent plane through the point of reflection the heights are h'_T and h'_R. Simple geometry gives

$$d_1^2 = [r_e + (h_T - h'_T)]^2 - r_e^2 = (h_T - h'_T)^2 + 2r_e(h_T - h'_T) \simeq 2r_e(h_T - h'_T) \qquad (2.12)$$

and similarly

$$d_2^2 \simeq 2r_e(h_R - h'_R) \qquad (2.13)$$

Using eqns. (2.12) and (2.13) we obtain

$$h'_T = h_T - \frac{d_1^2}{2r_e} \quad \text{and} \quad h'_R = h_R - \frac{d_2^2}{2r_e} \qquad (2.14)$$

The reflecting point, where the two angles marked ψ are equal, can be determined by noting that, providing $d_1, d_2 \gg h_T, h_R$, the angle ψ (radians) is given by

$$\psi = \frac{h'_T}{d_1} = \frac{h'_R}{d_2}$$

Hence

$$\frac{h'_T}{h'_R} \simeq \frac{d_1}{d_2} \qquad (2.15)$$

Using the obvious relationship $d = d_1 + d_2$ together with equations (2.14) and (2.15) allows us to formulate a cubic equation in d_1:

$$2d_1^3 - 3dd_1^2 + [d^2 - 2r_e(h_T + h_R)]d_1 + 2r_e h_T d = 0 \qquad (2.16)$$

The appropriate root of this equation can be found by standard methods starting from the rough approximation

$$d_1 \simeq \frac{d}{1 + h_T/h_R}$$

To calculate the field strength at a receiving point, it is normally assumed that the difference in path length between the direct wave and the ground-reflected wave is negligible in so far as it affects attenuation, but it cannot be neglected with regard to the phase difference along the two paths. The length of the direct path is

$$R_1 = d\left(1 + \frac{(h'_T - h'_R)^2}{d^2}\right)^{1/2}$$

and the length of the reflected path is

$$R_2 = d\left(1 + \frac{(h'_T + h'_R)^2}{d^2}\right)^{1/2}$$

The difference $\Delta R = R_2 - R_1$ is

$$\Delta R = d\left\{\left(1 + \frac{(h'_T + h'_R)^2}{d^2}\right)^{1/2} - \left(1 + \frac{(h'_T - h'_R)^2}{d^2}\right)^{1/2}\right\}$$

and if $d \gg h'_T, h'_R$ this reduces to

$$\Delta R = \frac{2h'_T h'_R}{d} \tag{2.17}$$

The corresponding phase difference is

$$\Delta\phi = \frac{2\pi}{\lambda}\Delta R = \frac{4\pi h'_T h'_R}{\lambda d} \tag{2.18}$$

If the field strength at the receiving antenna due to the direct wave is E_d, then the total received field E is

$$E = E_d[1 + \rho \exp(-j\,\Delta\phi)]$$

where ρ is the reflection coefficient of the Earth and $\rho = |\rho|\exp j\theta$. Thus,

$$E = E_d\{1 + |\rho|\exp[-j(\Delta\phi - \theta)]\} \tag{2.19}$$

This equation can be used to calculate the received field strength at any location, but note that the curvature of the spherical Earth produces a certain amount of divergence of the ground-reflected wave as Figure 2.3 shows. This effect can be taken into account by using, in eqn. (2.19), a value of ρ which is different from that derived in Section 2.3.1 for reflection from a plane surface. The appropriate modification consists of multiplying the value of ρ for a plane surface by a divergence factor D given by [3]:

$$D \simeq \left(1 + \frac{2d_1 d_2}{r_e(h'_T + h'_R)}\right)^{-1/2} \tag{2.20}$$

The value of D can be of the order of 0.5, so the effect of the ground-reflected wave is considerably reduced.

2.3.3 Propagation over a plane reflecting surface

For distances less than a few tens of kilometres, it is often permissible to neglect Earth curvature and assume the surface to be smooth and flat as shown in Figure 2.4. If we also assume grazing incidence so that $\rho = -1$, then eqn. (2.19) becomes

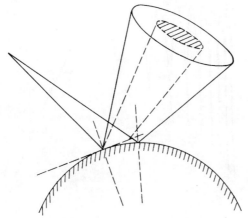

Figure 2.3 Divergence of reflected rays from a spherical surface.

$$E = E_d[1 - \exp(-j\Delta\phi)]$$
$$= E_d[1 - \cos\Delta\phi + j\sin\Delta\phi]$$

Thus,

$$|E| = |E_d|[1 + \cos^2\Delta\phi - 2\cos\Delta\phi + \sin^2\Delta\phi]^{1/2}$$
$$= 2|E_d|\sin\frac{\Delta\phi}{2}$$

and using eqn. (2.18), with $h'_T = h_T$ and $h'_R = h_R$,

$$|E| = 2|E_d|\sin\left(\frac{2\pi h_T h_R}{\lambda d}\right)$$

The received power P_R is proportional to E^2 so

$$P_R \propto 4|E_d|^2\sin^2\left(\frac{2\pi h_T h_R}{\lambda d}\right)$$
$$= 4P_T\left(\frac{\lambda}{4\pi d}\right)^2 G_T G_R \sin^2\left(\frac{2\pi h_T h_R}{\lambda d}\right) \tag{2.21}$$

If $d \gg h_T,\ h_R$ this becomes

$$\frac{P_R}{P_T} = G_T G_R \left(\frac{h_T h_R}{d^2}\right)^2 \tag{2.22}$$

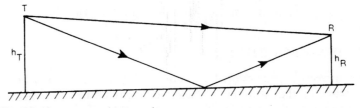

Figure 2.4 Propagation over a plane earth.

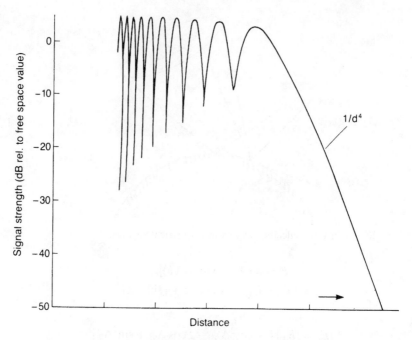

Figure 2.5 Variation of signal strength with distance in the presence of specular reflection.

Equation (2.22) is known as the *plane earth propagation equation*. It differs from the free space relationship (2.3) in two important ways. First, since we assumed that $d \gg h_T$, h_R, the angle $\Delta\phi$ is small and λ cancels out of eqn. (2.22), leaving it frequency independent. Secondly, it shows an inverse fourth-power law with range rather than the inverse square law of eqn. (2.3). This means a far more rapid decrease in received power with range, 12 dB for each doubling of distance in this case.

Note that eqn. (2.22) only applies at ranges where the assumption $d \gg h_T$, h_R is valid. Close to the transmitter, eqn. (2.21) must be used and this gives alternate maxima and minima in the signal strength as shown in Figure 2.5.

In convenient logarithmic form, eqn. (2.22) can be written

$$L_P \text{ (dB)} = 10 \log_{10}(P_T/P_R)$$
$$= -10 \log_{10} G_T - 10 \log_{10} G_R - 20 \log_{10} h_T - 20 \log_{10} h_R + 40 \log_{10} d$$

$$(2.23)$$

and by comparison with eqn (2.6) we can write a 'basic loss' L_B between isotropic antennas as

$$L_B \text{ (dB)} = 40 \log_{10} d - 20 \log_{10} h_T - 20 \log_{10} h_R \qquad (2.24)$$

2.4   GROUND ROUGHNESS

The previous section presupposed a smooth reflecting surface and the analysis was therefore based on the assumption that a specular reflection takes place at the point where the transmitted wave is incident on the Earth's surface. When the surface is rough the specular reflection assumption is no longer realistic since a rough surface presents many facets to the incident wave. A diffuse reflection therefore occurs and the mechanism is more akin to scattering. In these conditions characterisation by a single complex reflection coefficient is not appropriate since the random nature of the surface results in an unpredictable situation. Only a small fraction of the incident energy may be scattered in the direction of the receiving antenna, and the 'ground-reflected' wave may therefore make a negligible contribution to the received signal.

In these circumstances it is necessary to define what constitutes a rough surface. Clearly a surface that might be considered rough at some frequencies and angles of incidence may approach a smooth surface if these parameters are changed. A measure of roughness is needed to quantify the problem, and the criterion normally used is known as the *Rayleigh criterion*. The problem is illustrated in Figure 2.6(a) and an idealised rough surface profile is shown in Figure 2.6(b).

Consider the two rays A and B in Figure 2.6(b). Ray A is reflected from the upper part of the rough surface and ray B from the lower part. Relative to the wavefront AA′ shown, the difference in path length of the two rays when they reach the points C and C′ after reflection is

$$\Delta l = (AB + BC) - (A'B' + B'C')$$

$$= \frac{d}{\sin \psi}(1 - \cos 2\psi)$$

$$= 2d \sin \psi \qquad\qquad (2.25)$$

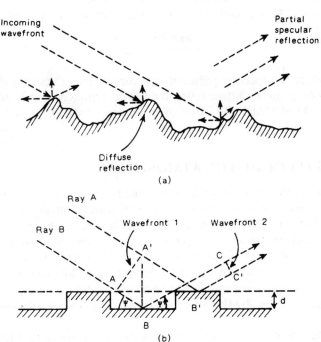

Figure 2.6   Reflections from a semi-rough surface: (a) practical terrain situation, (b) idealised model.

The phase difference between C and C' is therefore

$$\Delta\theta = \frac{2\pi}{\lambda}\Delta l = \frac{4\pi d \sin \psi}{\lambda} \tag{2.26}$$

If the height d is small in comparison with λ then the phase difference $\Delta\theta$ is also small. For practical purposes a specular reflection appears to have occurred and the surface therefore seems to be smooth. On the other hand, extreme roughness corresponds to $\Delta\theta = \pi$, i.e. the reflected rays are in antiphase and therefore tend to cancel. A practical criterion to delineate between rough and smooth is to define a rough surface as one for which $\Delta\theta \geqslant \pi/2$. Substituting this value into eqn. (2.26) shows that for a rough surface

$$d_R \geqslant \frac{\lambda}{8 \sin \psi} \tag{2.27}$$

In the mobile radio situation ψ is always very small and it is admissible to make the substitution $\sin \psi = \psi$. In these conditions eqn. (2.27) reduces to

$$d_R \geqslant \frac{\lambda}{8\psi} \tag{2.28}$$

In practice, the surface of the Earth is more like Fig. 2.6(a) than the idealised surface in Figure 2.6(b). The concept of height d is therefore capable of further interpretation and in practice the value often used as a measure of terrain undulation height is σ, the standard deviation of the surface irregularities relative to the mean height. The Rayleigh criterion is then expressed by writing eqn. (2.26) as

$$C = \frac{4\pi\sigma \sin \psi}{\lambda} \simeq \frac{4\pi\sigma\psi}{\lambda} \tag{2.29}$$

For $C < 0.1$ there is a specular reflection and the surface can be considered smooth. For $C > 10$ there is highly diffuse reflection and the reflected wave is small enough to be neglected. At 900 MHz the value of σ necessary to make a surface rough for $\psi = 1°$ is about 15 m.

2.5   THE EFFECT OF THE ATMOSPHERE

The lower part of the atmosphere, known as the *troposphere*, is a region in which the temperature tends to decrease with height. It is separated from the *stratosphere*, where the air temperature tends to remain constant with height, by a boundary known as the *tropopause*. In general terms the height of the tropopause varies from about 9 km at the Earth's poles to about 17 km at the equator. The height of the tropopause also varies with atmospheric conditions; for instance, at middle latitudes it may reach about 13 km in anticyclones and decline to less than about 7 km in depressions.

At frequencies above 30 MHz there are three effects worthy of mention:

- localised fluctuations in refractive index, which can cause scattering
- abrupt changes in refractive index as a function of height, which can cause reflection
- a more complicated phenomenon known as ducting (Section 2.5.1).

All these mechanisms can carry energy beyond the normal optical horizon and therefore have the potential to cause interference between different radio communication systems. Forward scattering of radio energy is sufficiently dependable that it may be used as a mechanism for long-distance communications, especially at frequencies between about 300 MHz and 10 GHz. Nevertheless, this *troposcatter* is not used for mobile radio communications and we will not consider it any further. Reflection and ducting are much less predictable.

Variations in the climatic conditions within the troposphere, i.e. changes of temperature, pressure and humidity, cause changes in the refractive index of the air. Large-scale changes of refractive index with height cause radio waves to be refracted, and at low elevation angles the effect can be quite significant at all frequencies, especially in extending the radio horizon distance beyond the optical horizon. Of all the influences the atmosphere can exert on radio signals, refraction is the one that has the greatest effect on VHF and UHF point-to-point systems; it is therefore worthy of further discussion. We start by considering an idealised model of the atmosphere and then discuss the effects of departures from that ideal.

An ideal atmosphere is one in which the dielectric constant is unity and there is zero absorption. In practice, however, the dielectric constant of air is greater than unity and depends on the pressure and temperature of the air and the water vapour; it therefore varies with weather conditions and with height above the ground. Normally, but not always, it decreases with increasing height. Changes in the atmospheric dielectric constant with height mean that electromagnetic waves are bent in a curved path that keeps them nearer to the Earth than would be the case if they truly travelled in straight lines. With respect to atmospheric influences, radio waves behave very much like light.

The refractive index of the atmosphere at sea level differs from unity by about 300 parts in 10^6 and it falls approximately exponentially with height. It is convenient to refer to the refractivity in N-units, where

$$N = (n - 1) \times 10^6$$

and n is the refractive index of the atmosphere expressed as

$$n \approx (1 + 300 \times 10^{-6})$$

A well known expression for N is [1, Ch. 4]:

$$N = \frac{77.6}{T} \left(P + \frac{4810e}{T} \right) \tag{2.30}$$

where P is the total pressure (mb)

e is the water vapour pressure (mb)

T is the temperature (K)

and as an example, if $P = 1000$ mb, $e = 10$ mb and $T = 290$ K then $N = 312$.

In practice P, e and T tend to fall exponentially with height, and therefore so does N. The value of N at height h can therefore be written in terms of the value N_s at the Earth's surface:

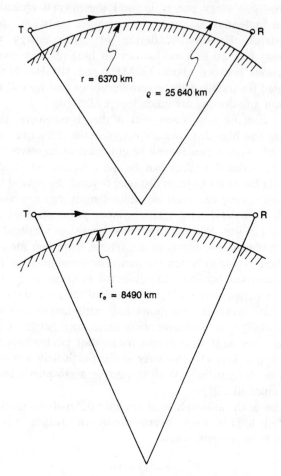

Figure 2.7 An effective Earth radius of 8490 km (6730×4/3) permits the use of straight-line propagation paths.

$$N(h) = N_s \exp(-h/H) \tag{2.31}$$

where H is a scale height (often taken as 7 km).

Over the first kilometre or so, the exponential curve can be approximated by a straight line and in this region the refractivity falls by about 39 N-units. Although this may appear to be a small change, it has a profound effect on radio propagation.

In a so-called standard exponential atmosphere, i.e. one in which eqn. (2.31) applies, the refractivity decreases continuously with height and ray paths from the transmitter are therefore curved. It can be shown that the radius of curvature is given by

$$\rho = -\frac{\mathrm{d}h}{\mathrm{d}n}$$

and that in a standard atmosphere $\rho = 10^6/39 = 25\,640$ km. This ray path is curved and so of course is the surface of the Earth. The geometry is illustrated in Figure 2.7, where it can be seen that a ray launched parallel to the Earth's surface is bent downwards but not enough to reach the ground. The distance d, from an antenna of height h to the optical horizon, can be obtained from the geometry of Figure 2.1. The maximum line-of-sight range d is given by

$$d^2 = (h + r)^2 - r^2 = h^2 + 2hr \simeq 2hr \qquad (2.32)$$

so that $d \approx \sqrt{2hr}$ when $h \ll r$.

The geometry of a curved ray propagating over a curved surface is complicated and in practical calculations it is common to reduce the complexity by increasing the true value of the Earth's radius until ray paths, modified by the refractive index gradient, become straight again. The modified radius can be found from the relationship

$$\frac{1}{r_e} = \frac{1}{r} + \frac{dn}{dh} \qquad (2.33)$$

where dn/dh is the rate of change of refractive index with height.

The ratio r_e/r is the effective Earth radius factor k, so the distance to the radio horizon is $\sqrt{2krh}\ (=\sqrt{2r_e h})$. The average value for k based on a standard atmosphere is 4/3 and use of this four-thirds Earth radius is very widespread in the calculation of radio paths. It leads to a very simple relationship for the horizon distance: $d = \sqrt{2h}$ where d is in miles and h is in feet.

In practice the atmosphere does not always behave according to this idealised model, hence the radio wave propagation paths are perturbed.

2.5.1 Atmospheric ducting and non-standard refraction

In a real atmosphere the refractive index may not fall continuously with height as predicted by eqn. (2.31) for a standard exponential atmosphere. There may be a general decrease, but there may also be quite rapid variations about the general trend. The relative curvature between the surface of the Earth and a ray path is given by eqn. (2.33) and if $dn/dh = -1/r_e$ we have the interesting situation of zero relative curvature, i.e. a ray launched parallel to the Earth's surface remains parallel to it and there is no radio horizon. The value of dn/dh necessary to cause this is -157 N-units per kilometre ($1/6370 = 157 \times 10^{-6}$). In certain parts of the world it is often found that the index of refraction has a rate of decrease with height over a short distance that is greater than this critical rate and sufficient to cause the rays to be refracted back to the surface of the Earth. These rays are then reflected and refracted back again in such a manner that the field is trapped or guided in a thin layer of the atmosphere close to the Earth's surface (Figure 2.8). This is the phenomenon known as *trapping* or *ducting*. The radio waves will then propagate over quite long distances with much less attenuation than for free space propagation; the guiding action is in some ways similar to the Earth–ionosphere waveguide at lower frequencies.

Ducts can form near the surface of the Earth (surface ducts) or at heights up to about 1500 m above the surface (elevated ducts). To obtain long-distance propagation, both the transmitting and the receiving antennas must be located within the duct in order to couple effectively to the field in the duct. The thickness of the duct may range from a few metres to several hundred metres. To obtain trapping or ducting, the rays must propagate in a nearly horizontal direction, so to satisfy

Figure 2.8 The phenomenon of ducting.

conditions for guiding within the duct the wavelength has to be relatively small. The maximum wavelength that can be trapped in a duct of 100 m thickness is about 1 m, (i.e. a frequency of about 300 MHz), so the most favourable conditions for ducting are in the VHF and UHF bands. For good propagation, the relationship between the maximum wavelength λ and the duct thickness t should be $t = 500\lambda^{2/3}$.

A simplified theory of propagation which explains the phenomenon of ducting can be expressed in terms of a modified index of refraction that is the difference between the actual refractive index and the value of $-157\,N$-units per kilometre that causes rays to remain at a constant height above the curved surface of the Earth [4, Ch. 6]. Under non-standard conditions the refractive index may change either more rapidly or less rapidly than $-157\,N$-units per kilometre. When the decrease is more rapid, the ray paths have a radius of curvature less than 25 640 km, so waves propagate further without getting too far above the Earth's surface. This is termed *super-refraction*. On the other hand, when the refractive index decreases less rapidly there is less downward curvature and *substandard refraction* is said to exist.

Figure 2.9 shows how changes in refractive index cause a surface duct to form and indicates some typical ray paths within the duct. Near the ground, $\mathrm{d}n/\mathrm{d}h$ is negative with a magnitude greater than $157\,N$-units per kilometre. Above height h_0 the gradient has magnitude less than 157. Below h_0 the radius of curvature of rays launched at small elevation angles is less than the radius of curvature of the Earth, and above h_0 it is greater. Rays 1, 2 and 3 are trapped between the Earth and an imaginary sphere at height h_0. Rays 2 and 3 are tangential to the sphere and represent the extremes of the trapped waves. Rays 4 and 5, at high angles, are only weakly affected by the duct and resume a normal path on exit. This kind of duct can cause anomalous propagation conditions, as a result of which the interference between radio services can be very severe.

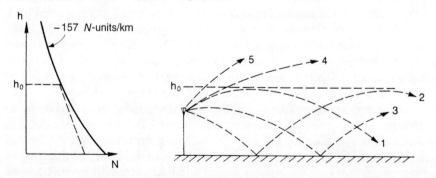

Figure 2.9 Refractive index variation and subsequent ray paths in a surface duct.

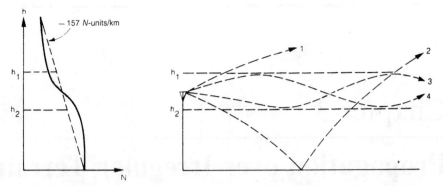

Figure 2.10   Refractive index variation and subsequent ray paths in an elevated duct.

Elevated ducts can also be formed as Figure 2.10 shows. An inversion (i.e. an increasing refractive index) exists up to height h_0 then there is a fast decrease up to height h_1. Rays launched over quite a wide range of angles can become trapped in this elevated duct; the mechanism of propagation is similar to that in a surface (or ground-based) duct.

The formation of ducts is caused primarily by the water vapour content of the atmosphere since, compared with the temperature gradient, this has a stronger influence on the index of refraction. For this reason, ducts commonly form over large bodies of water, and in the trade wind belt over warm seas there is often more or less permanent ducting; the thickness of the ducts is about 1.5 to 2 m. A quiet atmosphere is essential for ducting, hence the occurrence of ducts is a maximum in calm weather conditions over water or plains; there is too much turbulence over mountains. Ground ducts are produced in three ways:

- A mass of warm air arriving over a cold ground or the sea
- Night frosts which cause ducts during the second half of the night
- High humidity in the lower troposphere

Night frosts frequently occur in desert and tropical climates. Elevated ducts are caused principally by the subsidence of an air mass in a high-pressure area. As the air descends it is compressed and is thus warmed and dried. Elevated ducts occur mainly above the clouds and can interfere with ground–aircraft communications.

Anomalous propagation due to ducting can often cause television transmissions from one country to be received several hundred miles away in another country when atmospheric conditions are suitable. However, ducting is not a major source of problems to mobile radio systems in temperate climates.

REFERENCES

1. Jordan E.C. and Balmain K.G. (1968) *Electromagnetic Waves and Radiating Systems*. Prentice Hall, New York.
2. Friis H.T. (1946) A note on a simple transmission formula. *Proc. IRE*, **34**, 254–6.
3. Griffiths J. (1987) *Radio Wave Propagation and Antennas: An Introduction*. Prentice Hall, London.
4. Collin R.E. (1985) *Antennas and Radiowave Propagation*. McGraw-Hill, New York.

Chapter 3

Propagation over Irregular Terrain

3.1 INTRODUCTION

Land mobile radio systems are used in a wide variety of scenarios. At one extreme, county police and other emergency services operate over fairly large areas using frequencies in the lower part of the VHF band. The service area may be large enough to require several transmitters, operating in a quasi-synchronous mode, and is likely to include rural, suburban and urban areas. At the other extreme, in major cities, individual cells within a 900 or 1800 MHz cellular radio telephone system can be very small in size, possibly less than 1 km in radius, and service has to be provided to both vehicle-mounted installations and to hand-portables which can be taken inside buildings. It is clear that predicting the coverage area of any base station transmitter is a complicated problem involving knowledge of the frequency of operation, the nature of the terrain, the extent of urbanisation, the heights of the antennas and several other factors.

Moreover, since in general the mobile moves in or among buildings which are randomly sited on irregular terrain, it is unrealistic to pursue an exact, deterministic analysis unless highly accurate and up-to-date terrain and environmental databases are available. Satellite imaging and similar techniques are helping to create such databases and their availability makes it feasible to use prediction methods such as ray tracing (see later). For the present, however, in most cases an approach via statistical communication theory remains the most realistic and profitable. In predicting signal strength we seek methods which, among other things, will enable us to make a statement about the percentage of locations within a given, fairly small, area where the signal strength will exceed a specified level.

In practice, mobile radio channels rank among the worst in terrestrial radio communications. The path loss often exceeds the free space or plane earth path loss by several tens of decibels; it is highly variable and it fluctuates randomly as the receiver moves over irregular terrain and/or among buildings. The channel is also corrupted by ambient noise generated by electrical equipment of various kinds; this noise is impulsive in nature and is often termed man-made noise. All these factors will be considered in the chapters that follow; for now we will concentrate on methods of estimating the mean or average signal strength in a given small area. Several methods exist, some having specific applicability over irregular terrain, others in built-up areas, etc. None of the simple equations derived in Chapter 2 are suitable in unmodified form for predicting average signal strength in the mobile radio context, although as we will see, both the free space and plane earth equations are used as an underlying basis for several models that are used. Before going any further, we will deal with some further theoretical and analytical techniques that underpin many prediction methods.

3.2 HUYGENS' PRINCIPLE

Discussions of reflection and refraction are usually based on the assumption that the reflecting surfaces or refracting regions are large compared with the wavelength of the radiation. When a wavefront encounters an obstacle or discontinuity that is not large then Huygens' principle, which can be deduced from Maxwell's equations, is often useful in giving an insight into the problem and in providing a solution. In simple terms, the principle suggests that each point on a wavefront acts as the source of a secondary wavelet and that these wavelets combine to produce a new wavefront in the direction of propagation. Figure 3.1 shows a plane wavefront that has reached the position AA'. Spherical wavelets originate from every point on AA' to form a new wavefront BB', drawn tangential to all wavelets with equal radii. As an illustration, Figure 3.1 shows how wavelets originating from three representative points on AA' reach the wavefront BB'.

To explain the observable effect, i.e. that the wave propagates only in the forward direction from AA' to BB', it must be concluded that the secondary wavelets originating from points along AA' do not have a uniform amplitude in all directions and if α represents the angle between the direction of interest and the normal to the wavefront, then the amplitude of the secondary wave in a given direction is proportional to $(1 + \cos \alpha)$. Thus, the amplitude in the direction of propagation is proportional to $(1 + \cos 0) = 2$ and in any other direction it will be less than 2. In particular, the amplitude in the backward direction is $(1 + \cos \pi) = 0$. Consideration of wavelets originating from all points on AA' leads to an expression for the field at

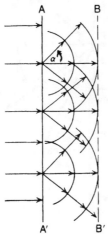

Figure 3.1 Huygens' principle applied to propagation of plane waves.

any point on BB' in the form of an integral, the solution of which shows that the field at any point on BB' is exactly the same as the field at the nearest point on AA', with its phase retarded by $2\pi d/\lambda$. The waves therefore appear to propagate along straight lines normal to the wavefront.

3.3 DIFFRACTION OVER TERRAIN OBSTACLES

The analysis in Section 3.2 applies only if the wavefront extends to infinity in both directions; in practice it applies if AA' is large compared to a wavelength. But suppose the wavefront encounters an obstacle so that this requirement is violated. It is clear from Figure 3.2 that beyond the obstacle (which is assumed to be impenetrable or perfectly absorbing) only a semi-infinite wavefront CC' exists. Simple ray theory would suggest that no electromagnetic field exists in the shadow region below the dotted line BC, but Huygens' principle states that wavelets originating from all points on BB', e.g. P, propagate into the shadow region and the field at any point in this region will be the resultant of the interference of all these wavelets. The apparent bending of radio waves around the edge of an obstruction is known as *diffraction*.

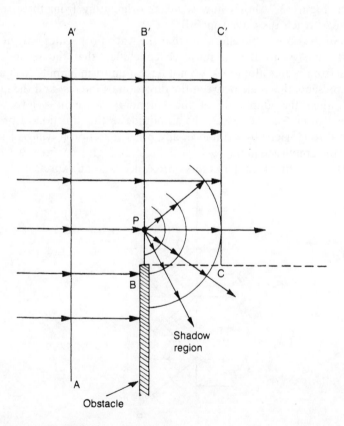

Figure 3.2   Diffraction at the edge of an obstacle.

To introduce some concepts associated with diffraction we consider a transmitter T and a receiver R in free space as in Figure 3.3. We also consider a plane normal to the line-of-sight path at a point between T and R. On this plane we construct concentric circles of arbitrary radius and it is apparent that any wave which has propagated from T to R via a point on any of these circles has traversed a longer path than TOR. In terms of the geometry of Figure 3.4 , the 'excess' path length is given by

$$\Delta \simeq \frac{h^2}{2}\left(\frac{d_1 + d_2}{d_1 d_2}\right) \tag{3.1}$$

assuming $h \ll d_1, d_2$. The corresponding phase difference is

$$\phi = \frac{2\pi\Delta}{\lambda} = \frac{2\pi}{\lambda}\left(\frac{h^2}{2}\right)\frac{d_1 + d_2}{d_1 d_2} \tag{3.2}$$

This is often written in terms of a parameter v, as

$$\phi = \frac{\pi}{2}v^2 \tag{3.3}$$

where

$$v = h\sqrt{\frac{2(d_1 + d_2)}{\lambda d_1 d_2}} \tag{3.4}$$

and is known as the *Fresnel–Kirchhoff diffraction parameter*.

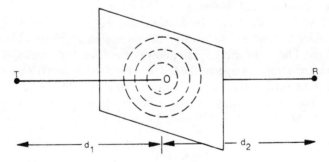

Figure 3.3 Family of circles defining the limits of the Fresnel zones at a given point on the radio propagation path.

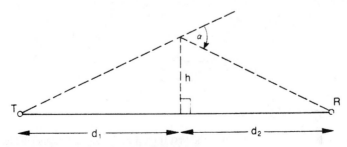

Figure 3.4 The geometry of knife-edge diffraction.

Alternatively, using the same approximation we can obtain

$$\phi = \frac{\pi\alpha^2}{\lambda}\left(\frac{d_1 d_2}{d_1 + d_2}\right) \tag{3.5}$$

and

$$v = \alpha\sqrt{\frac{2d_1 d_2}{\lambda(d_1 + d_2)}} \tag{3.6}$$

There is a need to keep a region known as the first Fresnel zone substantially free of obstructions, in order to obtain transmission under free space conditions (see Section 1.3.1). In practice this usually involves raising the antenna heights until the necessary clearance over terrain obstacles is obtained. However, if the terminals of a radio link path for which line-of-sight (LOS) clearance over obstacles exists, are low enough for the direct path to pass close to the surface of the Earth at some intermediate point, then there may well be a path loss considerably in excess of the free space loss, even though the LOS path is not actually blocked. Clearly we need a quantitative measure of the required clearance over any terrain obstruction and this may be obtained in terms of Fresnel zone ellipsoids drawn around the path terminals.

3.3.1   Fresnel-zone ellipsoids

If we return to Figure 3.3 then it is clear that on the plane passing through the point O, we could construct a family of circles having the specific property that the total path length from T to R via each circle is $n\lambda/2$ longer than TOR, where n is an integer. The innermost circle would represent the case $n = 1$, so the excess path length is $\lambda/2$. Other circles could be drawn for λ, $3\lambda/2$, etc. Clearly the radii of the individual circles depend on the location of the imaginary plane with respect to the path terminals. The radii are largest midway between the terminals and become smaller as the terminals are approached. The loci of the points for which the 'excess' path length is an integer number of half-wavelengths define a family of ellipsoids (Figure 3.5). The radius of any specific member of the family can be expressed in terms of n and the dimensions of Figure 3.4 as [1, Ch. 4]:

$$h = r_n = \sqrt{\frac{n\lambda d_1 d_2}{d_1 + d_2}} \tag{3.7}$$

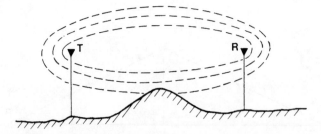

Figure 3.5   Family of ellipsoids defining the first three Fresnel zones around the terminals of a radio path.

and hence, $v_n = \sqrt{2n}$

This is an approximation which is valid provided $d_1, d_2 \gg r_n$ and is therefore realistic except in the immediate vicinity of the terminals. The volume enclosed by the ellipsoid defined by $n = 1$ is known as the first Fresnel zone. The volume between this ellipsoid and the ellipsoid defined by $n = 2$ is the second Fresnel zone, etc.

It is clear that contributions from successive Fresnel zones to the field at the receiving point tend to be in phase opposition and therefore interfere destructively rather than constructively. If an obstructing screen were actually placed at a point between T and R and if the radius of the aperture were increased from the value that produces the first Fresnel zone to the value that produces the second Fresnel zone, the third Fresnel zone, etc., then the field at R would oscillate. The amplitude of the oscillation would gradually decrease since smaller amounts of energy propagate via the outer zones.

3.3.2 Diffraction losses

If an ideal, straight, perfectly absorbing screen is interposed between T and R in Figure 3.4 then when the top of the screen is well below the LOS path it will have little effect and the field at R will be the 'free space' value E_0. The field at R will begin to oscillate as the height is increased, hence blocking more of the Fresnel zones below the line-of-sight path. The amplitude of the oscillation increases until the obstructing edge is just in line with T and R, at which point the field strength is exactly half the unobstructed value, i.e. the loss is 6 dB. As the height is increased above this value, the oscillation ceases and the field strength decreases steadily.

To express this in a quantitative way, we use classical diffraction theory and we replace any obstruction along the path by an absorbing plane placed at the same position. The plane is normal to the direct path and extends to infinity in all directions except vertically, where it stops at the height of the original obstruction. *Knife-edge diffraction* is the term used to describe this situation, all ground reflections being ignored.

The field strength at the point R in Figure 3.4 is determined as the sum of all the secondary Huygens sources in the plane above the obstruction and can be expressed as [2, Ch. 16]:

$$\frac{E}{E_0} = \frac{(1+j)}{2} \int_v^\infty \exp\left(-j\frac{\pi}{2}t^2\right) dt \tag{3.8}$$

This is known as the complex Fresnel integral and v is the value given by eqn. (3.4) for the height of the obstruction under consideration. We note that if the obstruction lies below the line-of-sight then h, and hence v, is negative. If the path is actually obstructed then h and v are positive, as in Figure 3.6.

An interesting and relevant insight into the evaluation of eqn. (3.8) can be obtained in the following way. We can write

$$\int_v^\infty \exp\left(-j\frac{\pi}{2}t^2\right) dt = \int_v^\infty \cos\left(\frac{\pi}{2}t^2\right) dt - j\int_v^\infty \sin\left(\frac{\pi}{2}t^2\right) dt$$

and

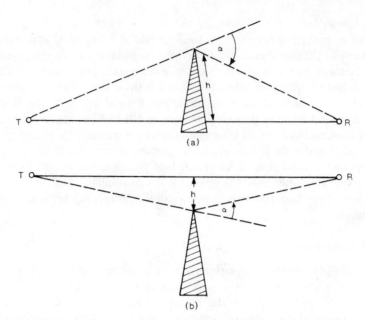

Figure 3.6 Knife-edge diffraction: (a) h and v positive, (b) h and v negative.

$$\int_v^\infty \cos\left(\frac{\pi}{2}t^2\right)\mathrm{d}t = \frac{1}{2} - \int_0^v \cos\left(\frac{\pi}{2}t^2\right)\mathrm{d}t$$

which is usually written as $\frac{1}{2} - C(v)$.

Similarly,

$$\int_v^\infty \sin\left(\frac{\pi}{2}t^2\right)\mathrm{d}t = \tfrac{1}{2} - S(v).$$

The complex Fresnel integral (3.8) can therefore be expressed as

$$\frac{E}{E_0} = \frac{(1+\mathrm{j})}{2}\{(\tfrac{1}{2} - C(v)) - \mathrm{j}(\tfrac{1}{2} - S(v))\} \tag{3.9}$$

Let us now consider the integral

$$C(v) - \mathrm{j}S(v) = \int_0^v \exp\left(-\mathrm{j}\frac{\pi}{2}t^2\right)\mathrm{d}t \tag{3.10}$$

Plotting this integral in the complex plane with C as the abscissa and S as the ordinate results in Figure 3.7, a curve known as *Cornu's spiral*. In this curve, positive values of v appear in the first quadrant and negative values in the third quadrant. The spiral has the following properties:

- A vector drawn from the origin to any point on the curve represents the magnitude and phase of eqn. (3.10).
- The length of arc along the curve, measured from the origin, is equal to v. As $v \to \infty$ the curve winds an infinite number of times around the points $(\tfrac{1}{2}, \tfrac{1}{2})$ or $(-\tfrac{1}{2}, -\tfrac{1}{2})$.

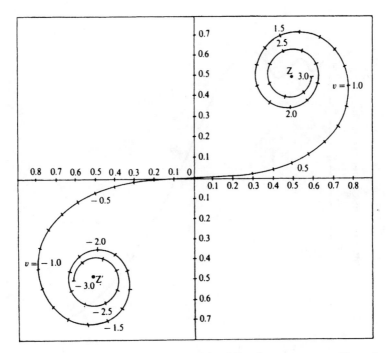

Figure 3.7 Plots of the Fresnel integral in terms of the diffraction parameter v (Cornu's spiral).

It is clear that $[\frac{1}{2} - C(v)]$ and $[\frac{1}{2} - S(v)]$ represent the real and imaginary parts of a vector drawn from the point $(\frac{1}{2}, \frac{1}{2})$ to a point on the spiral. Thus the value of $|E|$ corresponding to any particular value of v, say v_0, is proportional to the length of the vector joining $(\frac{1}{2}, \frac{1}{2})$ to the point on the spiral corresponding to v_0. Thus Cornu's spiral gives a visual indication of how the magnitude and phase of E varies as a function of the Fresnel parameter v.

Figure 3.8 shows the diffraction loss in decibels relative to the free space loss, as given by eqn. (3.9). In the shadow zone below the LOS path the loss increases smoothly; above the LOS path the loss oscillates about its free space value, the amplitude of oscillation decreasing as v becomes more negative. When there is grazing incidence over the obstacle there is a 6 dB loss, i.e. the field strength is $0.5E_0$; but Figure 3.8 shows that this loss can be avoided if $v \approx -0.8$, which corresponds to about 56% of the first Fresnel zone being clear of obstructions. In practice, therefore, designers of point-to-point links try to make the heights of antenna masts such that the majority of the first Fresnel zone is unobstructed.

As an alternative to using Figure 3.8, nomographs of the form shown in Figure 3.9 exist in the literature [3]. They enable the diffraction loss to be calculated to within about 2 dB. Alternatively, various approximations are available that enable the loss to be evaluated in a fairly simple way. Modified expressions as given by Lee [4] are

$$L(v) \text{ (dB)} = \begin{cases} -20 \, \log(0.5 - 0.62v) & -0.8 < v < 0 \\ -20 \, \log[0.5 \, \exp(-0.95v)] & 0 < v < 1 \\ -20 \, \log[0.4 - \{0.1184 - (0.38 - 0.1v)^2\}^{1/2}] & 1 < v < 2.4 \\ -20 \, \log(0.225/v) & v > 2.4 \end{cases} \tag{3.11}$$

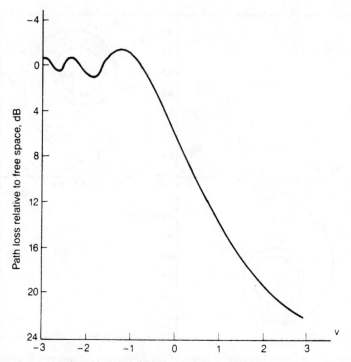

Figure 3.8   Diffraction loss over a single knife-edge as a function of the parameter v.

The approximation used for $v > 2.4$ arises from the fact that as v becomes large and positive then eqn. (3.8) can be written as

$$\left| \frac{E}{E_0} \right| \to \frac{2^{1/2}}{2\pi v}$$

an asymptotic result which holds with an accuracy better than 1 dB for $v > 1$, but breaks down rapidly as v approaches zero.

Ground reflections

The previous analysis has ignored the possibility of ground reflections either side of the terrain obstacle. To cope with this situation (Figure 3.10), four paths have to be taken into account in computing the field at the receiving point [5]. The four rays depicted in Figure 3.10 have travelled different distances and will therefore have different phases at the receiver. In addition the Fresnel parameter v is different in each case, so the field at the receiver must be computed from

$$E = E_0 \sum_{k=1}^{4} L(v_k) \exp(j\phi_k) \tag{3.12}$$

In any particular situation a ground reflection may exist only on the transmitter or receiver side of the obstacle, in which case only three rays exist.

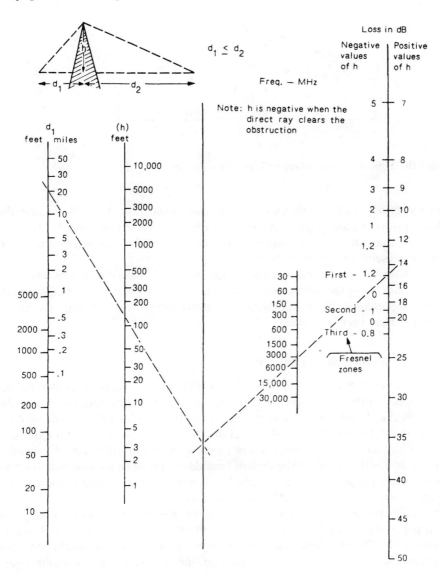

Figure 3.9 Nomograph for calculating the diffraction loss due to an isolated obstacle (after Bullington).

3.4 DIFFRACTION OVER REAL OBSTACLES

We have seen earlier that geometrical optics is incapable of predicting the field in the shadow regions, indeed it produces substantial inaccuracies near the shadow boundaries. Huygens' principle explains why the field in the shadow regions is non-zero, but the assumption that an obstacle can be represented by an ideal, straight, perfectly absorbing screen is in most cases a very rough approximation. Having said that, and despite the fact that the knife-edge approach ignores several

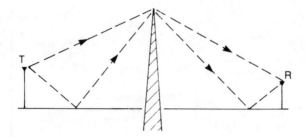

Figure 3.10   Knife-edge diffraction with ground reflections.

important effects such as the wave polarisation, local roughness effects and the electrical properties and lateral profile of the obstacle, it must be conceded that the losses predicted using this assumption are sufficiently close to measurements to make them useful to system designers.

Nevertheless, objects encountered in the physical world have dimensions which are large compared with the wavelength of transmission. Neither hills nor buildings can be truly represented by a knife-edge (assumed infinitely thin) and alternative approaches have been developed.

3.4.1   The uniform theory of diffraction

The original geometric theory of diffraction (GTD) was developed by Keller and his seminal paper on this subject [6] was published in 1962. By adding diffracted rays, the GTD overcame the principal shortcoming of geometrical optics, i.e. the prediction of a zero field in the shadow region. Keller developed his theory using wedge diffraction as a canonical problem but the theory remained incomplete because it predicted a singular diffracted field in the vicinity of the shadow boundaries, i.e. when the source, diffracting edge and receiving point lie in a straight line (earlier termed grazing incidence) and because it considered only perfectly conducting wedges.

These limitations were partially addressed by Kouyoumjian and Pathak in a classic paper published in 1974 [7] setting out the *uniform geometrical theory of diffraction (UTD)*. By performing an asymptotic analysis and multiplying the diffraction coefficients by a transition function, they succeeded in developing a ray-based uniform diffraction theory valid at all spatial locations. Even so, imperfections still remained and have prompted a very extensive volume of literature. Luebbers [8], for example, considered diffraction boundaries with finite conductivity and produced a widely used heuristic diffraction coefficient. More rigorous work on wedges with finite conductivity had been undertaken earlier by Maliuzhinets [9].

To illustrate the theory very briefly, we consider a two-dimensional diagram of a wedge with straight edges (Figure 3.11). It is conventional to label the faces of the wedge the *o*-face and the *n*-face. We measure angles from the *o*-face. The interior angle of the wedge is $(2 - n)\pi$ and is less than 180°. If E_0 is the field at the source, then the UTD gives the field at the receiving point as

$$E^{d}(s) = E_0 \bar{D} A(s', s) \exp(-jks) \tag{3.13}$$

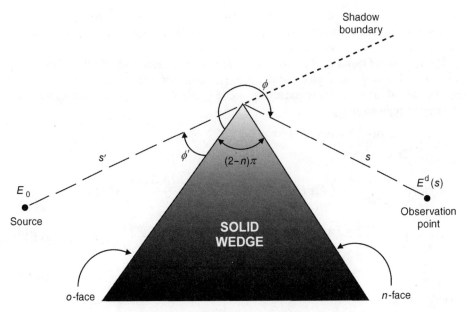

Figure 3.11 The geometry for wedge diffraction using UTD.

where $\bar{D}$ represents the dyadic diffraction coefficient of the wedge, s' and s are the distances along the ray path from the source to the edge and from the edge to the receiving point respectively, $A(s', s)$ is a spreading factor which describes the amplitude variation of the diffracted field and $\exp(-jks)$ is a phase factor ($k = 2\pi/\lambda$). The form of $A(s', s)$ depends on the type of wave being considered and is given by $1/\sqrt{s}$ for plane and conical wave incidence. For cylindrical incidence s is replaced by $s \sin \beta_0$, the perpendicular distance to the edge; β_0 is the angle between the incident ray and the tangent to the edge. For spherical wave incidence,

$$A(s', s) = \sqrt{\frac{s'}{s(s' + s)}} \tag{3.14}$$

If the receiving point is not close to a shadow or reflection boundary, then for all types of wave the scalar diffraction coefficient is [10]:

$$D_{h,s} = \frac{\exp(-j\pi/4) \sin(\pi/n)}{n\sqrt{2\pi k} \sin \beta_0}$$

$$\times \left[\frac{1}{\cos(\pi/n) - \cos\left(\dfrac{\phi - \phi'}{n}\right)} \pm \frac{1}{\cos(\pi/n) - \cos\left(\dfrac{\phi + \phi'}{n}\right)} \right] \tag{3.15}$$

The subscripts h and s represent the so-called *hard* polarisation (*H*-field parallel to both faces of the wedge) and *soft* polarisation (*E*-field parallel to both faces) and correspond to the $+$ and $-$ signs on the right-hand side of the equation. This expression becomes singular as shadow or reflection boundaries are approached, causing problems in these regions.

The regions of rapid field change adjacent to the shadow and reflection boundaries are termed transition regions and an expression for the dyadic edge diffraction coefficient of a perfectly conducting wedge, valid both inside and outside the transition regions is:

$$D_{s,h} = \frac{-\exp[-j(\pi/4)]}{2n\sqrt{2\pi k}\ \sin\beta_0}$$

$$\times \left[\cot\left(\frac{\pi+(\phi-\phi')}{2n}\right) F[kLa^+(\phi-\phi')] + \cot\left(\frac{\pi-(\phi-\phi')}{2n}\right) F[kLa^-(\phi-\phi')]\right.$$

$$\left. \pm \left\{\cot\left(\frac{\pi+(\phi+\phi')}{2n}\right) F[kLa^+(\phi+\phi')] + \cot\left(\frac{\pi-(\phi+\phi')}{2n}\right) F[kLa^-(\phi+\phi')]\right\}\right]$$

$$(3.16)$$

where $F[\cdot]$ is

$$F(X) = 2j\sqrt{X} \int_{\sqrt{X}}^{\infty} \exp(-j\tau^2)\, d\tau \tag{3.17}$$

in which the positive value of the square root is taken, and

$$a^{\pm}(\beta) = 2\ \cos^2\left(\frac{2n\pi N^{\pm} - \beta}{2}\right) \tag{3.18}$$

In eqn. (3.18) the N are the integers that most nearly satisfy the equations

$$2\pi n N^+ - \beta = \pi \text{ and } 2\pi n N^- - \beta = -\pi$$

$$\text{with } \beta = \phi \pm \phi'$$

It is apparent that N^+ and N^- each have two values. The distance parameter L is given by

$$L = \begin{cases} s\ \sin^2\beta_0 & \text{for plane wave incidence} \\ \dfrac{ss'}{s+s'}\ \sin^2\beta_0 & \text{for conical and spherical wave incidences} \end{cases} \tag{3.19}$$

The UTD method can easily cope with wedges which have curved faces and different internal angles, so reasonably accurate modelling of real terrain obstacles is fairly straightforward. Furthermore, a 90° wedge can be used to model the edge of a building, so diffraction losses around corners can also be handled [11]. Wedges with finite conductivity also fall within the scope of the method [10], so accurate diffraction calculations along a path profile depend on producing a series of models for the obstacles which are truly representative of their actual shape.

The UTD equations are easily implemented on a computer and the resulting subroutines are only marginally more demanding computationally than those for knife-edge diffraction. The advantages are that polarisation, local surface roughness and the electrical properties of the wedge material (natural or man-made) can be taken into account.

Other approaches

The problem of non-idealised obstacles has also been treated in other ways. Probably most notable are Pathak [12], who represented obstacles as convex surfaces, and Hacking [13], who had shown earlier that the loss due to rounded obstacles exceeds the knife-edge loss. If a rounded hilltop as in Figure 3.12 is replaced by a cylinder of radius r equal to that of the crest, then the cylinder supports reflections either side of the hypothetical knife-edge that coincides with the peak, and the Huygens wavefront above that point is therefore modified. This is similar to the mechanism in the four-ray situation described above. An excess loss (dB) can be added to the knife-edge loss to account for this; the value is given by [13]:

$$L_{ex} \approx 11.7 \left(\frac{\pi r}{\lambda} \right)^{1/3} \alpha \tag{3.20}$$

If the hilltop is rough, due to the presence of trees, then the diffraction loss is about 65% of the value given above.

An alternative solution [14] is available through a dimensionless parameter ρ defined as

$$\rho = \left(\frac{\lambda}{\pi} \right)^{1/6} r^{1/3} \left(\frac{d_1 + d_2}{d_1 d_2} \right)^{1/2} \tag{3.21}$$

The diffraction loss can then be represented by the quantity $A(v, \rho)$, normally expressed in decibels. It is related to the ideal knife-edge loss $A(v, 0)$ by

$$A(v, \rho) = A(v, 0) + A(0, \rho) + U(v\rho) \tag{3.22}$$

$U(v\rho)$ is a correction factor given by Figure 3.13 and $A(0, \rho)$ is shown in Figure 3.14. The knife-edge loss $A(v, 0)$ is given by Figure 3.8. Approximations are available for $A(0, \rho)$ and $U(v\rho)$ as [15]:

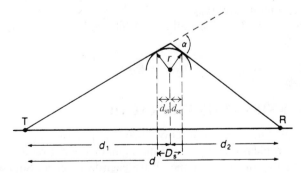

Figure 3.12 Diffraction over a cylinder.

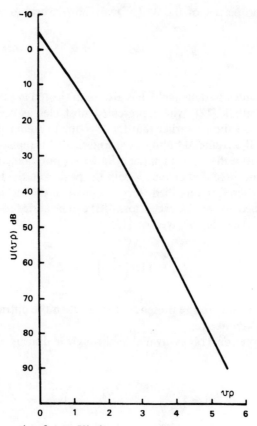

Figure 3.13   The correction factor $U(v\rho)$.

$$A(0, \rho) = 6 + 7.19\rho - 2.02\rho^2 + 3.63\rho^3 - 0.75\rho^4 \qquad\qquad \rho < 1.4 \qquad (3.23)$$

$$U(v\rho) = \begin{cases} (43.6 + 23.5v\rho)\log_{10}(1 + v\rho) - 6 - 6.7v\rho & v\rho < 1 \\ 22v\rho - 20\,\log_{10}(v\rho) - 14.13 & v\rho \geqslant 2 \end{cases} \qquad (3.24)$$

Strictly, both these methods are applicable only to horizontally polarised signals, but measurements [13] have shown that at VHF and UHF they can be applied to vertical polarisation with reasonable accuracy.

With reference to Figure 3.12, the radius of a hill crest may be estimated as

$$r = \frac{2D_s d_{st} d_{sr}}{\alpha(d_{st}^2 + d_{sr}^2)} \qquad (3.25)$$

3.5   MULTIPLE KNIFE-EDGE DIFFRACTION

The extension of single knife-edge diffraction theory to two or more obstacles is not an easy matter. The problem is complicated mathematically but reduces to a double integral of the Fresnel form over a plane above each knife-edge. Solutions for the case of two edges have been available for some time [16,17] and more recently an

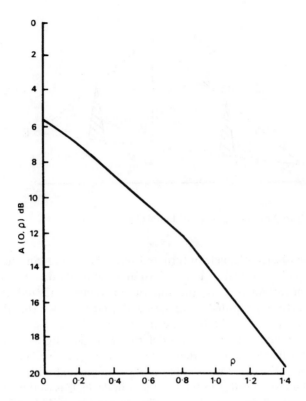

Figure 3.14 The rounded-hill loss $A(0, \rho)$.

expression for the attenuation over multiple knife-edges has been obtained by Vogler [18] using a computer program that handles up to 10 edges by making use of repeated integrals of the error function. Nevertheless, different approximations to the problem have been suggested, and because of the length and mathematical intricacy of the exact solution, their use has become widespread.

3.5.1 Bullington's equivalent knife-edge

In this early proposal [3] the real terrain is replaced by a single 'equivalent' knife-edge at the point of intersection of the horizon ray from each of the terminals as shown in Figure 3.15. The diffraction loss is then computed using the methods described in Section 3.3 using $L = f(d_1, d_2, h)$ where h is the height above the line-of-sight path between the terminals. Bullington's method has the advantage of simplicity but important obstacles below the paths of the horizon rays are sometimes ignored and this can cause large errors to occur. Generally, it underestimates path loss and therefore produces an optimistic estimate of field strength at the receiving point.

3.5.2 The Epstein–Peterson method

The primary limitation of the Bullington method – that important obstacles can be ignored – is overcome by the Epstein–Peterson method [19]; this computes the

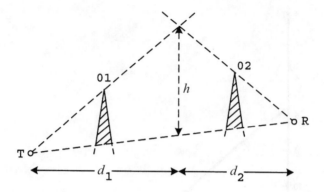

Figure 3.15   The Bullington 'equivalent' knife-edge.

attenuation due to each obstacle in turn and sums them to obtain the overall loss. A three-obstacle path is shown in Figure 3.16 and the method is as follows. A line is drawn from the terminal T to the top of obstruction 02 and the loss due to obstruction 01 is then computed using the standard techniques; the effective height of 01 is h_1, the height above the baseline from T to 02, i.e. $L_{01} = f(d_1, d_2, h_1)$. In a similar way the attenuation due to 02 is determined by joining the peaks of 01 and 03 and using the height above that line as the effective height of 02, i.e. $L_{02} = f(d_2, d_3, h_2)$. Finally, the loss due to 03 is computed with respect to the line joining 02 to the terminal R and the total loss in decibels is obtained as the sum. In the case illustrated, all the obstacles actually obstruct the path, but the technique can also be applied if one or more are subpath obstacles encroaching into the lower-numbered Fresnel zones.

For two knife-edges, comparison of results obtained using this method with Millington's rigorous solution [16] has revealed that large errors occur when the two obstacles are closely spaced. A correction has been derived [16] for the case when the v-parameters of both edges are much greater than unity. This correction is added to

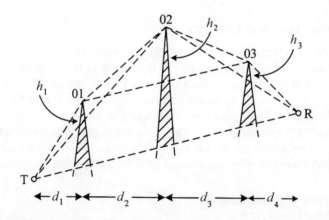

Figure 3.16   The Epstein–Peterson diffraction construction.

the loss originally calculated and is often expressed in terms of a spacing parameter α as

$$L' = 20 \log_{10}(\operatorname{cosec} \alpha) \qquad (3.26)$$

where, for edges 01 and 02,

$$\operatorname{cosec} \alpha = \left(\frac{(d_1 + d_2)(d_2 + d_3)}{d_2(d_1 + d_2 + d_3)} \right)^{1/2}$$

3.5.3 The Japanese method

The Japanese method [20] is similar in concept to the Epstein–Peterson method. The difference is that, in computing the loss due to each obstruction, the effective source is not the top of the preceding obstruction but the projection of the horizon ray through that point onto the plane of one of the terminals. In terms of Figure 3.17 the total path loss is computed as the sum of the losses L_{01}, L_{02} and L_{03}, where $L_{01} = f(d_1, d_2, h_1)$, $L_{02} = f([d_1 + d_2], d_3, h_2)$ and $L_{03} = f([d_1 + d_2 + d_3], d_4, h_3)$, the baseline for each calculation being as illustrated.

It has been shown [21] that the use of this construction is exactly equivalent to using the Epstein–Peterson method and then adding the Millington correction as given by eqn. (3.26). However, although these methods are generally better than Bullington's method, they too tend to underestimate the path loss.

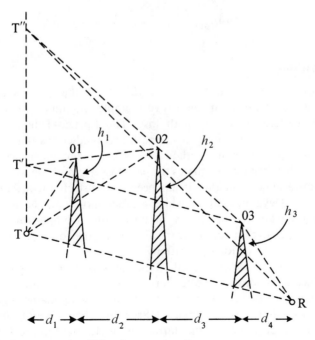

Figure 3.17 The Japanese atlas diffraction construction.

3.5.4 The Deygout method

The Deygout method is illustrated in Figure 3.18 for a three-obstacle path. It is often termed the *main edge* method because the first step is to calculate the v-parameter for each edge alone, as if all other edges were absent, i.e. we calculate the v-parameters for paths T–01–R, T–02–R and T–03–R. The edge having the largest value of v is termed the main edge and its loss is calculated in the standard way. If in Figure 3.18 edge 02 is the main edge, then the diffraction losses for edges 01 and 03 are found with respect to a line joining the main edge to the terminals T and R and are added to the main edge loss to obtain a total.

More generally, for a path with several obstacles, the total loss is evaluated as the sum of the individual losses for all the obstacles in order of decreasing v, as the procedure is repeated recursively. As an illustration, assume that two obstacles exist between the main edge 02 and terminal T. We then have to find which of them is the subsidiary main edge, evaluate its loss and then find the additional loss in the manner indicated above for the remaining obstacle. In practice it is common to compute the total loss as the sum of three components only: the main edge and the subsidiary main edges on either side.

Estimates of the path loss using this method [22] generally show very good agreement with the rigorous approach but they become pessimistic, i.e. overestimate the path loss, when there are multiple obstacles and/or if the obstructions are close together [15]. The accuracy is highest when there is one dominant obstacle. For the case of two comparable obstacles, corrections can be found in the literature [15] using the spacing parameter α described above.

When $v_1 \geqslant v_2$ and $v_1, v_2, (v_2 \operatorname{cosec} \alpha - v_1 \cot \alpha) > 1$ the required correction is

$$L' = 20 \log_{10} \left(\operatorname{cosec}^2 \alpha - \frac{v_2}{v_1} \operatorname{cosec} \alpha \cot \alpha \right) \tag{3.27}$$

3.5.5 Comparison

There are comparisons in the literature [23,24] of the various approximations described above. Bullington's method is very simple, but almost invariably produces results which underestimate the path loss. The Epstein–Peterson and Japanese methods are better but can also provide path loss predictions that are too low. On the other hand, the Deygout method shows good agreement with the rigorous theory for two edges but overestimates the path loss in circumstances where the other methods produce underestimates. It has been demonstrated [24] that the analytical superiority of the Deygout method, which is much more complicated to implement, lies in its relationship to the theory of diffraction. Complication, however, has ceased to be a problem in recent years and computer routines have been written [25] for evaluating the various algorithms.

The pessimism of the Deygout method increases as the number of obstructions is increased, hence calculations are often terminated after consideration of three edges. Giovaneli [26] has devised an alternative technique which remains in good agreement with the values obtained by Vogler [18] even when several obstructions are considered. Giovaneli considers the diffraction angles used in the Deygout method and reasons as follows. In Figure 3.18 the diffraction angle used in calculating the

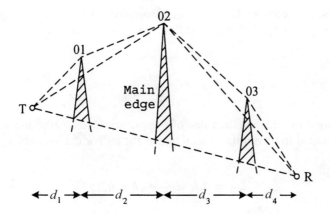

Figure 3.18 The Deygout diffraction construction.

loss due to 02 (the main edge) alone is larger than the angle through which a ray from 01 must actually be diffracted in order to reach the top of 03. The difference increases when the individual obstructions have similar losses, particularly when they are close together. A pessimistic value for the v-parameter is therefore obtained and hence too great a value for the diffraction loss. An approach using a different geometry which maintains the proper diffraction angles is proposed and this is illustrated in Figure 3.19 for the case of two obstacles.

An observation plane RR′ is considered, passing through the terminal R. A source is located at T and we assume that 01 is the principal obstacle (the main edge). A ray from T reaches the observation plane at R″ after diffraction through an angle α_1 at the top of 01. To obtain the parameter v for this obstruction an effective height h_1' is found, given by

$$h_1' = h_1 - \frac{d_1 H_1}{d_1 + d_2 + d_3}$$

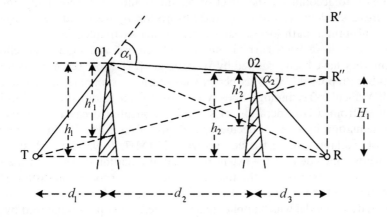

Figure 3.19 The Giovaneli diffraction construction.

and this is used in eqn. (3.4). The loss associated with 02 is then obtained by considering the path 01–02–R with a diffraction angle α_2 and an effective height h_2' given by

$$h_2' = h_2 - \frac{d_3 h_1}{d_2 + d_3}$$

which is also used to calculate the value of v appropriate to obstruction 02. As usual, the losses are added to obtain the overall figure, the individual losses being calculated from

$$L_{01} = f(d_1, d_2 + d_3, h = h_1')$$

and

$$L_{02} = f(d_2, d_3, h = h_2').$$

Giovaneli shows how the method can be extended to paths with several obstacles, including subpath obstacles; he also presents examples to illustrate the technique and demonstrates that this method retains its comparability with results from Vogler's computer program in conditions where the original Deygout method becomes pessimistic.

3.6   PATH LOSS PREDICTION MODELS

The prediction of path loss is a very important step in planning a mobile radio system, and accurate prediction methods are needed to determine the parameters of a radio system which will provide efficient and reliable coverage of a specified service area. Earlier in this chapter we showed that in order to make predictions we need a proper understanding of the factors which influence the signal strength and some of these have already been covered. Other factors exist however, for example in urban areas we have to account for the effect of buildings and other man-made obstacles. In rural areas, shadowing, scattering and absorption by trees and other vegetation can cause substantial path losses, particularly at higher frequencies.

Many studies have been carried out to characterise and model the effects of vegetation; they have been reviewed by Weissberger [27]. More recent measurements have also been reported [28]. Weissberger's conclusions, summarised very briefly by the IEEE Vehicular Technology Society Committee on Radio Propagation [29, p. 11], resulted from a consideration of several exponential decay models based on specific attenuation in terms of decibels per metre of path length and a comparison with sets of available data at frequencies from 230 MHz to 95 GHz. Most reported measurements conclude that the extent of signal attenuation depends on the season of the year, i.e. whether or not the trees are in leaf, the propagation distance within the vegetation and the frequency of the transmitted signal. Weissberger's modified exponential decay model which applies in areas where a ray path is blocked by dense, dry, in-leaf trees is

$$L\,(\text{dB}) = \begin{cases} 1.33F^{0.284}d_{\text{f}}^{0.588} & 14 < d_{\text{f}} \leqslant 400 \\ 0.45F^{0.284}d_{\text{f}} & 0 \leqslant d_{\text{f}} \leqslant 14 \end{cases} \tag{3.28}$$

where L is the loss, F is the frequency (GHz) and d_{f} is the depth of the trees (m).

Other well-known empirical models for the attenuation due to foliage are the ITU Recommendation [30] and the so-called COST235 model [31], which also includes an adjustment to account for seasonal variation in tree condition. The relationship in the ITU Recommendation is

$$L\,(\text{dB}) = 0.2F^{0.3}d_{\text{f}}^{0.6} \tag{3.29}$$

The COST235 model is

$$L\,(\text{dB}) = 26.6F^{-0.2}d_{\text{f}}^{0.5} \tag{3.30a}$$

for vegetation out of leaf, and

$$L\,(\text{dB}) = 15.6F^{-0.009}d_{\text{f}}^{0.26} \tag{3.30b}$$

for vegetation in leaf. In equations (3.29) and (3.30), F is in megahertz and d_{f} is in metres. The seasonal difference is of the order of 4–6 dB. Equation (3.29) has been shown to give good agreement with measurements at 1800 MHz.

Existing prediction models differ in their applicability over different terrain and environmental conditions; some purport to have general applicability, others are restricted to more specific situations. What is certain is that no one model stands out as being ideally suited to all environments, so careful assessment is normally required. Most models aim to predict the median path loss, i.e. the loss not exceeded at 50% of locations and/or for 50% of the time; knowledge of the signal statistics then allows estimation of the variability of the signal so it is possible to determine the percentage of the specified area that has an adequate signal strength and the likelihood of interference from a distant transmitter. The remainder of this chapter is a brief survey of some better-known methods; for details the reader will have to consult the original references.

3.6.1 The Egli model

Following a series of measurements over irregular terrain at frequencies between 90 and 1000 MHz, Egli [32] observed there was a tendency for the median signal strength in a small area to follow an inverse fourth-power law with range from the transmitter, so it was natural for him to produce a model based on plane earth propagation. However, he also observed firstly that there was an excess loss over and above that predicted by eqn. (2.22) and secondly that this excess loss depended upon frequency and the nature of the terrain. It was necessary to introduce a multiplicative factor to account for this, and Egli's model for the median (i.e. 50%) path loss is based on

$$L_{50} = G_{\text{b}}G_{\text{m}}\left(\frac{h_{\text{b}}h_{\text{m}}}{d^2}\right)^2 \beta \tag{3.31}$$

where the suffices b and m refer to base and mobile respectively. β is the factor included to account for the excess loss and is given by

$$\beta = \left(\frac{40}{f}\right)^2 \quad (f \text{ in MHz}) \tag{3.32}$$

from which it is apparent that 40 MHz is the reference frequency at which the median path loss reduces to the plane earth value, irrespective of any variations in the irregularity of the terrain.

In practice, Egli found that the value of β was a function of terrain irregularity, the value obtained from eqn. (3.32) being a median value. He then related the standard deviation of β to that of the terrain undulations by assuming the terrain height to be lognormally distributed about its median value. Hence he produced the family of curves given in Figure 3.20, showing how β departs from its median value at 40 MHz, as a function of terrain factor (dB) and the frequency of transmission.

Note that although Egli's method includes a terrain factor, this is derived empirically and the method does not explicitly take diffraction losses into account. Despite the obvious limitations of Egli's method, it does introduce two factors that will appear several times later. These are the fourth-power law relating path loss to range from the transmitter, and the lognormal variation in median path loss (or signal strength) over a small area.

3.6.2 The JRC method

A method that has been in widespread use for many years, particularly in the UK, is the terrain-based technique originally adopted by the Joint Radio Committee of the

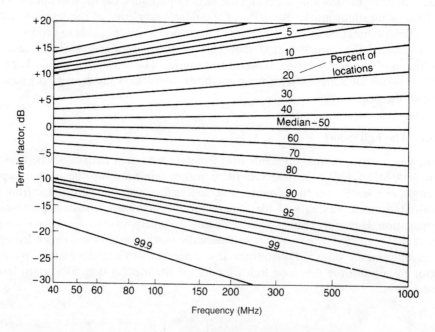

Figure 3.20 The terrain factor for base-to-mobile propagation (after Egli).

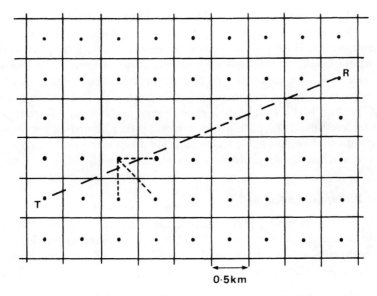

Figure 3.21 Matrix of terrain heights illustrating row, column and diagonal interpolation.

Nationalised Power Industries (JRC). It was described, at various stages of its development by Edwards and Durkin [33] and Dadson [34]. The method uses a computer-based topographic database which, in the original version, provided height reference points at 0.5 km intervals (Figure 3.21). The computer program uses this topographic data to reconstruct the ground path profile between the transmitter and a chosen receiver location using row, column and diagonal interpolation to improve accuracy. The heights and positions of obstructions (including subpath obstacles) are determined. The computer then tests for the existence of a line-of-sight path and whether adequate Fresnel zone clearance exists over that path. If both tests are satisfied, the larger of the free space and plane earth losses is taken, i.e. in these circumstances

$$L = \max(L_F, L_p) \tag{3.33}$$

If no line-of-sight path exists or if there is inadequate Fresnel zone clearance, the computer estimates the diffraction loss L_D along the path and computes the total loss as

$$L = \max(L_F, L_p) + L_D \tag{3.34}$$

In computing the diffraction loss, the computer uses the Epstein–Peterson construction (Section 3.5.2) for up to three edges. If more than three obstructions exist along the path, an equivalent knife-edge is constructed, in the manner suggested by Bullington, to represent all obstructions except the outer two.

In calculating the plane earth path loss, the reference plane for antenna heights is taken as that passing through the foot of the terminal with the lower ground height. This, however, can cause large prediction errors and an alternative was suggested by Fraser and Targett [35]. They determine the effective reflection plane as the line which best fits the terrain between the transmitter and receiver (least mean square error). However, in mobile communications, it is possible for the mobile antenna height to be small with respect to local terrain variations and a negative value of h_m can result. In this case the antenna height above local ground is used. A similar definition was used by Fouladpouri [25]. The principle embodied in the JRC method is still widely used, even though in its original form it generally tended to underestimate the path losses. Unless further databases are available, it has the limitation of being unable to account for losses due to trees and buildings, although approaches to this problem are available [35,36].

3.6.3   The Blomquist–Ladell model

The Blomquist–Ladell model [37] considers the same type of losses as the JRC method but combines them in a different way in an attempt to provide a smooth transition between points where the prediction is based on L_F and those where L_p is used. The basic formulation gives the path loss as

$$L \text{ (dB)} = L_F + [(L_p' - L_F)^2 + L_D^2]^{1/2} \qquad (3.35)$$

In this equation L_p' is a modified plane earth path loss which takes into account factors such as the effect of the troposphere and, over long paths, Earth curvature. The original publication gives an approximate expression for $(L_p' - L_F)$ which has been quoted by Delisle *et al.* [38]. Diffraction losses are estimated using the Epstein–Peterson method.

It is apparent from eqn. (3.35) that over highly obstructed paths, for which $L_D \gg (L_p' - L_F)$, the total loss can be approximated by

$$L = L_F + L_D \qquad (3.36)$$

Conversely, for unobstructed paths L_D approaches zero and the total losses become

$$L = L_p' \qquad (3.37)$$

It is clear that the computed path loss will never be less than L_F and the limiting cases represented by eqns. (3.36) and (3.37) appear intuitively reasonable. The similarity between these equations and eqns. (3.33) and (3.34) for the JRC model is obvious, but there is no theoretical justification whatsoever for combining the losses in the way indicated by eqn. (3.35).

3.6.4   The Longley–Rice models

The Longley–Rice models date from 1968 and the publication of an ESSA technical report [39] which introduced the methods and a computer program for predicting the median path loss over irregular terrain. The method may be used either with detailed terrain profiles for actual paths, or with profiles representative of median terrain characteristics for a given area. It includes estimates of variability with time and location, and a method of computing service probability. The range of system parameters over which the models are applicable are

Transmission frequency (MHz)   20 to 20 000
Range (km)   1 to 2000
Antenna heights (m)   0.5 to 3000
Polarisation   vertical or horizontal

Five further inputs are required by the program:

- Antenna heights above local ground
- Surface refractivity (250 to 400 N-units)
- Effective Earth radius
- Ground constants
- Climate

In addition it is necessary to provide a number of path-specific factors:

- Effective antenna heights
- Horizon distances of the antennas, d_{Lb} and d_{Lm}
- Horizon elevation angles, θ_{eb} and θ_{em}
- Angular distance for a transhorizon path, θ_e
- Terrain irregularity parameter, Δh

The definitions of some of these parameters are illustrated in Figure 3.22. Finally a deployment parameter is necessary to describe the antenna siting as random, careful or very careful. When the terminals of the radio system are on high ground and an effort is made to locate them where the signal is likely to be strong, this is regarded as very careful. If the sites are elevated, but no more than that, the siting is careful; but when the choice of antenna sites is dictated by factors other than radio reception, and there is an equal probability of good or bad reception, siting is said to be random.

If a terrain data map is available so that these parameters can be determined for any particular path, the prediction technique operates in a 'point-to-point' mode. However, if the terrain profile is not available, the report gives techniques for

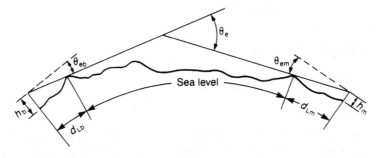

Figure 3.22   Geometry of a transhorizon radio path.

Table 3.1 Estimated values of Δh

Type of terrain	Δh
Water or very smooth plains	0–5
Plains	~ 30
Hills	80–150
Mountains	150–300
Rugged mountains	300–700

estimating these path-related parameters for use in an 'area' mode. The terrain irregularity parameter Δh (roughness indicator) is related to another parameter $\Delta h(d)$, the interdecile range of heights, evaluated at fixed distances along the path. The value of $\Delta h(d)$ increases with path length, and for long enough paths reaches an asymptotic value given by

$$\Delta h(d) = \Delta h[1 - 0.8 \exp(-0.02d)] \tag{3.38}$$

Estimates of Δh for different types of terrain are given in Table 3.1; for suggested values of ground constants see Table 2.1.

Longley and Rice emphasise the importance of effective antenna height in relation to their prediction method. They define these heights with respect to the dominant reflecting plane and discuss how this plane might be determined. In terms of effective height, the smooth earth horizon distance is given, in metric units, by $\sqrt{17h_e}$, so the total distance between the antennas and their respective horizons is $d_{LS} = d_{LSb} + d_{LSm}$.

In the area mode, the prediction technique calls for statistical estimation of the relevant parameters over irregular terrain, and the expressions given are

$$d_{Lb} = d_{LSb} \exp(-0.07\sqrt{\Delta h/h_e})$$
$$d_{Lm} = d_{LSm} \exp(-0.07\sqrt{\Delta h/h_e}) \tag{3.39}$$

where h_e is h_{eb} or h_{em} as appropriate, provided the value is greater than 5 m, otherwise a value of 5 m is used. The total distance between the antennas and their respective horizons is now $d_L = d_{Lb} + d_{Lm}$. The estimate for θ_{eb} (radians) is

$$\theta_{eb} = \frac{0.0005}{d_{LSb}}\left[1.3\left(\frac{d_{LSb}}{d_{Lb}} - 1\right)\Delta h - 4h_{eb}\right] \tag{3.40}$$

and similarly for θ_{em}.

The sum of the elevation angles is

$$\theta_e = \theta_{eb} + \theta_{em} \quad \text{or} \quad -\frac{d_L}{8495}$$

whichever is the larger.

For transhorizon paths, the path length d_i is greater than or equal to d_L, the sum of the horizon distances. The angular distance for a transhorizon path is always positive and is given by

$$\theta = \theta_e + \frac{d_i}{8495} \tag{3.41}$$

where d_i is the length of the transmission path in kilometres.

In computing diffraction loss using this technique it is necessary to express the distances d_1 and d_2 to two ideal (knife-edge) obstacles in terms of the horizon distances. The expressions used for the first and second obstacles are

$$d_1 = \begin{cases} d_{LS} & d' \leqslant d_{LS} \\ d'_1 & d'_1 > d_{LS} \end{cases} \tag{3.42}$$

where d'_1 and d_2 can be expressed in kilometres as:

$$d'_1 = d_L + 0.5(72\,165\,000/f_c)^{1/3}$$

and

$$d_2 = d_1 + (72\,165\,000/f_c)^{1/3} \tag{3.43}$$

The v-parameters appropriate to obstacles at distances d_1 and d_2 are then computed from

$$\begin{aligned} v_{b,i} &= 1.2915\theta_{ebi}[f_c d_{Lb}(d_i - d_L)/(d_i - d_{Lm})]^{1/2} \\ v_{m,i} &= 1.2915\theta_{emi}[f_c d_{Lm}(d_i - d_L)/(d_i - d_{Lb})]^{1/2} \end{aligned} \tag{3.44}$$

with $i = 1$ and 2.

The diffraction losses A_1 and A_2 are then estimated using

$$A_1 \text{ (dB)} = A(v_{b,1}) + A(v_{m,1})$$
$$A_2 \text{ (dB)} = A(v_{b,2}) + A(v_{m,2})$$

with the approximations for $A(v)$ given by eqn. (3.11).

The explicit expression used to determine the diffraction loss L_D to a mobile located a distance d from the base station is

$$L_D \text{ (dB)} = dm_d + A_0 \tag{3.45}$$

where

$$m_d = \frac{A_1 - A_2}{d_2 - d_1} \tag{3.46}$$

and

$$A_0 = A_{fo} + A_2 - d_2 m_d \tag{3.47}$$

Equation (3.47) includes an empirical clutter factor A_{fo}, estimated as

$$A_{fo} \text{ (dB)} = \min(A'_{fo}, 15) \tag{3.48}$$

where A'_{fo} is given by

$$A'_{fo} \text{ (dB)} = 5 \log_{10}[1 + h_m h_b f_c \sigma(d_{LS}) \times 10^{-5}] \tag{3.49}$$

and $\sigma(d_{LS})$ in metres is given by

$$\sigma(d_{LS}) = 0.78h(d) \exp\{-0.5[\Delta h(d)]^{1/4}\} \tag{3.50}$$

Delisle *et al.* [38] state that this model gives reasonably accurate predictions and is not restricted to short paths.

To predict the median transmission loss, the reference attenuation below free space is first calculated as a continuous function of distance from the transmitter: the free space loss at each distance is then added. The reference attenuation is computed in three different ways depending on the distance from the transmitter:

- For distances less than the smooth earth horizon distance d_{LSb}, the computation is based on two-ray reflection theory (plane earth) and an extrapolated value of diffraction loss.
- For distances just beyond the horizon from d_{LS} to d_X – d_X being the distance where diffraction and scatter losses are equal – the reference attenuation is a weighted average of knife-edge and smooth earth diffraction calculations. The weighting factor is a function of frequency, terrain irregularity and antenna heights. For highly irregular terrain the horizon obstacles as seen from the terminals are considered as sharp ridges and the diffraction loss is calculated over a double knife-edge path using the Epstein–Peterson approximation.
- For transhorizon paths where the range is greater than d_X, the reference attenuation is calculated either as a diffraction loss or as a forward scatter loss, whichever is the smaller.

Since the original publication there have been several revisions and modificaions of the Longley–Rice model and some corrections have been made. They are described in a 1982 report [40] and a subsequent memorandum [41]. An extensive summary of the method is given in reference 29. One significant development, relevant to mobile radio propagation, has been the introduction of an urban factor (UF) used to make predictions in urban areas [42]. It has been derived by comparing predictions from the original model with a curve given by Okumura (Chapter 4) for urban areas. The value of UF is given by

$$\text{UF (dB)} = 16.5 + 15 \log_{10}(f/100) - 0.12d \tag{3.51}$$

where f is in megahertz and d in kilometres.

The model allows a small adjustment for changes in climate and further allowances can be made to introduce time and location variability. Apart from equation (3.51) it does not contain any specific provision for corrections due to buildings or foliage which may exist in the immediate vicinity of the mobile.

3.6.5   CCIR methods

Figure 3.23 shows an example of the field strength prediction curves published in CCIR literature. These curves are based on statistical analysis of a considerable amount of experimental data collected in many countries. They are applicable over the kind of rolling hilly terrain found in many parts of Europe and North America for which the interdecile terrain irregularity parameter Δh is typically 50 m. Reference field strength curves are given, and to determine the field strength in

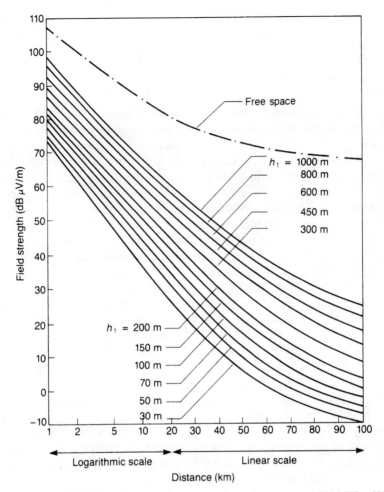

Figure 3.23 CCIR field strength prediction curves for urban areas at 900 MHz: 50% of the time, 50% of locations. Field strength (dBμV/m) for 1 kW ERP, $h_2 = 1.5$ m.

any specific situation it is necessary to look up another curve which gives a correction related to the value of Δh. Values are given for 50% of locations and 50% of the time, although in mobile radio applications it is only the spatial variability that is usually relevant. The reference curves are given for a mobile antenna height of 1.5 m and base station antenna heights between 30 and 1000 m. It is implicitly assumed that the values of field strength measured in a small area will be lognormally distributed around the predicted median value, i.e. the field strength in decibels follows a Gaussian (normal) distribution. Standard deviations, expressed as functions of distance and terrain irregularity, can be used to estimate values for other quantiles of interest, e.g. 5%, 10%, 90% and 95%.

Although CCIR curves are generally regarded as authoritative, it may be deduced from the literature [39,43,44] that the single parameter Δh is inadequate to define the required correction factor with sufficient accuracy. In addition, terrain variations in the immediate vicinity of the mobile are not explicitly taken into account and for any specific location it is not unusual to find a prediction error of about 10 dB. A more accurate method is therefore required.

The clearance angle method

This method, first proposed by the European Broadcasting Union (EBU) and later adopted by the CCIR, is an attempt to provide the extra precision in a small area. The principle is to retain the CCIR reference field strength curves, hence the simplicity of application, but to improve the prediction accuracy by taking into account the terrain variations in a small area surrounding the receiver, since they are not adequately reflected in the global value of Δh for the path concerned. These terrain effects are taken into account by a correction based on a terrain clearance angle. This angle is meant to be representative of those angles in the receiving area which are measured between the horizontal through the receiver and a line that clears all obstacles within 16 km in a direction towards the transmitter. Figure 3.24(a) shows the idea and indicates the sign convention.

The two curves in Figure 3.24(b) give values for the required correction factor in terms of the clearance angle; the correction should be added to the field strength obtained from the CCIR reference curves [45]. The correction curves have been derived by reference to calculations of field strength made for over 200 paths in Europe using a rather involved, but precise point-to-point method developed in the UK [46] and their comparison with results from the CCIR method. Because of the relatively small number of paths involved, correction factors for clearance angles outside the range $-5°$ to $+0.5°$ are not given by Figure 3.24(b). Values can be obtained, however, by linear interpolation from Figure 3.24(b) to limiting values of 30 dB (VHF) and 40 dB (UHF) at $+1.5°$ and to -40 dB (VHF and UHF) at $-15°$, subject always to the condition that the free space field strength is not exceeded. The clearance angle method has been shown by experiment [44] to compare favourably with several other methods, including Longley–Rice.

3.6.6   Other methods

Within a limited space it is impossible to cover all the methods for predicting propagation losses over irregular terrain; we have only covered a selection of the more popular or better-known methods. But other methods do exist, some of them extremely simple, others much more complicated.

The Longley–Rice model, for example, is an outgrowth of NBS Technical Note 101 [47] which consists of curves, theoretical equations and empirical formulas for predicting cumulative probability distributions of propagation loss to small areas, for a wide range of frequencies over various types of terrain in different climates. Application of the original technique requires very careful and detailed calculations, having found the curves that are applicable to the situation under consideration.

The terrain-integrated rough earth model (TIREM) also uses ideas presented in Technical Note 101. It exists as a computer program developed by the Electromagnetic Compatibility Analysis Center (ECAC) and it predicts propagation loss between two points, using as inputs the transmission frequency, atmospheric

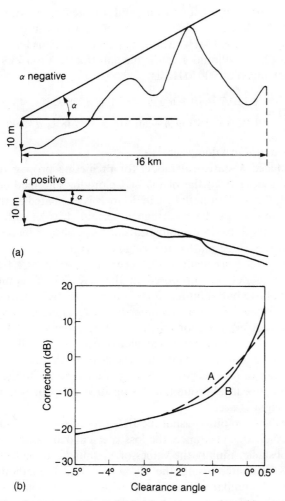

Figure 3.24 The clearance angle method: (a) sign convention, (b) correction factors for VHF (A) and UHF (B).

and ground constants and characteristics of the terrain profile [48]. A digitised terrain database is used to provide profile information and the program then selects an appropriate algorithm to calculate the path loss. TIREM is one of a series of point-to-point propagation models in the Master Propagation System (MPS11) developed in the USA and is the preferred model. An extended description is given in reference 29 and a program tape is available.

The Federal Communications Commission (FCC) uses a set of curves giving field strength F as a function of distance for propagation under average terrain conditions. Those described by Carey [49] give $F(50,50)$ and $F(50,10)$ for a mobile antenna height of 1.8 m, base station antenna heights in the range 30–1500 m above average terrain height and distances up to 130 km and 240 km for the $F(50,50)$ and $F(50,10)$ curves respectively. Note that $F(a,b)$ represents the field strength exceeded at a% of locations for b% of the time, but at distances involved in the land mobile service only spatial variation is significant and the temporal variation can be considered as $b = 100\%$. The curves have been based on CCIR information relevant to the frequency range 450–1000 MHz. Mathematical expressions have been fitted to the published curves [29, p. 20] and give the median transmission loss (presumably between isotropic antennas) at 900 MHz as

$$L \text{ (dB)} = \begin{cases} 110.7 - 19.1 \log h + 55 \log d & 8 \leqslant d < 48 \\ 91.8 - 18 \log h + 66 \log d & 48 \leqslant d < 96 \end{cases} \qquad (3.52)$$

where h is in metres and d in kilometres.

Murphy [50] produced a statistical model for predicting propagation loss over irregular terrain using data from the plains of Colorado. The method follows the Egli pattern in which the median path loss in decibels is estimated as the sum of the plane earth loss L_p and a loss due to irregular terrain. Kessler and Wiggins [51] developed a further method, based on statistical curves published by the FCC, which allows coverage contours to be plotted for irregular terrain.

Following a series of measurements at distances up to 40 km at a frequency near 140 MHz, a different model was proposed in the UK [25]. It combines several techniques described above but in outline it is similar to the JRC method. Firstly the terrain profile is determined from a computer-based terrain map and the terrain heights along the path are adjusted for Earth curvature using $k = 4/3$. Secondly the effective base station antenna height is calculated from the median deviation of heights along the path about a straight line drawn through the bases of the two terminals. If this median value is negative, its magnitude is added to the height of the base station antenna above local ground; if it is positive, its value is ignored and the structural antenna height is used.

Diffraction loss is then computed using the JRC procedure: if an LOS path exists with inadequate Fresnel zone clearance, the loss is calculated as the loss due to the dominant subpath obstacle. Finally the values of L_F and L_p are calculated and the overall path loss is estimated using the Blomquist–Ladell method (3.35) with $L'_p = L_p$. This method compares favourably with the JRC and Longley–Rice methods. Another method using a computer program in a manner similar to the JRC method has been published by Palmer of the Canadian Communications Research Centre (CRC) as part of a more comprehensive computer program for coverage prediction [52,53].

3.7 DISCUSSION

Several prediction methods have been described in this chapter. They all aim to predict the median signal strength either at a specified receiving point or in a small area. Receiving point methods are needed for point-to-point links whereas small area methods are more useful for base-to-mobile paths where the precise location of the receiver is not known. Some of the methods have been available for many years and have stood the test of time, possibly with modification and updating. They differ widely in approach, complexity and accuracy, and quite often the application of two different methods to precisely the same problem will yield results which differ by a wide margin, thereby producing a degree of uncertainty and lack of confidence on the part of the user.

One fact is quite clear even after many years of research and development: when it comes to accuracy, no one method outperforms all others in all conditions. In any case the engineer may well be prepared to trade accuracy for simplicity and ease of application. At some locations, especially close to the transmitter, accurate prediction of signal strength is a secondary consideration; often the primary concern is to predict the limits of the coverage area of a given base site and to identify the inevitable 'black spots' that occur; other objectives may be to predict the probability of interference between services or to plan a frequency assignment strategy for radio channels. Choosing a method appropriate to the specific problem under consideration is a vital step in reaching a valid prediction.

In general, the models described are a mixture of empiricism and the application of propagation theory. The empirical approach relies on fitting curves or analytical expressions to sets of measured data and has the advantage of implicitly taking into account all factors (known and unknown). However, a purely empirical model must always be subjected to stringent validation by testing it on data sets collected at locations and transmission frequencies other than those used to produce the model in the first place. Theoretical equations such as those for the free space or plane earth propagation loss often underpin models which include additional empirical (or semi-empirical) factors to account for diffraction loss, Earth curvature, atmospheric effects or vegetation loss.

In deriving prediction models, and in considering the applicability of a particular model to a specific problem, it is prudent to consider the input data required by the model, the availability and accuracy of that data and the effect on the prediction if only partial (or crudely defined) input data is available. In this context it is useful to take terrain data as an example. It has recently become possible to obtain highly accurate terrain data over large areas from satellite imaging and stereo aerial photography, and this can readily be presented in a form suitable for propagation prediction. A detailed path profile can always be derived from a map for a specific path, but in general it is necessary to rely on topographic databases which give terrain height information at regular intervals (Section 3.6.2).

Interpolation between these points is required if calculation of diffraction loss along the profile forms part of the prediction procedure. Older databases with heights based on 500 m intervals are now being replaced with much more accurate 50 m or even 10 m information.

Inherent in many methods of predicting path loss is the assumption that only the terrain directly between the two points concerned is relevant to the calculation, and off-path obstacles play no part. This was always known to be invalid, and when basing calculations on a representative point chosen within a small area, variations about the calculated value within the area concerned should be expected.

The use of diffraction calculations based on knife-edge theory to account for losses caused by real obstacles and the empirical methods of estimating losses over paths with many obstructions are demonstrably unjustifiable on any grounds other than that they provide a reasonably accurate, simple and efficient solution. More sophisticated techniques such as UTD, based on representing obstacles by wedges or cylinders, can be used to improve accuracy. In this context also, the diffraction loss should theoretically be added to the free space path loss L_F; nevertheless, some models call for it to be added to the plane earth path loss L_p if $L_p > L_F$.

A detailed path profile is not always required. In the Egli model, for example, only the interdecile height Δh is required and although an estimated value of Δh can be used (Table 3.1) a more accurate value for a given region can always be found if a computerised terrain database is available. However, in all cases, lack of detailed knowledge about other features of the terrain, such as buildings and trees which can influence the signal, cause uncertainty and lead to the introduction of empirical clutter factors. Nowadays, integrated ray tracing methods using topographical and environmental databases are being developed [54] and these allow significant off-path obstacles (which act as scatterers or reflectors) to be identified and their effect taken into account.

In view of this, it is interesting to speculate on the relationship between the availability of more detailed terrain databases and improved prediction accuracy. In many cases the database is used only to locate the heights and positions of the peaks of obstacles, after which the terrain is often represented by a series of knife-edges of suitable height located at the positions defined. This is crude to say the least, but while it might be tempting to suggest that only marginal improvement would come from using a database with terrain heights at say 50 m intervals rather than 500 m, this is not always the case.

Over long paths, and using this particular prediction methodology, the argument might hold true, but for short paths where only a few terrain height points are available it becomes much less convincing. In such cases it is clearly advantageous to have more detail and in particular, it must be conceded that near the mobile, when the antenna is low, accurate terrain height information can be very important. Short paths are of increasing importance with the growth in cellular radio and the interest in microcells, but small cells exist mainly in built-up areas where losses due to buildings are often just as important, if not more so, than losses caused by terrain variations. Propagation models specifically intended for built-up areas have not been discussed in this chapter, but it is clear that a detailed terrain database and accurate environmental information are required to deal with this situation.

As far as specific models are concerned, some (e.g. Egli) were originally intended for manual use, whereas others (e.g. JRC, Longley–Rice) were developed as computer programs. Some use terrain databases to produce detailed path profiles, others merely use them to derive a terrain parameter such as the interdecile height Δh. All the methods, however, can easily be implemented as computer programs and can make use of detailed terrain data when available. Most of them are capable of producing plots of field strength contours that can be used as map overlays and can deal with system performance calculations for specific applications since they can cope with problems of interference as well as problems of coverage. Using a computer it is easy to compare signal strengths at a given location from more than one transmitter site.

Some of the methods discussed in this chapter, together with others that will be described later, were compared a few years ago [55] for the type of terrain covered, the form of prediction, the ease of implementation and the accuracy. Accuracy may be determined principally from the extent to which the particular method concerned accounts for terrain irregularities. By this criterion the CRC [52] and JRC methods, which were deemed to have a high degree of complexity, came out best, followed by Longley–Rice and Kessler–Wiggins. On the other hand, the Egli and Murphy techniques were simple but they tended to take an 'average' picture of the terrain over a wide area and therefore their accuracy suffered. These were tentative conclusions; although complexity and accuracy generally increase together, the real criterion is often how well the prediction procedure deals with the terrain features in the area of interest.

It was implied at the end of Chapter 1 that there was little point in constructing highly accurate databases if available propagation methods were unable to use the information to good effect. Nevertheless, the availability of accurate geographical data is a precursor to the development of improved propagation models, and in recent years substantial strides have been made in this respect. Moving away from the traditional way of extracting geographical information from maps and replacing it with the use of high-resolution data obtained by remote-sensing techniques, and the extensive use of digital storage and dissemination methods, is just one example of the progress that has been made. The relationship between accuracy of data and accuracy of prediction is clear in general, but unestablished in any mathematical sense.

It is obvious that high-resolution databases have the potential to improve predictions but the extent of the improvement remains to be determined. There is a clear need to develop new prediction techniques which can use the available data effectively. In this context, semideterministic ray tracing methods threaten to dominate in the immediate future, especially for the shorter paths that are so relevant to cellular radio systems. An exact calculation of the signal strength at a specific receiving point in the mobile radio scenario is not a realistic endeavour; the scattering, reflecting and diffracting surfaces that dominate the propagation mechanism can never be described with sufficient accuracy.

Finally, it is worth expanding slightly on the question of signal variability with location. This has been alluded to several times without any quantitative description, although it is intuitively obvious that the median signal will vary from place to place within any small area for which a prediction has been made by taking a representative point. Observations show that, statistically, this variability follows a lognormal distribution with a standard deviation which depends on the roughness of the terrain but is typically of the order of 4–10 dB. This will be discussed later.

In general, the prediction methods give only the median value of the path loss and do not deal with the subject of variability either implicitly or explicitly. The notable exception is the Egli technique, which contains variability as a built-in but empirical feature. In practice, however, a quantitative measure of signal variability is essential. It allows us to estimate the percentage of a given area that has an adequate signal strength and the probability of interference from a distant transmitter. It could be claimed that an estimate of the variability is no less important than a prediction of the median signal strength itself.

Measurements in a wide variety of terrain situations have indicated that the signal variability increases with frequency and terrain irregularity. Egli [32] suggested that in rural areas the standard deviation of received signal level is related to the transmission frequency by

$$\sigma \text{ (dB)} = 5 \log_{10} f_c + 2 \tag{3.53}$$

where f_c is the frequency in megahertz.

Longley [56] cited the results of Hufford and Montgomery [57] and hence suggested a slightly different relationship:

$$\sigma \text{ (dB)} = 3 \log_{10} f_c + 3.6 \tag{3.54}$$

There seems to be no real evidence that σ is a function of either path length or antenna height.

The CCIR [46,58] has recognised that variability is a function of frequency and terrain irregularity. It is recommended that for frequencies below 250 MHz a value of $\sigma = 8$ dB should be used but that above 450 MHz values of 10, 15 and 18 dB should be used in average, hilly and mountainous terrain respectively. In the context of terrain irregularity, Longley [56] suggests the use of a parameter that combines the terrain irregularity factor Δh with the transmission wavelength or frequency, and which increases if Δh and/or f_c increase. The dimensionless parameter $\Delta h/\lambda$ fits this requirement, and using measured results Longley has determined best-fit expressions as

$$\sigma \text{ (dB)} = \begin{cases} 6 + 0.55(\Delta h/\lambda)^{1/2} - 0.004(\Delta h/\lambda) & \text{for } \Delta h/\lambda < 4700 \\ 24.9 & \text{for } \Delta h/\lambda > 4700 \end{cases} \tag{3.55}$$

REFERENCES

1. Griffiths J. (1987) *Radio Wave Propagation and Antennas: An Introduction*. Prentice Hall, London.
2. Jordan E.C. and Balmain K.G. (1968) *Electromagnetic Waves and Radiating Systems*. Prentice Hall, New York.
3. Bullington K. (1947) Radio propagation at frequencies above 30 Mc. *Proc IRE*, **35**(10), 1122–36.
4. Lee W.C.-Y. (1983) *Mobile Communications Engineering*. McGraw Hill, New York.
5. Anderson L.J. and Trolese L.G. (1958) Simplified method for computing knife-edge diffraction in the shadow region. *IRE Trans.*, **AP6**, 281–6.
6. Keller J.B. (1962) Geometrical theory of diffraction. *J. Opt. Soc. Am.*, **52**, 116–30.
7. Kouyoumjian R.G. and Pathak P.H. (1974) A uniform geometrical theory of diffraction for an edge in a perfectly conducting surface. *Proc. IEEE*, **62**(11), 1448–61.
8. Luebbers R.J. (1984) Finite conductivity uniform GTD versus knife-edge diffraction in the prediction of propagation path loss. *IEEE Trans.*, **AP32**(1), 70–6.
9. Maliuzhinets G.D. (1958) Excitation, reflection and emission of surface waves from a wedge with given face impedances. *Sov. Phys. Dokl.*, **3**, 752–5.
10. McNamara D.A., Pistorius C.W.I. and Malherbe J.A.G. (1990) *Introduction to the Uniform Geometrical Theory of Diffraction*. Artech House, London.
11. Anderson H.R. (1998) Building corner diffraction measurements and predictions using UTD. *IEEE Trans.*, **AP46**, 292–3.
12. Pathak P.H. *et al.* (1988) A uniform GTD analysis of the diffraction of electromagnetic waves by a smooth convex surface. *IEEE Trans.*, **AP28**, 631–42.
13. Hacking K. (1968) *Propagation over rounded hills*. BBC Research Report RA-21.
14. Dougherty H.T. and Maloney L.J. (1964) Applications of diffraction by convex surfaces to irregular terrain situations. *Radio Science*, **68D**(2), 284–305.
15. Causebrook J.H. and Davies B. (1971) *Tropospheric radio wave propagation over irregular*

terrain: the computation of field strength for UHF broadcasting. BBC Research Report 43.

16. Millington G., Hewitt R. and Immirzi F.S. (1963) Double knife-edge diffraction in field-strength prediction. *IEE Monograph 507E*, pp. 419–29.

17. Furutzu K. (1963) On the theory of radiowave propagation over inhomogeneous earth. *J. Res. NBS*, **67D**, 39–62.

18. Vogler L.E. (1981) *The attenuation of electromagnetic waves by multiple knife-edge diffraction.* NTIA Report 81–86. Available as PB82-139239, National Technical Information Service, Springfield VA.

19. Epstein J. and Peterson D.W. (1953) An experimental study of wave propagation at 850 MC. *Proc. IRE*, **41**(5), 595–611.

20. *Atlas of radio wave propagation curves for frequencies between 30 and 10,000 Mc/s* (1957) Radio Research Lab, Ministry of Postal Services, Tokyo, Japan, pp. 172–9.

21. Hacking K. (1966) Approximate methods for calculating multiple-diffraction losses. *Electron. Lett.*, **2**(5), 179–80.

22. Deygout J. (1966) Multiple knife-edge diffraction of microwaves. *IEEE Trans.*, **AP14**(4), 480–9.

23. Wilkerson R.E. (1966) Approximations to the double knife-edge attenuation coefficient. *Radio Science*, **1**(12), 1439–43.

24. Pogorzelski R.J. (1983) A note on some common diffraction link loss models. *Radio Science*, **17**, 1536–40.

25. Fouladpouri S.A.A. (1988) An investigation of computerised prediction models for mobile radio propagation over irregular terrain. PhD thesis, University of Liverpool, UK.

26. Giovaneli C.L. (1984) An analysis of simplified solutions for multiple knife-edge diffraction. *IEEE Trans.*, **AP32**(3), 297–301.

27. Weissberger, M.A. (1983) *An initial critical summary of models for predicting the attenuation of radio waves by trees.* ESD-TR-81-101. EMC Analysis Center, Annapolis MD.

28. Vogel, W.J. and Goldhirsch J. (1986) Tree attenuation at 869 MHz derived from remotely piloted aircraft measurements. *IEEE Trans.*, **AP34**, 1460–4.

29. IEEE Vehicular Technology Society Committee on Radio Propagation (1988) Coverage prediction for mobile radio systems operating in the 800/900 MHz frequency range. *IEEE Trans.* **VT37**(1); special issue on mobile radio propagation.

30. CCIR (1986) *Influences of terrain irregularities and vegetation on tropospheric propagation.* CCIR Report 235-6, Geneva.

31. *Radiowave propagation effects on next generation terrestrial telecommunication services* (1996) COST235, final report.

32. Egli J.J. (1957) Radio propagation above 40 Mc over irregular terrain. *Proc, IRE*, **45**(10), 1383–91.

33. Edwards R. and Durkin J. (1969) Computer prediction of service area for VHF mobile radio networks. *Proc. IEE*, **116**(9), 1493–500.

34. Dadson C.E. (1979) Radio network and radio link surveys derived by computer from a terrain data base. *NATO-AGARD Conference Publication CPP-269.*

35. Frazer E.L. and Targett D.J. (1985) A comparison of models for the prediction of service area of cellular radio telephone sites. *Proc. ICAP'85 (IEE Conference Publication 248)*, pp. 390–94.

36. Ibrahim M.F., Parsons J.D. and Dadson C.E. (1983) Signal strength prediction in urban areas using a topographical and environmental data base. *Proc. ICC'83*, pp. A3.5.1 to A3.5.4.

37. Blomquist A. and Ladell L. (1974) Prediction and calculation of transmission loss in different types of terrain. *NATO-AGARD Conference Publication CP-144*, Res. Inst. Nat. Defense Dept 3, S-10450, Stockholm 80, pp. 32/1 to 32/17.

38. Delisle G.Y., Lefevre J.P., Lecours M. and Chouin J.Y. (1985) Propagation loss prediction: a comparative study with application to the mobile radio channel. *IEEE Trans.*, **VT34**(2), 86–95.

39. Longley A.G. and Rice P.L. (1968) *Prediction of tropospheric radio transmission over irregular terrain; a computer method – 1968.* ESSA Technical Report ERL 79-ITS67.

40. Hufford G.A., Longley A.G. and Kissick W.A. (1982) *A guide to the use of the ITS irregular terrain model in the area prediction mode.* NTIA Report 82-100.

41. Hufford G.A. Memorandum to users of the ITS irregular terrain model, 30 January 1985.

42. Longley A.G. (1978) *Radio propagation in urban areas.* Office of Telecommunications Report OT78-144.

43. Thelot B. (1981) Method of calculating the propagation parameters used in the VHF and UHF bands. *EBU Review*, **186**, 76–81.

44. Paunovic D.S., Stojanovic Z.D. and Stojanovic I.S. (1984) Choice of a suitable method for the prediction of the field strength in planning land mobile radio systems. *IEEE Trans.*, **VT33**(4), 259–65.

45. CCIR (1983) Methods and statistics for estimating field strength values in the land mobile services using the frequency range 30 MHz to 1 GHz. *CCIR XV Plenary Assembly*, Geneva, Report 567, Vol. 5.

46. Causebrook J.H. and King R.W. (1974) *Computer programs for UHF co-channel interference prediction using a terrain data bank.* BBC Research Report 1974/6.

47. Rice P.L., Longley A.G., Norton K.A. and Barsis A.P. (1965) *Transmission loss predictions for tropospheric communication circuits.* NBS Technical Note 101; issued May 1965, revised May 1966 and Jan 1967. US Government Printing Office, Washington DC.

48. *Master Propagation System (MPS11) User's Manual.* US Department of Commerce, NTIS Accession No. PB83-178624.

49. Carey R. (1964) *Technical factors affecting the assignment of frequencies in the domestic public land mobile radio service.* Federal Communications Commission, Washington DC, Report R-6406.

50. Murphy J.P. (1970) Statistical propagation model for irregular terrain paths between transportable and mobile antennas. *NATO-AGARD Conf. Proc.*, **70**, 49/1 to 49/20.

51. Kessler W.J. and Wiggins M.J. (1977) A simplified method for calculating UHF base-to-mobile statistical coverage contours over irregular terrain. *Proc. 27th IEEE Veh. Tech. Conf.*, pp. 227–36.

52. Palmer F.H. (1978) The CRC VHF/UHF propagation prediction program: description and comparison with field measurements. *NATO-AGARD Conf. Proc.*, **238**, 49/1 to 49/15.

53. Palmer F.H. (1979) VHF/UHF path-loss calculations using terrain profiles deduced from a digital topographic data base. *NATO-AGARD Conf. Proc.*, **269**, 26/1 to 26/11.

54. Tameh E.K. (1999) The development and evaluation of a deterministic mixed cell propagation model based on radar cross-section theory. PhD thesis, University of Bristol.

55. Aurand J.F. and Post R.E. (1985) A comparison of prediction methods for 800 MHz mobile radio propagation. *IEEE Trans.*, **VT34**(4), 149–53.

56. Longley A.G. (1976) *Location variability of transmission loss for land mobile and broadcast systems.* Office of Telecommunications Report OT76-87, NTIS Accession No. PB–254472.

57. Hufford G.A. and Montgomery J.L. (1966) *On the statistics of VHF field strength measurements using low antenna heights.* NBS Report 9223, NTIS Accession No. AD487672.

58. CCIR (1978) VHF and UHF propagation curves for the frequency range from 30 MHz to 1000 MHz. *CCIR XIV Plenary Assembly*, Kyoto, Recommendation 370-3, Vol. 5.

Chapter 4

Propagation in Built-up Areas

4.1 INTRODUCTION

Having looked at how irregular terrain affects VHF and UHF radio wave propagation and the effects of multipath, we are now in a position to discuss propagation in built-up areas. This chapter will deal principally with propagation between base stations and mobiles located at street level; propagation into buildings and totally within buildings will be discussed later. Although losses due to buildings and other man-made obstacles are of major concern, terrain variations also play an important role in many cases.

Within built-up areas, the shadowing effects of buildings and the channelling of radio waves along streets make it difficult to predict the median signal strength. Often the strongest paths are not the most obvious or direct ones and the signal strength in streets that are radial or approximately radial with respect to the direction of the base station often exceeds that in streets which are circumferential. Figure 4.1 is a recording of the signal envelope measured in a vehicle travelling along two city streets. For the first 65 m the street is radial; the Rayleigh fading is clearly observed along with the increase in mean level at intersections. The vehicle then turned into a circumferential street, where the mean signal strength is a little lower and the fading pattern is somewhat different. In suburban areas there are fewer large buildings and the channelling effects are less apparent. However foliage effects, often negligible in city centres, can be quite important. Generally, the effects of trees are similar to those of buildings, introducing additional path losses and producing spatial fading.

Estimation of the received mobile radio signal is a two-stage process which involves predicting the median signal level in a small region of the service area and describing the variability about that median value. Quantifying the extent to which the signal fluctuates within the area under consideration is also a problem in which there are two contributing factors. Short-term variations around the local mean value will be discussed in Chapter 5 and are commonly termed *multipath, fast fading* or *Rayleigh fading*. Longer-term variations in the local mean are caused by gross variations in the terrain profile between the mobile and the base station as the mobile moves from place to place and by changes in the local topography. They are often

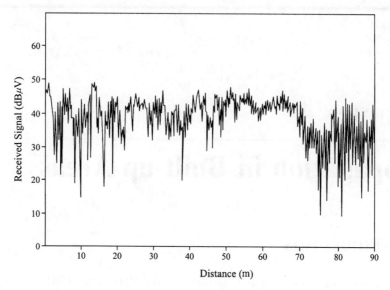

Figure 4.1   Recording of signal strength in an urban area.

termed *slow fading* and, as mentioned in Chapter 3, the characteristics can be described by a lognormal statistical distribution.

4.2   BUILT-UP AREAS: A CLASSIFICATION PROBLEM

The propagation of radio waves in built-up areas is strongly influenced by the nature of the environment, in particular the size and density of buildings. In propagation studies for mobile radio, a qualitative description of the environment is often employed using terms such as rural, suburban, urban and dense urban. Dense urban areas are generally defined as being dominated by tall buildings, office blocks and other commercial buildings, whereas suburban areas comprise residential houses, gardens and parks. The term 'rural' defines open farmland with sparse buildings, woodland and forests. These qualitative descriptions are open to different interpretations by different users; for example, an area described as urban in one city could be termed suburban in another. This leads to doubts as to whether prediction models based on measurements made in one city are generally applicable elsewhere. There is an obvious need to describe the environment quantitatively to surmount the unavoidable ambiguity embodied in the qualitative definitions which can arise from cultural differences and subjective judgement.

To illustrate the argument, Figure 4.2 shows building height histograms for two 500 m Ordnance Survey (OS) map squares in central London. In qualitative terms both areas would be classed as dense urban. It is obvious that the percentage of square A occupied by tall buildings is much greater than the percentage of square B, so a higher path loss value would be expected. In practice it is higher by 8–10 dB [1].

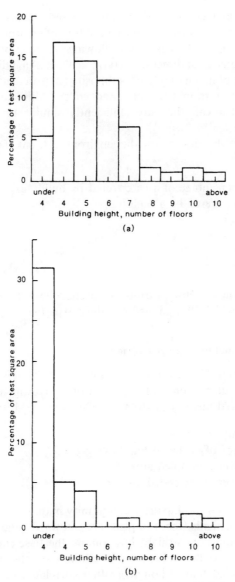

Figure 4.2 Building height histograms for central London: (a) Soho area, (b) Euston area.

4.2.1 A classification approach

In situations of practical interest, the environment can be regarded as composed of many different mutually independent scatterer classes or types. Features such as buildings and trees are common and a town might appear as a random collection of buildings, each building being a scatterer. Likewise a forest appears as a random collection of trees. If the statistical properties of groups or clusters of individual scatterers are known, as well as the scatterer population per group, then it is possible to derive quantitative descriptions of the environment using the statistics [2].

An environment classification method can be based on this approach. Any given mobile radio service area can be viewed as a mixture of environments (e.g. a mixture of urban, suburban and rural localities). Following OS descriptions, the service area can be divided into squares of dimension 500 m × 500 m. An individual square is then regarded as a sample of an ensemble of composite environments with the ensembles described by different terrain type and land cover. Although sample cells in an ensemble are not identical, they are sufficiently similar to allow a meaningful statistical description.

When considering the effects of the environment, six factors are useful in classifying land usage:

- Building density (percentage of area covered by buildings)
- Building size (area covered by a building)
- Building height
- Building location
- Vegetation density
- Terrain undulations

Using some or all of these factors, various researchers have devised classifications for the environments in which they carried out their experiments.

4.2.2   Classification methods: a brief review

Kozono and Watanabe [3] working in Tokyo in 1977 attempted a quantitative description of the urban environment as part of their investigation into the influence of buildings on received mean field strength. They proposed four parameters:

- Area factor of occupied buildings, α
- Extended area factor of occupied buildings, α'
- Building volume over a sampled area, β
- Building volume over an extended area, β'

A sampled area, based on the Japanese community map, is a circle of radius 250 m. The extended area extends the sampled area towards the base station by a 500 m × 500 m area along the straight line joining the base station to the sampled area. In their study into the influence of buildings on the mean received signal strength, they concluded that although β often correlated better with the median received signal, α was more suitable since it is easier to extract from the maps.

Ibrahim and Parsons [4], characterising the test areas for their experiments in inner London, introduced two parameters: land usage factor L and degree of urbanisation factor U. Land usage factor L is defined as the percentage of the 500 m × 500 m test square that is covered by buildings, regardless of their height. This is essentially the same as the factor α used by Kozono and Watanabe. Good correlation was observed between the path loss value and L. Degree of urbanisation factor U is defined as the percentage of building site area, within the test square, occupied by buildings having four or more floors. The decision to use four floors as the reference was taken after plotting the cumulative frequency distribution of the building area against the number of floors, for a large number of OS map squares. Comparison with the propagation loss from a base station to a mobile moving in the square revealed that the percentage of buildings having four or more floors correlated best with the measured propagation data. The factor U may vary between zero and 100%; a value approaching zero indicates a suburb whereas a value approaching 100% indicates a highly developed urban area.

British Telecom [5] proposed a ten-point land usage categorisation based on qualitative descriptions. This scale is shown in Table 4.1. These categories, though comprehensive, can be interpreted differently by other service providers. Table 4.2 shows how the BT categories compare to those employed by other organisations [6–9].

The comparisons in Table 4.2 clearly indicate the fallibility of employing mainly qualitative descriptions in classifying land use within mobile radio service areas. In Germany, built-up areas are classified under one category, whereas in Britain and Japan they come under three broad classes: suburban, urban and dense urban. Experiments have shown, however, that these three categories do not cause the same level of signal attenuation and it would therefore be inappropriate to compare results obtained in built-up areas in Germany with those collected in the UK. A more detailed description of land use in Germany would be required, and this would be

Table 4.1 British Telecom categories of land usage

Category	Description
0	Rivers, lakes and seas
1	Open rural areas, e.g. fields and heathlands with few trees
2	Rural areas similar to the above but with some wooded areas, e.g. parkland
3	Wooded or forested rural areas
4	Hilly or mountainous rural areas
5	Suburban areas, low-density dwellings and modern industrial estates
6	Suburban areas, higher-density dwellings, e.g. council estates
7	Urban areas with buildings of up to four storeys, but with some open space between
8	Higher-density urban areas in which some buildings have more than four storeys
9	Dense urban areas in which most of the buildings have more than four storeys and some can be classed as skyscrapers (this category is restricted to the centre of a few large cities)

Table 4.2 Comparisons of BT and other land use categories

BT (UK)	Germany	BBC (UK)	Denmark	Okumura (Japan)
0	4	–	–	Land or sea
1	2	1	0, 1, 2	–
2	3	1	1, 2	–
3	2	1	4	–
4	2, 3	1	–	Undulating
5	1	2	3	Suburban
6	1	2	6	Suburban
7	1	3	7	Urban
8	1	3	8	Urban
9	1	4	9	Urban

more expensive in terms of cost and time. The need for a more accurate and universal standard of categorisation is therefore very apparent, particularly now that the pan-European mobile radio system GSM is in widespread use and third-generation systems have been planned.

Some years ago the derivation of land usage data involved costly and time-consuming manual procedures. Now it is possible to use geographic information systems (GIS) where digital database technology indexes items to a coordinate system for storage and retrieval [10]. Digitised maps are now generally available and for the future it seems most appropriate to adopt some standard categories of land use which relate to a GIS and which will be applicable worldwide.

In association with a computer-based simulation, a more refined method of categorisation has been proposed [11]. From a digitised map it is possible to extract the following land usage parameters:

- Building location (with respect to some reference point)
- Building size, or base area
- Total area occupied by buildings
- Number of buildings in the area concerned
- Terrain heights
- Parks and/or gardens with trees and vegetation

When this information is available it becomes possible to develop further parameters:

- The building size distribution (BSD): a probability density function defined by a mean and standard deviation. The standard deviation is an indication of homogeneity. A small value indicates an area where the buildings are of a fairly uniform size; a large value implies a more diverse range.
- Building area index (BAI): similar to α [3] or L [4].
- Building height distribution (BHD): a probability density function of the heights of all buildings within the area concerned.
- Building location distribution: a probability density function describing the location of buildings with the area.
- Vegetation index (VI): the percentage of the area covered by trees, etc.
- Terrain undulation index: similar to Δh.

Three classifications of environment are also proposed, with subclasses as appropriate:

- Class 1 (rural)

  (A) Flat
  (B) Hilly
  (C) Mountainous

- Class 2 (suburban)

  (A) Residential with some open spaces
  (B) Residential with little or no open space
  (C) High-rise residential

- Class 3 (urban and dense urban)

  (A) Shopping area
  (B) Commercial area
  (C) Industrial area

Digitised maps, in the form of computer tape, are supplied with software that enables the user to create an output file for plotting the map. Further software has been developed to extract the information needed to calculate the parameters for an appropriate area classification. Based on the observed statistics of the extracted data, values have been proposed for the parameters associated with the subclasses in Class 2 and Class 3 environments (Table 4.3).

Table 4.3 Descriptive parameters for Class 2 and 3 environments

Class	BAI (%)	BSD (m^2)		BHD (no. of storeys)		VI (%)
		μ_s	σ_s	μ_H	σ_H	
2A	12–20	95–115	55–70	2	1	$\geqslant 2.5$
2B	20–30	100–120	70–90	2–3	1	< 5
2C	$\geqslant 12$	$\geqslant 500$	> 90	$\geqslant 4$	1	$\leqslant 2$
3A	$\geqslant 45$	200–250	$\geqslant 180$	$\geqslant 4$	1	0
3B	30–40	150–200	$\geqslant 160$	3	1	0
3C	35–45	$\geqslant 250$	$\geqslant 200$	2–3	1	$\leqslant 1$

4.3 PROPAGATION PREDICTION TECHNIQUES

Some of the techniques in Chapter 3 can be applied to propagation in urban areas, but Chapter 3 did not cover methods specifically developed for application in urban areas, i.e. methods primarily intended to predict losses due to buildings rather than losses due to terrain undulations. We now review a further selection of models but there is no suggestion that the two sets are mutually exclusive. Just as some of the 'irregular terrain' methods have factors that can be used to account for buildings, some of the techniques described here are applicable in a wider range of scenarios than built-up areas.

Before describing the better-known techniques, it is worth re-emphasising that there is no single method universally accepted as the best. Once again the accuracy of any particular method in any given situation will depend on the fit between the parameters required by the model and those available for the area concerned. Generally, we are concerned with predicting the mean (or median) signal strength in a small area and, equally importantly, with the signal variability about that value as the mobile moves.

4.3.1 Young's measurements

Young [12] did not develop a specific prediction method but he reported an important series of measurements in New York at frequencies between 150 and 3700 MHz. His findings proved to be influential and have been widely quoted. The experimental results of some field trials in which the signal from a base station was received at a vehicle moving in the city streets confirmed that the path loss was much greater than predicted by the plane earth propagation equation. It was clear that the path loss increased with frequency and there was clear evidence of strong correlation between path losses at 150, 450 and 900 MHz. The sample size at 3700 MHz was not large enough to justify a similar conclusion.

In fact, Young did not compare his measured results with the theoretical plane earth equation, but an investigation of some of his results (Figure 4.3) strongly suggests the existence of high correlation. In other words, Young's results show that an inverse fourth-power law relates the loss to distance from the transmitter, and in terms of the Egli model (Section 3.6.1) the relationship can be expressed as

$$L_{50} = G_b G_m \left(\frac{h_b h_m}{d^2} \right)^2 \beta \tag{4.1}$$

In this case the clutter factor β represents losses due to buildings rather than terrain features, and Figure 4.3 shows that at 150 MHz in New York β is approximately 25 dB. From his experimental results, Young also plotted the path loss not exceeded at 1, 10, 50, 90 and 99% of locations within his test area and these are also shown in Figure 4.3. They reveal that the variability in the signal can be described by a lognormal distribution, although Young himself did not make such an assertion.

Finally, Young observed that the losses at ranges greater than 10 miles (16 km) were 6–10 dB less than might have been expected from the trend at shorter ranges. He reasoned, convincingly, that this was because the measurements at longer ranges were representative of suburban New York, whereas those nearer the transmitter

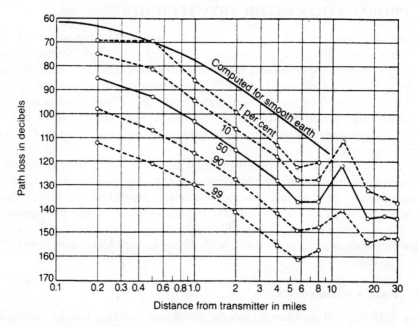

Figure 4.3 Measured path loss at 150 MHz in Manhattan and the Bronx and suburbs (after Young).

represented losses in urban Bronx and Manhattan. In summary we can say that as early as 1952 it could have been inferred from Young's results that the propagation losses were proportional to the fourth power of the range between transmitter and receiver, that the mean signal strength in a given area was lognormally distributed, and that the losses depended on the extent of urban clutter.

4.3.2 Allsebrook's method

A series of measurements in British cities at frequencies between 75 and 450 MHz were used by Allsebrook and Parsons [13] to produce a propagation prediction model. Two of the cities, Birmingham and Bath, were such that terrain features were negligible; the third, Bradford, had to be regarded as hilly.

Figure 4.4 shows results at 167 MHz, from which it is apparent that the fourth-power range law provides a good fit to the experimental data. Equation (4.1)

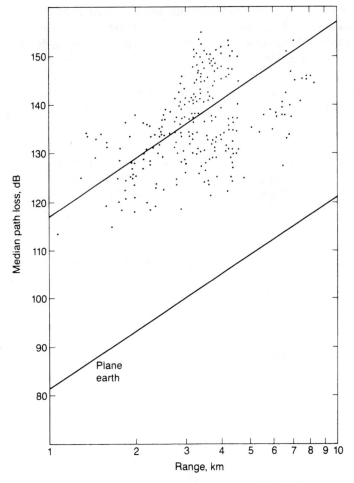

Figure 4.4 Median path loss between half-wave dipoles at 167.2 MHz.

therefore provides a basis for prediction, with an appropriate value of β. Where terrain effects are negligible the flat city model can be used:

$$L_{50} \text{ (dB)} = L_p + L_B + \gamma \tag{4.2}$$

where L_p is the plane earth path loss, L_B is the diffraction loss due to buildings and γ is an additional UHF correction factor intended for use if $f_c > 200\,\text{MHz}$. Effectively, in this model, $\beta = L_B + \gamma$.

For a hilly city it was necessary to add terrain losses, and following extensive analysis of the experimental results it was proposed to determine the diffraction loss using the Japanese method (Section 3.5.3) and to combine this with the other loss components in the manner suggested by Blomquist and Ladell. The hilly city model, which reduces to the flat city model if $L_D \rightarrow 0$, is

$$L_{50} \text{ (dB)} = L_F + [(L_p - L_F)^2 + L_D^2]^{1/2} + L_B + \gamma \tag{4.3}$$

It was shown that the diffraction loss due to buildings could be estimated by considering the buildings close to the mobile using the geometry in Figure 4.5. The receiver is assumed to be located exactly at the centre of the street, which has an effective width W'. This assumption is not exactly true but it is simple. It obviates the need to know the direction of travel and on which side of the street the vehicle is located. Figure 4.6 shows calculations based on knife-edge diffraction in an average street, compared with measured values of β. The calculations were based on the existence of coherent reflection on the base station side of the buildings, although it

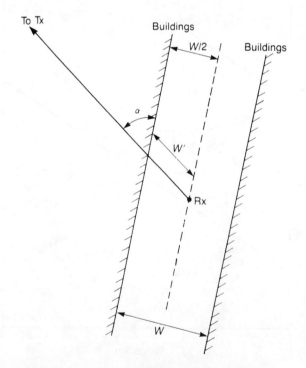

Figure 4.5   The geometry used by Allsebrook to calculate diffraction loss.

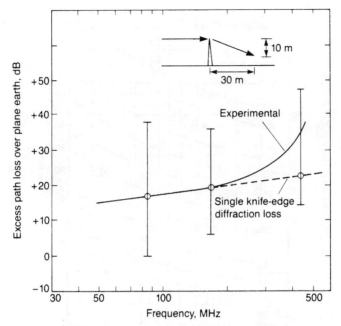

Figure 4.6 Experimental results compared with calculated losses based on the diffraction geometry in Figure 4.5; $h_0 = 10$ m, $h_m = 2$m, $W' = 30$ m.

was not suggested that this represents the true mechanism of propagation. The calculations and measurements are in good agreement at frequencies up to 200 MHz but the losses are underestimated above that frequency. This was atributed to the fact that at UHF the thickness of the buildings is several wavelengths and the difference between the two curves in Figure 4.6 represents the UHF correction factor γ.

In a paper comparing various propagation models, Delisle *et al.* [14] approximated L_B by

$$L_B \text{ (dB)} = 20 \log_{10} \left(\frac{h_0 - h_m}{548 \sqrt{W' f_c \times 10^{-3}}} \right) + 16 \qquad (4.4)$$

but they pointed out that it is very sensitive to the value of h_0, the average height of buildings in the vicinity of the mobile. Although their comparison ignored the factor γ, they admitted the need for a UHF correction factor, albeit not necessarily as large as suggested by Allsebrook and Parsons, i.e. increasing from 0 to 15 dB as f_c increases from 200 to 500 MHz.

4.3.3 The Okumura method

Following an extensive series of measurements in and around Tokyo at frequencies up to 1920 MHz, Okumura *et al.* [6] published an empirical prediction method for signal strength prediction. The basis of the method is that the free space path loss between the points of interest is determined and added to the value of $A_{mu}(f, d)$ obtained from Figure 4.7. A_{mu} is the median attenuation, relative to free space in an

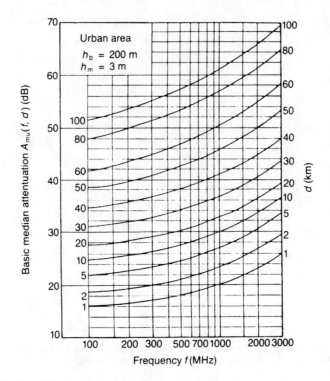

Figure 4.7 Basic median path loss relative to free space in urban areas over quasi-smooth terrain (after Okumura).

urban area over quasi-smooth terrain (interdecile range $< 20\,\text{m}$) with a base station effective antenna height h_{te} of $200\,\text{m}$ and a mobile antenna height h_{re} of $3\,\text{m}$. It is expressed as a function of frequency ($100\text{–}3000\,\text{MHz}$) and distance from the base station ($1\text{–}100\,\text{km}$). Correction factors have to be introduced to account for antennas not at the reference heights, and the basic formulation of the technique can be expressed as

$$L_{50}\ (\text{dB}) = L_F + A_{mu} + H_{tu} + H_{ru} \qquad (4.5)$$

H_{tu} is the base station antenna height gain factor; it is shown in Figure 4.8 as a function of the base station effective antenna height and distance. H_{ru} is the vehicular antenna height gain factor and is shown in Figure 4.9. Figure 4.8 shows that H_{tu} is of order $20\,\text{dB/decade}$, i.e. the received power is proportional to h_{te}^2, in agreement with the plane earth equation. From Figure 4.9 it is apparent that the same relationship applies in respect of H_{ru} if $h_{re} > 3\,\text{m}$; however, H_{ru} only changes by $10\,\text{dB/decade}$ if $h_{re} < 3\,\text{m}$.

Further correction factors are also provided, in graphical form, to allow for street orientation as well as transmission in suburban and open (rural) areas and over irregular terrain. These must be added or subtracted as appropriate. Irregular terrain is further subdivided into rolling hilly terrain, isolated mountain, general sloping terrain and mixed land–sea path. The terrain-related parameters that must be evaluated to determine the various correction factors are:

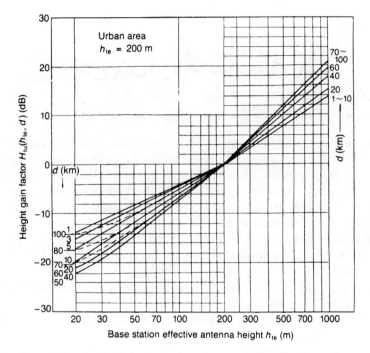

Figure 4.8 Base station height/gain factor in urban areas as a function of range (reference height=200 m).

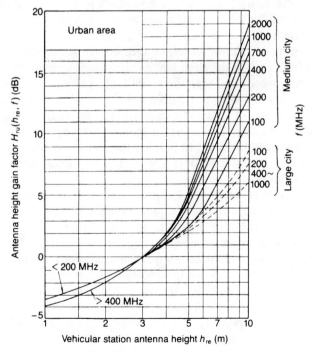

Figure 4.9 Vehicular antenna height/gain factor in urban areas as a function of frequency and urbanisation (reference height=3 m).

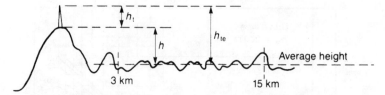

Figure 4.10   Method of calculating the effective base station antenna height.

- *Effective base station antenna height (h_{te})*: this is the height of the base station antenna above the average ground level calculated over the range interval 3–15 km (or less if the range is below 15 km) in a direction towards the receiver (Figure 4.10).
- The *terrain undulation height (Δh)*: this is the terrain irregularity parameter, defined as the interdecile height taken over a distance of 10 km from the receiver in a direction towards the transmitter.
- *Isolated ridge height:* if the propagation path includes a single obstructing mountain, its height is measured relative to the average ground level between it and the base station.
- *Average slope:* if the ground is generally sloping, the angle θ (positive or negative) is measured over 5–10 km.
- *Mixed land–sea path parameter:* this is the percentage of the total path length covered with water.

The Okumura model probably remains the most widely quoted of the available models. It has come to be used as a standard by which to compare others, since it is intended for use over a wide variety of radio paths encompassing not only urban areas but also different types of terrain.

The model can be made suitable for use on a computer by reading an appropriate number of points from each of the given graphs into computer memory and using an interpolation routine when accessing them. In some cases a correction factor is expressed as a function of another parameter by a number of prediction curves intended for various values of a second parameter, e.g. H_{tu} is given as a function of h_{te} for various values of range. Two consecutive interpolations are then necessary to derive the required correction factor. In practice the correction curves are contained as subprograms and a correction factor can be obtained by accessing the appropriate program with the required parameters.

There are two modes of operation. In quasi-smooth terrain the required input parameters include frequency, antenna heights, range, type of environment, size of city and street orientation. For irregular terrain a number of terrain-related parameters, as defined above, may also be required. If a terrain database is also stored in the computer then a computer routine can determine the type of irregularity from the path profile and hence derive the appropriate terrain parameters.

The wholly empirical nature of the Okumura model means that the parameters used are limited to specific ranges determined by the measured data on which the model is based. If, in attempting to make a prediction, the user finds that one or more parameters are outside the specified range then there is no alternative but to extrapolate the appropriate curve. Whether this is a reasonable course of action depends on the circumstances, e.g. how far outside the specified range the parameter is, and the smoothness of the curve in question. Simple extrapolation can sometimes lead to unrealistic results and care must be exercised.

Some constraints also exist in deriving the terrain-related parameters. For example, if the transmission range is less than 3 km it does not seem possible, or indeed reasonable, to use the definition of h_{te} given by the model. If the average terrain height along the path is greater than the height of the base station antenna then h_{te} is negative. In both these cases it seems sensible to ignore Okumura's definition and enter h_{te} as the actual height of the antenna above local ground level. Other problems can also occur, such as a possible ambiguity in how the terrain should be defined if there is one dominant obstruction in terrain which would otherwise be described as rolling hilly. It seems prudent to have the computer output a flag whenever a given parameter is out of range, so the user can decide whether extrapolation is appropriate or whether some other action needs to be taken.

Hata's formulation

In an attempt to make the Okumura method easy to apply, Hata [15] established empirical mathematical relationships to describe the graphical information given by Okumura. Hata's formulation is limited to certain ranges of input parameters and is applicable only over quasi-smooth terrain. The mathematical expressions and their ranges of applicability are as follows.

Urban areas

$$L_{50} \text{ (dB)} = 69.55 + 26.16 \log f_c - 13.82 \log h_t - a(h_r)$$
$$+ (44.9 - 6.55 \log h_t) \log d \tag{4.6}$$

where

$$150 \leqslant f_c \leqslant 1500 \quad (f_c \text{ in MHz})$$
$$30 \leqslant h_t \leqslant 200 \quad (h_t \text{ in m})$$
$$1 \leqslant d \leqslant 20 \quad (d \text{ in km})$$

$a(h_r)$ is the correction factor for mobile antenna height and is computed as follows. For a small or medium-sized city,

$$a(h_r) = (1.1 \log f_c - 0.7)h_r - (1.56 \log f_c - 0.8) \tag{4.7}$$

where $1 \leqslant h_r \leqslant 10 \, \text{m}$.
    For a large city,

$$a(h_r) = \begin{cases} 8.29(\log 1.54h_r)^2 - 1.1 & f \leqslant 200 \text{ MHz} \\ 3.2(\log 11.75h_r)^2 - 4.97 & f \geqslant 400 \text{ MHz} \end{cases} \tag{4.8}$$

Suburban areas

$$L_{50} \text{ (dB)} = L_{50}(\text{urban}) - 2[\log(f_c/28)]^2 - 5.4 \tag{4.9}$$

Open areas

$$L_{50} \text{ (dB)} = L_{50}(\text{urban}) - 4.78(\log f_c)^2 + 18.33 \log f_c - 40.94 \tag{4.10}$$

In quasi-open areas the loss is about 5 dB more than indicated by equation (4.10).

These expressions have considerably enhanced the practical value of the Okumura method, although Hata's formulations do not include any of the path-specific corrections available in the original model. A comparison of predictions given by these equations with those obtained from the original curves (with interpolation as necessary) reveals negligible differences that rarely exceed 1 dB. Hata's expressions are very easily entered into a computer. In practice the Okumura technique produces predictions that correlate reasonably well with measurements, although by its nature it tends to average over some of the extreme situations and not respond sufficiently quickly to rapid changes in the radio path profile.

Allsebrook found that an extended version of the Okumura technique produced prediction errors comparable to those of his own method. The comparisons made by Delisle *et al.* [14] and by Aurand and Post [16] also showed the Okumura technique to be among the better models for accuracy, although it was also rated as 'rather complex'. Generally the technique is quite good in urban and suburban areas, but not as good in rural areas or over irregular terrain. There is a tendency for the predictions to be optimistic, i.e. suggesting a lower path loss than actually measured.

The extended COST231–Hata model

The Hata model, as originally described, is restricted to the frequency range 150–1500 MHz and is therefore not applicable to DCS1800 and other similar systems operating in the 1800–1900 MHz band. However, under the European COST231 programme the Okumura curves in the upper frequency range were analysed, and an extended model was produced [17]. This model is

$$L_{50} \text{ (dB)} = 46.3 + 33.9 \log f_c - 13.82 \log h_t - a(h_r)$$
$$+ (44.9 - 6.55 \log h_t) \log d + C \tag{4.11}$$

In this equation $a(h_r)$ is as defined previously, with $C = 0$ dB for medium-sized cities and suburban centres with medium tree density and 3 dB for metropolitan centres.

Equation (4.11) is valid for the same range of values of h_t, h_r and d as eqn (4.6), but the frequency range is now $1500 < f_c$ (MHz) < 2000. The application of this model is restricted to macrocells where the base station antenna is above the rooftop levels of adjacent buildings. Neither the original nor the extended models are applicable to microcells where the antenna height is low.

Akeyama's modification

The Okumura technique adopts curves for urban areas as the datum from which other predictions are obtained. This presentation was adopted not because urban areas represent the most common situation, but to meet computational considerations and because the highest prediction accuracy was obtained if the urban curves were used as the 'standard'. In many countries the urban situation is far from being the most common.

Caution must be exercised in applying the environmental definitions as described by Okumura to locations in countries other than Japan. Okumura's definition of urban, for example, is based on the type and density of buildings in Tokyo and it may not be directly transferable to cities in North America or Europe. Indeed, experience with measurements in the USA has shown that the typical US suburban environment lies somewhere between Okumura's definition of suburban and open areas. Since the CCIR has adopted the Okumura urban curve as its basic model for 900 MHz propagation, it is also prudent to exercise caution when using these curves.

One other problem encountered in using the Okumura model is that the correction factor used to account for environments other than urban (suburban, quasi-open and open) is a function only of the buildings in the immediate vicinity of the mobile. It is often more than 20 dB, is discrete and cannot be objectively related to the height and density of the buildings. There is uncertainty over how the factors suggested by Okumura can be applied to cities other than Tokyo, particularly those where the architectural style and construction materials are quite different.

Some attempts have been made to expand the concept of degree of urbanisation to embrace a continuum of values [3,4] although others [5,11] prefer a finer, but discrete, categorisation along the lines proposed by Okumura. A ground cover (degree of urbanisation) factor has been proposed by Akeyama *et al.* [18] to account for values of α less than 50% in a continuous way. Figure 4.11 shows some experimental points together with a regression line drawn to produce a best fit. The value of S – the deviation from Okumura's reference median curve at 450 MHz – is given by

$$S \text{ (dB)} = \begin{cases} 30 - 25 \log \alpha & 5\% < \alpha < 59\% \\ 20 + 0.19 \log \alpha - 15.6(\log \alpha)^2 & 1\% < \alpha \leqslant 5\% \\ 20 & \alpha < 1\% \end{cases} \qquad (4.12)$$

where α is the percentage of the area covered by buildings.

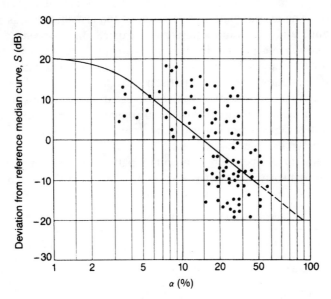

Figure 4.11 Deviation from median field strength curve due to buildings surrounding the mobile terminal.

4.3.4 The Ibrahim and Parsons method

Propagation models were produced by analysis of measured data collected principally in London with base station antennas at a height of 46 m above local ground [4]. The frequencies used were 168, 445 and 896 MHz and the signal from the base station transmitter was received in a vehicle that travelled in the city streets. Samples were taken every 2.8 cm of linear travel using positional information derived from a 'fifth-wheel' towed by the vehicle; these samples were digitised and recorded onto a tape recorder.

The measured data was collected in batches, each batch representing a 500 m × 500 m square as delineated on an OS map. This size of square was judged suitable as it was not so large that the type of environment varied substantially or so small that the propagation data became unrepresentative. The mobile route within each test square was carefully planned to include a random mixture of wide and narrow roads of as many orientations as possible, and the average route length within each square was 1.8 km. A total of 64 squares were selected in three arcs around the base station at ranges of approximately 2, 5 and 9 km. The total length of the measurement route was about 115 km. The same route was used for the two sets of trials at 168 and 455 MHz. At 900 MHz the test routes were limited to a range of 5 km due to the high path loss at this frequency and the limited transmitter power.

The value of the median path loss between two isotropic antennas was extracted from the data collected in each of the test squares and compared with the various factors likely to affect it, such as the range from the transmitter, the urban environment, the transmission frequency and terrain parameters. These factors act simultaneously and some lack of precision has to be accepted when trying to identify their individual contributions.

In general, the median received signal decreased as the mobile moved away from the base station. The median path loss for each of the test squares was plotted as a function of range and regression analysis was carried out to produce the best-fit straight line through the points; this was subsequently repeated forcing a fourth-power range law fit. The results are summarised in Table 4.4. The rather limited data at 900 MHz did not allow a valid comparison with data at 168 and 455 MHz.

It was evident that the rate at which the received signal attenuates with range increased with an increase in the transmission frequency. It also appeared that the fourth-power range dependence law is a good approximation at the two frequencies for ranges up to 10 km from the transmitter.

At all ranges and for all types of environment the path loss increased with an increase in the transmission frequency. For the test squares at 2 km range, the median path loss at

Table 4.4 Range dependence regression equations at 168 and 455 MHz

Frequency	Median path loss (dB)		RMS prediction error (dB)
168 MHz	Best fit	$1.6 + 36.2 \log d$	5.30
	Fourth law	$-12.5 + 40 \log d$	5.50
455 MHz	Best fit	$-15.0 + 43.1 \log d$	6.18
	Fourth law	$-4.0 + 40 \log d$	6.25

900 MHz was found to exceed the loss at 455 MHz by an average of 9 dB, and to exceed the loss at 168 MHz by an average of 15 dB. At 5 km range, the excess loss at 900 MHz relative to 455 and 168 MHz appeared to increase slightly, suggesting that as the transmission frequency increases, the signal attenuates faster with the increase in range.

Strong correlation was evident between the path loss at the three frequencies; this is evident from Figure 4.12 which shows the median path loss to each of the test squares at 2 km range. The correlation coefficient was 0.93 when the measurements at 168 and 455 MHz were considered, and 0.97 between the measurements at 455 and 900 MHz. When the 'local mean' of the received signal at the three frequencies along the test routes within test squares was compared, high correlation was again evident. It was therefore concluded that the propagation mechanism at the three frequencies is essentially similar.

The way in which urbanisation was treated has already been discussed in Section 4.2.2. The factors L and U were determined from readily available data, although the information necessary to calculate U was, at that time, only available for city-centre areas. In developing the prediction models this was taken into account and U was employed as an additional parameter to be used only in highly urbanised areas.

Two approaches to modelling were taken: the first was to derive an empirical expression for the path loss based on multiple regression analysis; the second was to start from the theoretical plane earth equation and to correlate the excess path loss (the clutter factor) with the parameters likely to influence it. The main difference between the first empirical method and the second semi-empirical method is that a fourth-power range dependence law is assumed, a priori, in the second approach – a not unreasonable assumption, as shown previously. A multiple regression analysis taking all factors into account, in decreasing order of importance, produced the following empirical equation for the path loss:

$$L_{50} \text{ (dB)} = -20 \log(0.7h_b) - 8 \log h_m + \frac{f}{40} + 26 \log \frac{f}{40} - 86 \log \left(\frac{f+100}{156} \right)$$

$$+ \left[40 + 14.15 \log \left(\frac{f+100}{156} \right) \right] \log d + 0.265L - 0.37H + K \qquad (4.13)$$

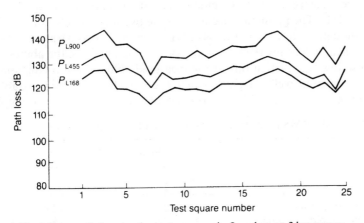

Figure 4.12 Mean path loss to the test squares in London at 2 km range.

where $K = 0.087U - 5.5$ for highly urbanised areas, otherwise $K = 0$.

In this equation the symbols have their usual meanings, H being the difference in average ground height between the OS map squares containing the transmitter and receiver. The value of $h_r \leqslant 3\,$m and $0 \leqslant d \leqslant 10\,$km.

The semi-empirical model is based on the plane earth equation. It suggests that the median path loss should be expressed as the sum of the theoretical plane earth loss and an excess clutter loss β. The values of β at 168, 455 and 900 MHz were computed for each test square. They were then related to the urban environment factors and a best-fit equation for β was found. Accordingly, the following model was proposed:

$$L_{50}\ (\text{dB}) = 40\ \log\ d - 20\ \log(h_t h_r) + \beta \tag{4.14}$$

where

$$\beta = 20 + \frac{f}{40} + 0.18L - 0.34H + K \tag{4.15}$$

and

$$K = 0.094U - 5.9$$

Here again K is applicable only in the highly urbanised areas, otherwise $K = 0$. The RMS prediction errors produced by the two models are summarised in Table 4.5.

Application of the model requires estimates for L, U and H of the test squares under consideration. Parameter H can be easily extracted from a map; L and U can sometimes be obtained from other stored information but they may have to be estimated either because the information is not readily available or simply to save time. As an illustration, the value of β given by equation (4.15) in a flat city ($H = 0$) at a frequency of 900 MHz is

$$\beta\ (\text{dB}) = 42.5 + 0.18L \tag{4.16}$$

and if L lies in the range 0 to 80% then β lies between 42.5 and 57 dB. This agrees well with some independently measured results shown in Figure 4.13 for which $\beta = 49\,$dB.

The models were compared with the independent data collected by Allsebrook at 85, 167 and 441 MHz. The prediction accuracy of the two models at 85 and 167 MHz was excellent, though it was only fair at 441 MHz. Even with two parameters (L and H) set to their mean values for the area in question – thus limiting the ability of the predictions to follow the fluctuations of the measured values – the predictions and measurements compared well, suggesting that the models are indeed suitable for global application. Comparing the performance of the two models, the empirical

Table 4.5   RMS prediction errors produced by the two models

	Frequency (MHz)		
	168	455	900
Empirical model	2.1	3.2	4.19
Semi-empirical model	2.0	3.3	5.8

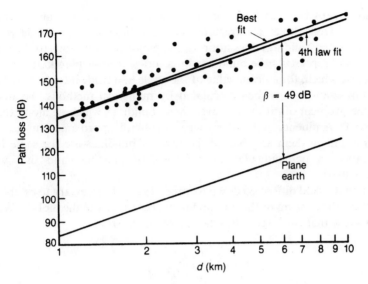

Figure 4.13 Experimental results in a city at 900 MHz compared with a best-fit regression line and an inverse fourth-power law line.

model seemed to perform slightly better at 85 and 167 MHz, whereas the semi-empirical model was markedly better at 441 MHz.

4.3.5 The Walfisch–Bertoni method

Walfisch and Bertoni [19] pointed out that although measurements have shown that in quasi-smooth terrain the average propagation path loss is proportional to (range)^n where n lies between 3 and 4, the influence of parameters such as building height and street width are poorly understood and are often accounted for by ad hoc correction factors [3,6,15]. They therefore developed a theoretical model based on the path geometry shown in Figure 4.14. The primary path to the mobile shown lies over the tops of the buildings in the vicinity [20,21], with the buildings closest to the mobile being the most important, as shown by path 1. Other possible propagation mechanisms exist, but although the total field at the receiving point may have components due to multiple reflections and diffractions (path 4) and building penetration (path 3), these are generally negligible.

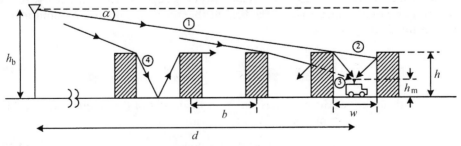

Figure 4.14 Geometry of rooftop diffraction.

Furthermore, gaps between buildings are randomly located and are not aligned with each other from street to street or with the base station–mobile path, so propagation between buildings does not produce major contributions to the received signal. The model represents the buildings by a series of absorbing knife-edges and establishes those which, for a given value of the angle α, intrude into the first Fresnel zone. The diffraction loss is then calculated by numerical methods. To obtain a solution to the problem of diffraction over many buildings which lie along the path and which have an influence, particularly when α is small, approximations have to be made. Central among them are that all the rows of buildings are the same height, that propagation is perpendicular to the rows of buildings and that vertical polarisation is used.

To determine the field diffracted down to street level, it is necessary to establish the field incident on the rooftop of the building immediately before the mobile. Walfisch and Bertoni show that for large n this can be obtained from

$$Q(\alpha) \approx 0.1 \left(\frac{\alpha \sqrt{b/\lambda}}{0.03} \right)^{0.9} \tag{4.17}$$

This is in addition to the d^{-1} dependence of the radiated field, giving an overall dependence of $d^{-1.9}$. This yields a $d^{-3.8}$ law for the received signal power, very close to the d^{-4} law for propagation over plane earth that is commonly used in empirical models. The overall path loss then consists of three factors: the path loss between the antennas in free space, the multiple-edge diffraction loss up to the rooftop closest to the mobile, and the diffraction and scatter loss from that point to the mobile at street level. Assuming isotropic antennas, the first of these is the basic path loss given by eqn. (2.6), i.e.

$$L_B \text{ (dB)} = 32.4 + 20 \log f_{MHz} + 20 \log d_{km}$$

Equation (4.17) is used to find $Q(\alpha)$ and it is possible to account for terrain slope at the mobile and for the Earth's curvature, provided the range d does not approach the radio horizon. For level terrain, α (radians) is given by

$$\alpha = \frac{h_b - h}{d} - \frac{d}{2r_e} \tag{4.18}$$

where r_e is the effective Earth radius ($\approx 8.5 \times 10^3$ km).

The additional loss due to the final diffraction down to street level is estimated by assuming the row of buildings to act as an absorbing edge located at the centre of the row. In this case the amplitude of the field at the mobile is obtained by multiplying the rooftop field by the factor

$$\frac{\sqrt{\lambda}}{2\pi} \left[\left(\frac{b}{2} \right)^2 + (h - h_m)^2 \right]^{-1/4} \left[\frac{-1}{\gamma - a} + \frac{1}{2\pi + \gamma - \alpha} \right] \tag{4.19}$$

where h is the height of the buildings and h_m is the height of the mobile antenna. The angles α and γ are both measured in radians with

$$\gamma = \tan^{-1}[2(h - h_m)/b] \tag{4.20}$$

Equation (4.19) can be simplified by neglecting $1/(2\pi + \gamma - \alpha)$ compared with $1/(\gamma - \alpha)$ and assuming that $\alpha \ll \gamma$.

The fact there is deep fading in the signal received by the mobile indicates that the field component received by reflection from the buildings next to the mobile is of a similar amplitude to that received by rooftop diffraction. These two components have random phases, however, so the RMS value of the total field is the sum of the RMS values of the individual components. In this case it is $\sqrt{2} \times$ the RMS value of the diffracted field. Using eqns (2.6) and (4.18) to (4.20) and including the factor $\sqrt{2}$ yields an expression for the reduction in the field over that experienced by the same antennas separated by a distance d in free space. This is the 'excess loss' over the free space path loss and is given by

$$L_{\text{ex}} \text{ (dB)} = 57.1 + A + \log f_{\text{c}} + 18 \log d - 18 \log(h_{\text{b}} - h) - 18 \log\left(1 - \frac{d^2}{17(h_{\text{b}} - h)}\right)$$

$$(4.21)$$

The final term in this expression accounts for Earth curvature and can often be neglected. The building geometry is incorporated in the term

$$A = 5 \log\left[\left(\frac{b}{2}\right)^2 + (h - h_{\text{m}})^2\right] - 9 \log b + 20 \log\{\tan^{-1}[2(h - h_{\text{m}})/b]\} \quad (4.22)$$

The total path loss is found by adding L_{ex} to the free space path loss L_{B} for isotropic antennas. Walfisch and Bertoni tested their model against published measurements [6,22] and found good agreement.

The COST–Walfisch–Ikegami model

During the COST231 project the subgroup on propagation models proposed a combination of the Walfisch–Bertoni method with the Ikegami model [21], to improve path loss estimation through the inclusion of more data. Four factors are included:

- Heights of buildings
- Width of roads
- Building separation
- Road orientation with respect to the LOS path

The model distinguishes between LOS and non-LOS paths as follows. For LOS paths the equation for L_{b} is

$$L_{\text{b}} \text{ (dB)} = 42.6 + 26 \log d + 20 \log f_{\text{c}} \quad (d \geqslant 20 \text{ m})$$

This was developed from measurements taken in Stockholm, Sweden. It has the same form as the free space path loss equation, and the constants are chosen such that L_{b} is equal to the free space path loss at $d = 20$ m. In the non-LOS case the basic transmission loss comprises the free space path loss L_{B} (2.6), the multiple-screen diffraction loss L_{msd} and the rooftop-to-street diffraction and scatter loss L_{rts}. Thus

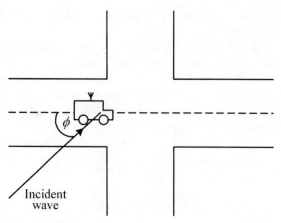

Figure 4.15 Defining the street orientation angle φ.

$$L_b = \begin{cases} L_B + L_{rts} + L_{msd} & L_{rts} + L_{msd} > 0 \\ L_B & L_{rts} + L_{msd} < 0 \end{cases} \qquad (4.23)$$

The determination of L_{rts} is based on the principle given in the Ikegami model [21], but with a different street orientation function. The geometry is shown in Figure 4.15 and the values of L_{rts} are as follows:

$$L_{rts} = -16.9 - 10 \log w + 10 \log f_c + 20 \log(h - h_m) + L_{ori} \qquad (4.24)$$

$$L_{ori} = \begin{cases} -10 + 0.354\varphi & 0° \leqslant \varphi < 35° \\ 2.5 + 0.075(\varphi - 35) & 35° \leqslant \varphi < 55° \\ 4.0 - 0.114(\varphi - 55) & 55° \leqslant \varphi < 90° \end{cases} \qquad (4.25)$$

Note that L_{ori} is a factor which has been estimated from only a very small number of measurements.

The multiple-screen diffraction loss was estimated by Walfisch and Bertoni for the case when the base antenna is above the rooftops, i.e. $h_b > h$. This has also been extended by COST to the case when the antenna is below rooftop height, using an empirical function based on measurements. The relevant equations are

$$L_{msd} = L_{bsh} + k_a + k_d \log d + k_f \log f_c - 9 \log b \qquad (4.26)$$

where

$$L_{bsh} = \begin{cases} -18 \log[1 + (h_b - h)] & h_b > h \\ 0 & h_b \leqslant h \end{cases} \qquad (4.27)$$

$$k_a = \begin{cases} 54 & h_b > h \\ 54 - 0.8(h_b - h) & h_b \leqslant h \text{ and } d \geqslant 0.5 \,\text{km} \\ 54 - 0.8(h_b - h)\dfrac{d}{0.5} & h_b \leqslant h \text{ and } d < 0.5 \,\text{km} \end{cases} \qquad (4.28)$$

$$k_d = \begin{cases} 18 & h_b > h \\ 18 - 15\dfrac{(h_b - h)}{h} & h_b \leqslant h \end{cases} \tag{4.29}$$

$$k_f = -4 + \begin{cases} 0.7\left(\dfrac{f_c}{925} - 1\right) & \text{for medium-sized cities and suburban centres with medium tree density} \\ 1.5\left(\dfrac{f_c}{925} - 1\right) & \text{for metropolitan centres} \end{cases} \tag{4.30}$$

The term k_a represents the increase in path loss when the base station antenna is below rooftop height. The terms k_d and k_f allow for the dependence of the diffraction loss on range and frequency, respectively. If data is unavailable the following default values are recommended:

$$h = 3\,\text{m} \times (\text{number of floors}) + \text{roof height}$$

$$\text{roof height} = \begin{cases} 3\,\text{m} & \text{for pitched roofs} \\ 0\,\text{m} & \text{for flat roofs} \end{cases}$$

$$b = 20 \text{ to } 50\,\text{m}$$
$$w = b/2$$
$$\varphi = 90°$$

The COST model is restricted to the following range of parameters:

f_c	800 to 2000 MHz
h_b	4 to 50 m
h_m	1 to 3 m
d	0.02 to 5 km

It gives predictions which agree quite well with measurements when the base station antenna is above rooftop height, producing mean errors of about 3 dB with standard deviations in the range 4–8 dB. However, the performance deteriorates as h_b approaches h_r and is quite poor when $h_b \ll h_r$. The model, as it stands, might therefore produce large errors in the microcellular situation.

Other solutions [23–25] have been published for evaluating L_{msd} and several papers [26–28] compare the different approaches with measurements. As might be expected, the results differ markedly depending on the situation where the models are applied. An adaptive combination of the different approaches has been used in urban macrocells at 1800 MHz [27] and yields better results than any single model.

4.3.6 Other models

A propagation model described by Lee [29, Ch. 3] is intended for use at 900 MHz and operates in two modes, an area-to-area mode and a point-to-point mode. In the first case the prediction is based on three parameters:

- The median transmission loss at a range of 1 km, L_0
- The slope of the path loss curve, γ dB/decade

- An adjustment factor, F_0

Hence the median loss at a distance d is given by

$$L_{50} \text{ (dB)} = L_0 + \gamma \log d + F_0 \tag{4.31}$$

Values of L_0 and γ derived from experiments are listed in Table 4.6 and in making predictions it is necessary to select values from the table by comparing the environment under consideration with the reference environment that most closely resembles it. It is interesting to note that with the exception of Tokyo, the value of γ for urban and suburban areas is always quite close to 40 dB/decade.

The experimental results on which Table 4.6 is based were obtained using a transmission system with the following parameters:

- Carrier frequency = 900 MHz
- Base station antenna height = 30.48 m (100 ft)
- Transmitter power = 10 W
- Base station antenna gain with respect to $\lambda/2$ dipole = 6 dB
- Mobile antenna height = 3 m (10 ft)

The adjustment factor F_0 is intended to compensate for using different values of these parameters and is expressed as

$$F_0 = F_1 F_2 F_3 F_4$$

The values of these various factors are given by

$$F_1 = \left(\frac{\text{actual base station antenna height}}{\text{reference base station height}} \right)^2$$

$$= \left(\frac{\text{actual base station antenna height (m)}}{30.5} \right)^2$$

$$F_2 = \frac{\text{actual transmitter power (W)}}{10}$$

$$F_3 = \frac{\text{actual gain of base station antenna}}{4}$$

Table 4.6 Propagation parameters for Lee's model

Environment	L_0 (dB)	γ
Free space	91.3	20
Open (rural) space	91.3	43.5
Suburban	104.0	38.5
Urban area		
Philadelphia	112.8	36.8
Newark	106.3	43.1
Tokyo	128.0	30

For F_3 the gain is measured with respect to a $\lambda/2$ dipole and the reference antenna gain is 6 dB ($= 4$). F_4 compensates for changes in the mobile antenna height. It can be determined in the same manner as F_1 for heights above 10 m, but for heights below 10 m the ratio of heights is used rather than the ratio of (heights)^2; compare with Okumura. Lee also suggested that a change in transmission frequency can be accommodated by using a factor of the form $(f/f_0)^n$ but was not specific about the value of n. However, the work of Okumura [6] and Young [12] suggests a value between 2 and 3.

In the more refined point-to-point mode, Lee's model [30] takes some account of the terrain. Over unobstructed paths the major task is to determine an appropriate value for the effective base station antenna height. In hilly terrain there are two possible reflection points shown in Figure 4.16; the effective one is the point closer to the mobile. An effective height can be found by extending the ground plane in this region back to the base station location. This may be greater than the actual height above the local ground, as in Figure 4.16, or less. The effective height is then used to correct eqn (4.31) using the expression

$$L'_{50} = L_{50} + 20 \log(h_e/30) \tag{4.32}$$

As the mobile moves, the effective base station antenna height changes and Figure 4.17(a) illustrates some of the possibilities. A separate evaluation of eqn (4.32) has to be made in each of these situations, hence the term 'point-to-point prediction'. Figure 4.17(b) shows the difference between the point-to-point model and predictions for a flat suburban area with $\gamma = 38.6$ dB/decade. The trend is as expected: for positions C to G the value of h_e is greater than the physical height above local ground, so the predicted loss is smaller; for positions H and I the value of h_e is less than the actual height. For obstructed paths, a terrain loss can be added using any of the methods discussed in Chapter 3. Incorporating terrain information can make a substantial difference to the predictions (Figure 4.17(b)) and the accuracy is generally increased. The standard deviation of error in the area mode is claimed to be 8 dB but in the point-to-point mode this is reduced to less than 3 dB.

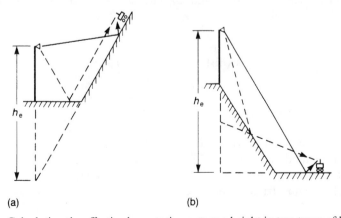

(a)   (b)

Figure 4.16   Calculating the effective base station antenna height in two types of hilly terrain.

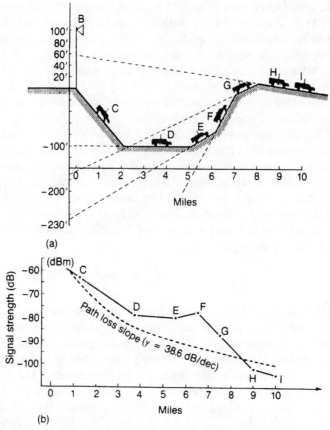

Figure 4.17 Influence of terrain on effective antenna height for different positions (after Lee): (a) hilly terrain contour, (b) point-to-point prediction.

There have been other approaches to formulating a simple prediction equation suitable for use in built-up areas. For example, McGeehan and Griffiths [31] started from the plane earth equation and introduced an environmental factor given empirically by

$$A \text{ (dB)} = A^* - 30 \log f_{\text{MHz}}$$

A^* is given for various environments as follows:

- 45 ± 5 dB for older cities with narrow, twisting streets
- 55 ± 5 dB for modern cities with long, straight, wide streets
- 65 ± 5 dB for typical suburban areas and some rural areas
- 75 ± 5 dB for unobstructed open areas

The received power P_R is then given by

$$P_R \text{ (dBm)} = \{P_T + 10 \log G_b + 10 \log G_m + 20 \log(h_b h_m)$$
$$- 30 \log f - 120\} + A^* - 40 \log d \qquad (4.33)$$

In this equation P_T is in dBm, heights h_b and h_m are in metres, and d is in kilometres. For any given radio system the terms within the curly brackets are constants, so eqn (4.33) can be simplified to

$$P_R \text{ (dBm)} = K + A^* - 40 \log d \qquad (4.34)$$

Following a series of measurements in London at 900 MHz, Atefi and Parsons [32] produced several equations of the form

$$L_{50} \text{ (dB)} = A + B \log d_{km}$$

Base station antenna heights varied between 22 and 88 m above sea level but the factor B was always fairly close to 40 (fourth-power law) and A represented the plane earth path loss plus the clutter factor β, with β lying in the range 38–44 dB. The variations of A (dB) and B (dB/decade) as a function of base station antenna height were examined and found to be

$$A = 140.1 - 12.2 \log h_b$$
$$B = 49.3 - 6.8 \log h_b \qquad (4.35)$$

These equations are very similar to those derived from Okumura's measurements:

$$A = 146 - 14 \log h_b$$
$$B = 45 - 6.5 \log h_b \qquad (4.36)$$

and this was considered to be very encouraging. Over the range of heights used for the experiments, the value of β was given by

$$\beta = 48.1 - 0.12 h_b \qquad (4.37)$$

which indicates that increasing h_b by a factor of 10 decreases β by 12 dB.

The model, as finally developed, needed to include the effects of carrier frequency and terrain loss. The former was obtained from Okumura's data and the latter was obtained from a terrain database and the Epstein–Peterson diffraction construction. The final model was expressed as

$$L_{50} \text{ (dB)} = 82 + 26.16 \log f + 38 \log d - 21.8 \log h_b - 0.15 \log h_m + L_D \quad (4.38)$$

where L_D represents the diffraction loss. The model was tested against the independent data collected by Allsebrook [13] and Ibrahim [4] and the results are summarised in Table 4.7. The proposed model is quite successful and gives results that are reasonably accurate over a wide range of transmission frequencies.

There have been several approaches to the modelling of large-scale terrain variations. Hviid *et al.* [33] have based an approach on the magnetic field integral equation (MFIE) with no transverse variations, no backscattering and a perfectly magnetically conducting surface, whereas the electric field integral equation (EFIE) has been used by others [34]. The EFIE approach includes both forward and back scattering and embodies a moment method using a novel set of complex basis

Table 4.7 Comparison of prediction models in urban areas

Frequency (MHz)	Prediction from (5.23)	Allsebrook's best-fit line	Ibrahim's best-fit line
85.87	$97.4 + 38 \log d$	$98 + 38 \log d$	
167.2	$105 + 38 \log d$	$101 + 38 \log d$	$106 + 38 \log d$
441.0	$116 + 38 \log d$	$117 + 39 \log d$	$115 + 38 \log d$

functions which reduces the computation time significantly. In principle there seems no reason why this more efficient approach should not be applied to any method using integral equations.

Parabolic integral equations have also been used to build macrocell models [35,36]. There is a heuristic method based on the parabolic heat or diffusion equation. Here a multiple knife-edge method is used to estimate the path loss over irregular terrain and the equation is solved using the simple explicit forward difference method. A 100 m grid is used in the propagation direction and the height grid has a 5 m resolution. In the second method, the FFT multiple half-screen diffraction model [35], the terrain profile is replaced by a number of absorbing half-screens similar to the method used by Walfisch and Bertoni, hence the propagation is modelled as a multiple diffraction mechanism with reflections neglected.

A more empirical approach has been proposed based on training a neural network [37]. The training can be done by theoretical methods or using measurements. If the training is done directly from measurements, the system becomes very flexible and readily adaptable to any environment. Although the training takes quite a long time, once it has been completed, results are available immediately.

Ray tracing methods (Chapter 7) have also received some attention [38]. In the uniform theory of diffraction (UTD) obstacles along the propagation path have to be modelled by simple geometrical objects such as wedges and convex surfaces, and expressions exist to calculate the diffraction coefficients for these shapes. Recent research [39] has produced a three-dimensional deterministic model which combines key features from many previous models. It uses four databases to characterise the principal features of the environment, i.e. terrain, buildings, foliage and land usage. The various channel effects such as reflection, scattering, diffraction and foliage loss, are taken into account and the outputs include signal strength, time dispersion, spatial dispersion and fading statistics. Diffraction losses can be calculated using either UTD or knife-edge approximations.

A feature of the model is that it uses a radar imaging technique, with each terrain pixel and wall treated as a radar target which scatters a portion of the incident energy in the direction of the receiver. Significantly this allows the model to take into account not only the direct path between the transmitter and receiver but also propagation paths that involve scattering from terrain features or buildings in any plane. In other words, it can take off-axis obstacles into account, i.e. it can consider significant scatterers which do not lie on the direct path between transmitter and receiver. The user can specify the extent of the area surrounding the transmit and receive locations that is to be considered in any prediction, and the model incorporates algorithms which establish whether a particular terrain pixel or other feature is a significant contributor in any given situation.

If the area considered is reduced to zero, the model degenerates into a two-dimensional one, which only takes account of obstacles that lie on the direct transmitter–receiver path. The number of diffractions to be considered in any off-axis scattered path is also under user control, and those paths which contain more than the specified maximum are ignored. A sensitivity analysis has established the effects of various factors such as terrain database resolution on the output. Comparison with measurements has shown that the three-dimensional approach, i.e. taking off-axis obstacles into account, produces a much improved prediction in almost all cases. Models of this type which are deterministic in nature have the potential to predict parameters of interest in the design of wideband systems.

4.4 MICROCELLULAR SYSTEMS

The type of cellular structure employed in a mobile radio-telephone system can have a profound effect on its performance. Cell shapes in the current generations of systems are often depicted as hexagonal [40], although this is a rather inflexible strategy. In a move towards providing communications to and from any person or terminal, irrespective of their location, it is necessary to accommodate a wide variety of scenarios, including motorways, urban and suburban areas, parks, rural areas and even locations remote from any centre of population. To cope with these requirements it is necessary to tailor the size and shape of cells to suit the local geography and the amount of teletraffic envisaged. The diverse nature of the required cell shapes can be accommodated through the use of different antennas, transmitter locations, power levels and carrier frequencies [41].

Although a number of approaches are available, the use of microcells [42] has become a popular solution in heavily built-up areas. A commonly used criterion to define a microcell is related to the base station antenna height, and a typical microcell may have dimensions of only a few hundred metres with a base station antenna mounted at street-lamp level. Even so, signal strength prediction may be necessary for distances of several kilometres for interference calculations. A fixed base transceiver is associated with each microcell, and in order to provide adequate coverage, a partial overlap with other microcells or macrocells is designed into the system. The extent of coverage from a given base station can be adjusted by using different antenna patterns and heights and by controlling the transmitted power. Microcells have the advantage that delay spreads are small and, together with intensive frequency reuse, this has the potential to allow high-capacity systems with high data rates.

The fact that propagation is required only over very short ranges suggests that much higher frequencies can be used in microcellular systems. Microwave frequencies offer a much enhanced bandwidth potential and the wavelengths are such that space diversity can be accommodated even within the dimensions of hand-portable equipment. A suggestion that was much discussed a few years ago is the use of frequencies where the radiation is partially absorbed by oxygen molecules in the atmosphere. Such resonant absorption lines are found in the frequency range from about 50 to 70 GHz (Figure 1.2) and the portion of the radio spectrum from 51.4 to 66 GHz has been designated absorption band A1. Some parts of this band have been provisionally allocated for mobile communications. The attenuation due to oxygen in the atmosphere at ground level has been measured [43] as 16 dB/km and this is in addition to the normal spatial attenuation and losses due to rain.

The 60 GHz band has been advocated for use in microcellular systems because of the oxygen absorption effects, the potentially large available bandwidths and the prospect of affordable transceivers for the public. However, while stand-alone 60 GHz systems might be a possibility for specialist scenarios, the question of compatibility and integration with existing 900 MHz and 1.8 GHz cellular radio systems poses enormous problems. Microcellular systems, however, are not restricted in principle and investigations of short-range propagation characteristics at frequencies which are of interest in current systems have also been carried out. We briefly review some of the findings.

4.4.1 Microwave measurements

A substantial research effort has been concentrated on characterising the transmission path at 60 GHz for point-to-point links under good and adverse weather conditions, and it is well-documented [44–47] that moderate rainfall rates can cause severe degradation even over relatively short paths. One solution is simply to overdesign the system by increasing the transmitter power or by employing some form of diversity reception [48]; alternatively a tracking antenna can be used to alleviate losses due to abnormal propagation conditions [49].

Experiments under stationary conditions [50] have shown that signal attenuation follows a $d^{-2.2}$ law, indicating that free space conditions are fairly representative of the propagation mechanism. Figure 4.18 shows that the mean signal strength closely

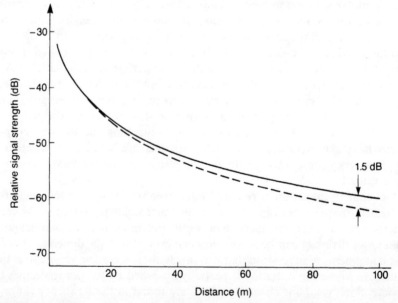

Figure 4.18 Variation of mean signal strength with distance: (- - - -) measured loss and (——) calculated free space loss.

follows the free space law over a distance of 100 m where a line-of-sight path exists. Obstruction of the path by trees causes the signal strength to fall by about 12 dB. Where there is no evident line-of-sight, or in urban areas where there is significant scattering from buildings, the short-term signal fading can often be represented by Rayleigh statistics, and Figure 4.19 shows experimental results taken in London over two paths of length 100 and 190 m [51].

The data at 190 m is a very good fit to a Rayleigh distribution, but closer to the transmitter the characteristics are somewhat different. These results are as expected, since the closer together the transmitter and receiver, the greater the chance that line-of-sight conditions will exist. By the same token, when the distance is greater, it is more likely that multipath transmission will occur with random fluctuations that conform to Rayleigh statistics. In more open environments, where the road surface is the only significant scatterer, the observed interference pattern is much simpler, the statistics are not Rayleigh, and the spectrum of the signal has a much smaller high-frequency content.

Figure 4.20 shows that severe diffraction losses occur at millimetre wavelengths. As the receiver moves into the shadow region, the signal strength drops very rapidly and the quality of reception in this region is primarily determined by the signal strength at the edge of the zone. Moving only a few metres into the shadow region causes an additional attenuation of up to 40 dB, which can be disastrous for a communication link. It seems likely that the amount of energy diffracted around obstacles is too small to be usable in the mobile environment, and although the use of diversity techniques can enhance performance, system designers will probably have to ensure that a direct line-of-sight path exists between the transmitter and receiver.

In addition to 60 GHz there have also been measurements at lower frequencies, particularly around 11 GHz [52,53] in both urban and rural areas. Reudink's results at 11.2 GHz in New Jersey and New York City were not obtained with microcellular systems in mind and the base antenna heights were typical of large-cell systems. Suburban measurements in New Jersey showed an excess loss of 37 dB relative to

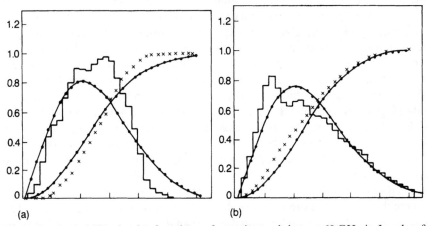

(a)   (b)

Figure 4.19 Probability density functions of experimental data at 60 GHz in London for (a) path length 100 m, (b) path length 190 m: (——) experimental PDF, (×) experimental CDF, (●) best-fit Rayleigh PDF and CDF.

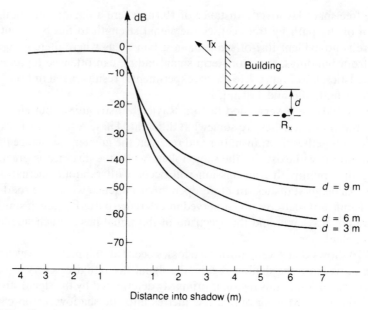

Figure 4.20 Signal attenuation in a shadow zone: *d* is distance from the wall.

the free space value, and in New York City this value increased to 46 dB. The distribution of the excess loss was closely represented by lognormal statistics with a standard deviation of about 10 dB. It was found that in urban areas the standard deviation decreased to about 8 dB within 1 mile of the base station, but this was principally due to changes in the local building density and height rather than being a function of range.

Interesting results have been presented for rural and semi-urban areas [53]. Figure 4.21 shows variations in the mean signal strength for a propagation path along a straight road where a line-of-sight path exists. Experimental conditions were such that the receiver (housed in a vehicle) could traverse a path of about 300 m on either side of a base site. The regular notch pattern is quite clear and it is apparent that the spatial frequency of the notches decreases with increased base-to-mobile distance. It was further observed that the notch separation increases with a decrease in base antenna height. The dotted curve in Figure 4.21 represents the calculated received power in a plane earth propagation mode and can be compared with Figure 2.5. It provides strong evidence that the propagation mechanism is due to direct and ground-reflected waves. The exact locations of some of the notches do not match the theoretical predictions but this is because the road surface is not perfectly smooth and flat. Small undulations of the road can easily shift the location of the notches.

In semi-urban areas the fit between measurements and theory is not quite so obvious as Figure 4.22 shows. A notch pattern is observable but the measurements are more random in nature, possibly due to the presence of more than one reflected (or scattered) ray. Where there are deep notches, one reflected ray dominates. The mean signal power at the receiver does not appear to decay faster than d^{-2} with base-to-mobile distance, and rapid fluctuations of the type normally associated with Rayleigh statistics are only apparent at very low signal levels.

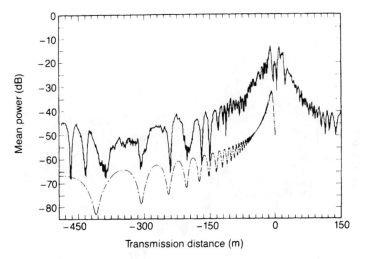

Figure 4.21 Measured and calculated signal power values at 11.2 GHz measured along a straight road in a rural area; $h_b = 9$ m and $h_m = 1.8$ m: (——) measured and ($-\cdot-$) calculated.

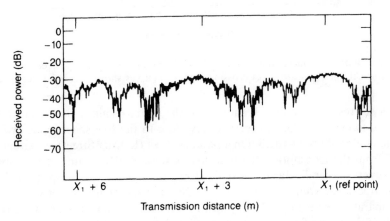

Figure 4.22 Worst-case variations in received signal power at 11.2 GHz measured over a 6 m length of road in a semi-urban area.

4.4.2 UHF measurements

Research workers such as Young and Okumura made measurements in urban areas at frequencies up to 3.7 GHz but they were not short-range measurements with relatively low base station antennas, designed with microcellular systems in mind. Measurements of that kind show somewhat different results, principally because a line-of-sight path often exists over some of the coverage area.

Whitteker [54], conducted measurements in Ottawa, Canada, using a transmitting antenna 8.5 m above the ground, the signal being received using a vehicle with an antenna at 3.65 m above ground level. Both antennas were vertical monopoles (nominally omnidirectional). The measurements were designed to determine representative signal levels, not to resolve the fast fading pattern. Figure 4.23 is

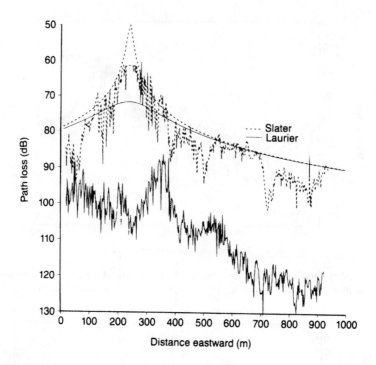

Figure 4.23 Path loss measurements at 910 MHz along two parallel streets in Ottawa, Canada; vertical lines indicate street crossings, smooth curves show calculated free space loss.

typical of the results obtained. It shows that the path loss along the street where the transmitter is located (Slater Street) is very close to the free space value except for occasional fades of about 10 dB. On a parallel street (Laurier Street), apparently well shielded from the transmitter, the path loss is some 25 dB greater. It was observed that on streets perpendicular to the main street, the signal level dropped very rapidly from the free space value – approximately 20 dB in less than a street width.

A comparative study at 900 MHz and 1.8 GHz has been undertaken in Melbourne, Australia [55]. At the fixed site, the antenna heights were between 5 and 20 m, well below the height of the surrounding buildings. The signal strength was measured at distances up to 1 km, the measuring vehicle having an antenna mounted 1.5 m above street level. A line-of-sight path always existed. It was found that the attenuation could be modelled by two straight lines, one representing the region between the base station and a point about 150 m away, the other representing greater distances. Close to the base station the slope is inverse square, suggestive of a ducting mode; beyond the so-called turning point (or breakpoint) the slope is greater. Other authors have also reported a turning point in the path loss curve. Results reported by Kaji and Akeyama [56] show a change at about 350 m and a similar trend was observed at 900 MHz in New Zealand [57]. There is a high correlation between the measurements made at 900 MHz and 1.8 GHz, although there is often a larger spread in the results at the higher frequency. The 1.8 GHz signal was more sensitive to shadowing when other vehicles moved in front of the measurement vehicle, and this could be the main reason for the increased variability.

Kaji and Akeyama offered no physical explanation for the existence of a breakpoint, but compared their experimental results with predictions made using extrapolated versions of the models proposed by Okumura [6] and Hata [15]. Although not intended for use at distances less than 1 km, these models are often used for predicting signal strength in cellular radio systems. The slope of the path loss curve at distances less than 1 km was lower than predicted by Okumura – 20–25 dB/decade compared with 31–35 dB/decade – and this is reasonable if a line-of-sight component exists in the signal. It was concluded that the extrapolated models do not adequately cover the microcellular situation, particularly with reference to the observed turning point.

It is often assumed that the size of a microcell will be controlled by adjusting the transmitter power, the shape being influenced by the antenna radiation pattern. Antenna height, however, also has the potential to be a powerful factor. Measurements have been made in the UK [58] at frequencies of 900, 1500 and 1800 MHz using transmitting antenna heights of 7, 14 and 21 m. Signal strength was measured using transportable equipment, mounted on a trolley and wheeled along the nearby streets. Signal variability was well predicted by the two-ray model of Rustako [53]. Because a line-of-sight path rarely exists to the side streets, the variability is greater than on the main streets. The fast fading component of the signal was extracted by normalising using the running mean (or moving average) technique with a window width of 20λ and analysis revealed that the data fitted well with a Rician distribution (Section 5.11). This is not surprising since, in the main streets at least, there will usually be a dominant component.

For distances up to about 100 m from the transmitter the signal level is almost independent of antenna height. Beyond 100 m height effects become apparent but the power/distance factor is of the order of 2, as suggested by Rustako [53]. The signal strength drops rapidly when the receiver moves into a side street. If the transmission to the side street is blocked by multi-storey buildings (a typical case in city centres) then, compared with the level at the intersection, it drops by about 20 dB over a distance of 25–30 m. This distance increases to 45–50 m if the transmission is blocked by one- or two-storey buildings. Similar results were presented by Chia *et al.* [59]. Over the short distances used in these tests it appeared that the transmission frequency did not affect either the cell size or the variability of the received signal.

4.4.3 Microcellular modelling

There have been several approaches to the modelling of propagation in a microcellular environment. Both theoretical and empirical methods have been used and ray tracing techniques have also been investigated. Ray tracing techniques require the availability of a detailed database but produce a site-specific model which gives both path loss and time dispersion. The multipath effects, so important in urban areas, depend on the relative height of the base station antenna and the surrounding buildings, and models therefore have to take into account diffraction over and around buildings if they aim to include all dominant paths. Models which assume infinitely high buildings can only consider diffraction around buildings.

Models based on ray tracing yield a very site-specific, three-dimensional path loss and time dispersion prediction, provided the database includes information about building height and shape. Computation time is often a problem, however, and in any case the prediction has to be rerun from the beginning for each and every location of the mobile, even if the base station is fixed. Generally, existing microcellular models are valid only in flat urban areas and little attention has been given to the influence of terrain variations; the effects of vegetation have also been largely ignored. Both these aspects are important, however, and need to be incorporated specifically as models are refined in the future.

It appears that the propagation mechanisms in a microcellular scenario differ from those of a large-cell system in that a street-guided wave exists, at least in the immediate vicinity of the transmitter. As a result of their measurements, Kaji and Akeyama [56] suggested an empirical model in the form

$$S = -20 \log_{10}\{d^a(1 + d/g)^b\} + c \tag{4.39}$$

for base antenna heights in the range 5–20 m and for distances between 200 m and 1 km. In this equation,

   S is the signal level (dBμV)
   d is the distance from the transmitter
   a is the basic attenuation rate for short distances (approximately 1)
   b is the additional attenuation rate for distances beyond the turning point
   g is the distance corresponding to the turning point
   c is the offset factor

The model has two limiting cases. For distances significantly less than g the attenuation is such that

$$S = -20 \log_{10} d^a + c \tag{4.40}$$

For distances greater than g the attenuation fall-off rate tends to be such that

$$S = -20 \log_{10} d^{(a+b)} + c + constant \tag{4.41}$$

This model was fitted by the authors to their measured results at 900 MHz; the values for the various coefficients are given in Table 4.8.

Several other models have been developed more recently for propagation below rooftop level. The dual-slope behaviour in the LOS situation has been confirmed [60], and the non-LOS situation has also received much attention. Close to the base station a line-of-sight condition exists but further away, and always in side streets, traffic, buildings and other obstacles intervene to cause shadowing and hence a different propagation law exponent. One approach is to recognise that as distance from the base station increases, the first Fresnel zone gradually becomes obstructed by the ground and there comes a point when the clearance is less than that required for free space transmission. The breakpoint distance can then be calculated approximately by assuming the breakpoint marks the position where the first Fresnel zone just touches the ground surface.

In several approaches, the Walfisch–Bertoni model or a variation of it [61,62], is used to model over-rooftop diffraction. An analytical, semideterministic approach has been described [63] in which ground reflections are considered and street-corner diffraction is modelled using the UTD; the finite conductivity of the walls is dealt

Table 4.8 Results of fitting the proposed propagation model to experimental results at 900 MHz using a least squares regression procedure

Antenna height (m)	a	b	g	C	Sum of errors squared
5	1.15	−0.14	148.6	94.5	309
9	0.74	0.27	151.8	79.8	246
15	0.20	1.05	143.9	55.5	577
19	−0.48	2.36	158.3	37.3	296

with by introducing heuristic coefficients. The model can cope with arbitrary angles of street crossings. Several ray tracing approaches have been described with various levels of sophistication. One uses a two-dimensional building layout database in which the positions of the walls are defined in terms of vectors [64]. The permittivity and conductivity of the walls can be taken into account but in practice these electrical characteristics are assumed to be the same for all buildings. The software computes all reflected and diffracted rays up to a specified order and the path loss is determined by the phasor addition of all rays that reach the receiving point. The channel impulse response can be evaluated by considering the magnitude, phase and delay of each received ray. A similar principle is used in another model [65] which also takes into account multiple reflected and multiple wall-penetrating rays as well as combinations of these rays.

Models which incorporate over-rooftop diffraction are often based on the Walfisch–Bertoni approach [66], although the Deygout diffraction construction has also been used [67]. An interesting extension to Walfisch–Bertoni is through the consideration of building heights within the vertical propagation plane. Some empiricism is always necessary, however, because even the most sophisticated building databases do not provide information about the shape of rooftops. For base stations which are well above rooftop height it is almost invariably assumed that rooftop-to-street diffraction over the last building is the dominant loss mechanism affecting the radio link.

Three-dimensional models have also been developed and the transition region between LOS and non-LOS, which is very important for system design, has been studied in some detail [68]. Again the primary approach is through ray tracing or ray launching techniques with a variety of methods used to describe the environment, to account for the electrical characteristics of the building materials and to estimate the diffraction losses [69–72].

Comparison with measurements

A path loss measurement campaign was undertaken in Munich, Germany, to provide data for model testing. The tests were carried out at 947 MHz using transmitting and receiving antenna heights of 13 m and 1.5 m respectively. The transmitting antenna height was below the rooftop level of most of the buildings in the test area, so the vast majority of the receiver locations were in a non-LOS situation. Taken overall, the generation of models outlined above produced an average standard deviation of the prediction error in the range 7–9 dB with the three-dimensional models among the best. Over-rooftop and around-building diffraction losses appear to be of the same order of magnitude, so models which incorporate only one of them are at risk of seriously underestimating the total loss. The reasons for the discrepancies between predictions and measurements almost certainly include inaccuracies in the database and the fact that the walls of buildings are represented as smooth surfaces. The real environment also includes features like lamp posts, cars and other vehicles which are not taken into account.

None of the models incorporate any losses due to vegetation, although there appears to be no reason why one of the models presented in Chapter 3 should not be included. Although his work was not specifically directed to the microcellular scenario, Tameh [39] has shown that vegetation is a significant contributor to loss in urban areas.

4.5   DISCUSSION

Although there appears to be a huge potential for improved prediction methods based on deterministic or semideterministic processes through the availability of improved databases of various kinds and the ready availability of small, powerful computers, the fact remains that currently, for macrocells, the Okumura–Hata model is still the most used. This is undoubtedly due to its simplicity and its proven reliability. However, many variations of the original approach have been proposed [73,74] where the basic loss has been combined with further losses calculated using various knife-edge diffraction models. Alternative methods of defining parameters such as the effective base station antenna height and establishing the correction factor in irregular terrain have also been investigated.

It is clear that accurate ways of representing geographical data and efficient methods of extracting information from the database are essential for the development of improved and computationally efficient propagation tools. Predictions in microcellular environments, for example, could benefit greatly from the availability of accurate high-resolution data, and aerial photography which can produce 1–2 m resolution is potentially very useful in this context. For urban areas in general, it is desirable to have reliable information about the average height of individual buildings and groups of buildings when making diffraction calculations or when antennas are likely to be close to rooftop height.

Clutter databases which include vegetation are also very important as the signal is highly sensitive to scatterers in the immediate vicinity of the antenna. As an example, it has been shown that the average difference in predictions for over-rooftop propagation estimates in Munich, Germany, obtained by using mean, rather than actual, rooftop height data is about 4 dB, and that errors of up to 15 dB in microcells are attributable to database inaccuracies arising from the omission of vegetation data and poor resolution of terrain data height. This is useful information, indicating what could be done if better data were available.

Nevertheless, the strength of the chain is that of its weakest link and nowhere is that more obvious than here. There is little point in having highly accurate methods for calculating diffraction loss if they depend on obstacle shapes still only coarsely defined. A strategic, integrated approach is surely needed.

REFERENCES

1. Ibrahim M.F. (1982) Signal strength prediction for mobile radio communication in built-up areas. PhD thesis, University of Birmingham.
2. Jao J.K. (1984) Amplitude distribution of composite terrain radar clutter and the *K*-distribution. *IEEE Trans.*, **AP32**(10), 1049–62.
3. Kozono S. and Watanabe K. (1977) Influence of environmental buildings on UHF land mobile radio propagation. *IEEE Trans.*, **COM25**(10), 1133–43; see also **COM26**, 199–200 (1978).
4. Ibrahim M.F. and Parsons J.D. (1983) Signal strength prediction in built-up areas. Part 1: median signal strength. *Proc. IEE Part F*, **130**(5), 377–84.
5. Huish P.W. and Gurdenli E. (1988) Radio channel measurements and prediction for future mobile radio systems. *Br. Telecom. Tech. J.*, **6**(1), 43–53.
6. Okumura Y., Ohmori E., Kawano T. and Fukuda K. (1968) Field strength and its variability in the VHF and UHF land mobile radio service. *Rev. Elec. Commun. Lab.*, **16**(9/10), 825–73.
7. King R.W. and Causebrook J.H. (1974) *Computer programs for UHF co-channel interference prediction using a terrain data bank*. BBC Research Report RD 1974/6.
8. Lorenz R.W. (1986) *Field strength prediction program DBPR1*. COST207 Technical Document TD(86) #02.
9. Juul-Nyholm G. (1986) *Land usage figures for use in field strength prediction as measured in Denmark*. COST207 Technical Document TD(86) #28.
10. Lord Chorley (1987) *Handling Geographic Information*. HMSO, London.
11. Kafaru, O.O. (1989) An environment-dependent approach to wideband modelling and computer simulation of UHF mobile radio propagation in built-up areas. PhD thesis, University of Liverpool.
12. Young, W.R. (1952) Comparison of mobile radio transmission at 150, 450, 900 and 3700 MC. *Bell Syst. Tech J.*, **31**, 1068–85.
13. Allsebrook, K. and Parsons, J.D. (1977) Mobile radio propagation in British cities at frequencies in the VHF and UHF bands. *Proc. IEE*, **124**(2), 95–102. See also *IEEE Trans.*, **VT26**, 313–22 (1977).
14. Delisle G.Y., Lefevre J.-P., Lecours M. and Chourinard J.-Y. (1985) Propagation loss prediction: a comparative study with application to the mobile radio channel. *IEEE Trans.*, **VT34**(2), 86–95.
15. Hata M. (1980) Empirical formula for propagation loss in land mobile radio services. *IEEE Trans.*, **VT29**(3), 317–25.
16. Aurand J.F. and Post R.E. (1985) A comparison of prediction models for 800 MHz mobile radio propagation. *IEEE Trans.*, **VT34**(4), 149–53.
17. COST231 (1999) *Digital mobile radio: towards future generation systems*. Final Report EUR18957, Ch. 4.
18. Akeyama A., Nagatsu T. and Ebine Y. (1982) Mobile radio propagation characteristics and radio zone design method in local cities. *Rev. Elec. Commun. Lab.*, **30**, 308–17.
19. Walfisch J. and Bertoni H.L. (1988) A theoretical model of UHF propagation in urban environments. *IEEE Trans.*, **AP36**(12), 1788–96.
20. Ikegami F. and Yoshida S. (1984) Analysis of multipath propagation structure in urban mobile radio environments. *IEEE Trans.*, **AP28**, 531–7.
21. Ikegami F., Yoshida S., Tacheuchi T. and Umehira M. (1984) Propagation factors controlling mean field strength on urban streets. *IEEE Trans.*, **AP32**(8), 822–9.
22. Ott G.D. and Plitkins A. (1978) Urban path-loss characteristics at 820 MHz. *IEEE Trans.*, **VT27**, 189–97.
23. Saunders S.R. and Bonar F. (1994) Prediction of mobile radio wave propagation over buildings of irregular heights and spacings. *IEEE Trans.*, **AP42**(2), 137–44.
24. Maciel L.R., Bertoni H.L. and Xia H.H. (1993) Unified approach to prediction of propagation over buildings for all ranges of base station antenna height. *IEEE Trans.*, **VT43**(1), 35–41.

25. Maciel L.R., Bertoni H.L. and Xia H.H. (1992) Propagation over buildings for paths oblique to the street grid. *Proc. PIMCR'92*, Boston MA, pp. 75–9.
26. Low K. (1992) Comparison of urban propagation models with CW-measurements. *Proc IEEE VTC'92 Conference*, pp. 936–42.
27. Kurner T., Fauss R. and Wasch A. (1996)A hybrid propagation modelling approach for DCS1800 macro cells. *Proc IEEE VTC'96 Conference*, pp. 1628–32.
28. Cardona N., Möller P. and Alfonso F. (1995) Applicability of Walfisch-type urban propagation models. *Electron. Lett.*, **31**(23), 1971–2.
29. Lee W.C.-Y. (1982) *Mobile Communications Engineering.* McGraw-Hill, New York.
30. Lee W.C.-Y. (1986) *Mobile Communication Design Fundamentals.* Sams & Co., Indianapolis.
31. McGeehan J.P. and Griffiths J. (1985) Normalised prediction chart for mobile radio reception. *Proc. ICAP'85 (IEE Conference Publication 248)*, pp. 395–9.
32. Atefi A. and Parsons J.D. (1986) Urban radio propagation in mobile radio frequency bands. *Proc. Comms 86 (IEE Conference Publication 262)*, pp. 13–18.
33. Hviid J.T., Anderson J.B., Toftgard J. and Bojer J. (1995) Terrain-based propagation model for rural areas – an integral equation approach. *IEEE Trans.*, **AP43**(1), 41–6.
34. Moroney D. and Cullen P. (1995) A fast integral equation approach to UHF coverage estimation. In *Mobile and Personal Communications* (ed. del Re). Elsevier, New York, pp. 343–50.
35. Berg J.E. and Holmquist H. (1994) An FFT multiple half-screen diffraction model. *Proc. IEEE VTC'94 Conference*, Stockholm, Sweden, pp. 195–9.
36. Berg J.E. (1994) A macrocell model based on the parabolic differential equation. *Proc. Virginia Tech 4th Symposium on Wireless Personal Communications*, pp. 9.1 to 9.10.
37. Stocker K.E., Gschwendtner B.E. and Landstorfer F.M. (1993) An application of neural networks to prediction of terrestrial wave propagation for mobile radio. *Proc. IEE Part H*, **140**(4), 315–20.
38. Lebherz M., Weisbeck W. and Krank W. (1992) A versatile wave propagation model for the VHF/UHF range considering three-dimensional terrain. *IEEE Trans.*, **AP40**(10), 1121–31.
39. Tameh E.K. (1999) The development and evaluation of a deterministic mixed cell propagation model based on radar cross–section theory. PhD thesis, University of Bristol.
40. McDonald V.H. (1979) The cellular concept. *Bell Syst. Tech. J.*, **58**(1), 15–41.
41. Steele R. (1985) Towards a high capacity digital cellular mobile radio system. *Proc. IEE Part F*, **132**(5), 405–15.
42. Steele R. and Prabhu V.K. (1985) Mobile radio cell structures for high user density and large data rates. *Proc. IEE Part F*, **132**(5), 396–404.
43. King B.G. (1977) A 60 GHz short-hop 50 Megabit digital radio system. *Proc. Eascon 77*, pp. 24-1A to 24-1D.
44. Bodtman W.F. and Ruthroff C.L. (1974) Rain attenuation and short radio paths: theory, experiment and design. *Bell Syst. Tech. J.*, **53**(7), 1329–49.
45. Delange O.E., Deitrich A.F. and Hogg D.C. (1975) An experiment on propagation of 60 GHz waves through rain. *Bell Syst. Tech. J.*, **54**(1), 165–76.
46. Straiton A.W. (1975) The absorption and re-radiation of radio waves by oxygen and water vapour in the atmosphere. *IEEE Trans.*, **AP23**, 595–7.
47. Liebe H.J., Gimnestad G.C. and Happonen J.D. (1977) Atmospheric oxygen microwave spectrum – experiment versus theory. *IEEE Trans.*, **AP25**(3), 327–35.
48. Zeimer R.E. (1984) An overview of millimetre wave communications. *Proc. 14th European Microwave Conference*, pp. 3–8.
49. Tharek A.R., Kanso A. and McGeehan J.P. (1986) Initial propagation measurements for 60 GHz mobile radio. *IEE Colloquium on Millimetre Wave Propagation, Digest*, 1986/17.
50. Hawkins N.D., Steele R., Reckard D.C. and Shepherd C.R. (1985) Path loss characteristics of 60 GHz transmissions. *Electron. Lett.*, **21**(22), 1054–5.
51. Thomas H.J,. Siqueira G.L. and Cole R.S. (1986) Millimetre-wavelength cellular radio: propagation measurements at 55 GHz. *Proc. 16th European Microwave Conf.*, Dublin, pp. 221–6.

52. Reudink D.O. (1972) Comparison of radio transmissions at X-band frequencies in suburban and urban areas *IEEE Trans.*, **AP20**, 470–3.
53. Rustako A.J., Amitay N., Owens G.J. and Roman R.S. (1989) Propagation measurements at microwave frequencies for microcellular mobile and personal communications. *Proc. 39th IEEE Vehicular Technology Conf.*, San Francisco CA, pp. 316–20.
54. Whitteker J.H. (1988) Measurements of path loss at 910 MHz for proposed microcell urban mobile systems. *IEEE Trans.*, **VT37**(3), 125–9.
55. Harley P. (1989) Short distance attenuation measurements at 900 MHz and 1.8 GHz using low antenna heights for microcells. *IEEE J. Selected Areas in Communications*, **7**(1), 5–11.
56. Kaji M. and Akeyama A. (1985) UHF-band propagation characteristics for land mobile radio. *AP-S International Symposium Digest*, **2**, 835–8.
57. Williamson A.G., Egan B. and Chester J.W. (1984) Mobile radio propagation in Auckland at 851 MHz. *Electron. Lett.*, **20**(12), 517–18.
58. Turkmani A.M.D., Parsons J.D., Ju Feng and Lewis D.G. (1989) Microcellular Radio Measurements at 900, 1500 and 1800 MHz. *Proc. 5th Int. Conf. on Land Mobile Radio*, Warwick, pp. 65–8.
59. Chia S.T.S., Steele R., Green E. and Baran A. (1987) Propagation and bit error ratio measurements for a microcellular system. *J. IERE.*, **57**(6), S255–66 (suppl.).
60. Bertoni H.L., Honcharenko L.R., Maciel L.R. and Xia H.H. (1994) UHF propagation prediction for wireless personal communications. *Proc. IEEE*, **82**(9), 1333–59.
61. Cardona N., Navarro F. and Möller P. (1994) Applicability of Walfisch-type urban propagation models. COST231 TD(94) 134.
62. Saunders S.R. and Boner F. (1994) Prediction of mobile radio propagation over buildings of irregular heights and spacing. *IEEE Trans.*, **AP42**(2), 137–44.
63. Wiart W. (1994) Microcellular modelling when base station is below rooftops. *Proc. IEEE VTC'94 Conference*, Stockholm, Sweden, pp. 200–4.
64. Rizk K., Wagen J.-F. and Gardiol F. (1994) Ray tracing based path loss prediction in two microcellular environments. *Proc. PIMCR'94*, The Hague, Netherlands, pp. 384–8.
65. Cichon D.J. and Wiesbeck W. (1994) Indoor and outdoor propagation modelling in picocells. *Proc. PIMCR'94*, The Hague, Netherlands, pp. 491–5.
66. Möller P., Cardona N. and Cichon D.J. (1994) *Investigations in a new developed urban propagation model*. COST231, TD(94)135.
67. Perucca M. (1994) *Small cells characterization using 3D database and Deygout diffraction approach*. COST231, TD(94)54.
68. Dersch U. and Zollinger E. (1994) Propagation mechanisms in microcell and indoor environments. *IEEE Trans.*, **VT43**(4), 1058–66.
69. Rossi J.P., Bic J.C. and Levy A.J. (1990) *A ray launching model in urban areas*. COST231, TD(90)78.
70. Daniell P., Degli-Espositi V., Falciasecca G. and Riva G. (1994) Field prediction tools for wireless communications in outdoor and indoor environments. *Proc IEEE MTT-Symposium on Technologies for Wireless Applications*, Turin, Italy.
71. Gschwendtner B.E., Wolfe G., Burk B. and Landsdorfer F.M. (1995) Ray tracing vs. ray launching in 3-D microcell modelling. *Proc. EPMCC'95*, Bologna, Italy, pp. 74–9.
72. Cichon D.J. and Wiesbeck W. (1995) Comprehensive ray optical propagation models for indoor and outdoor environments: theory and applications. *Proc. COMMSPHERE'95*, Eilat, Israel, pp. 201–8.
73. Kurner T. and Fauss R. (1995) Untersuchungen zur feldstarkepradition im 1800 MHz-bereich (Investigation of path-loss algorithms at 1800 MHz). *Nachrichtentechnik Elektronic, ne-Science*, **45**, 18–23.
74. Badsberg M., Andersen J.B. and Mogensen P. (1995) *Exploitation of the terrain profile in the Hata model*. COST231, TD(95)9.

Chapter 5

Characterisation of Multipath Phenomena

5.1 INTRODUCTION

In Chapter 3 we described some methods for predicting path losses, concentrating on those applicable to mobile communication systems. The discussion centred around techniques that deal principally with radio propagation over irregular terrain; methods of predicting signal strength in urban areas or in other environments, e.g. inside buildings, were deliberately left until Chapter 4. These propagation models are extremely important since the vast majority of mobile communication systems operate in and around centres of population. Having introduced them, we can now go into more detail about the propagation mechanism in built-up areas, not only qualitatively but also in terms of a mathematical model. In that way we can understand the full significance of the prediction techniques and indicate the ways forward towards a global model that includes the effects of topographic and environmental factors.

The major problems in built-up areas occur because the mobile antenna is well below the surrounding buildings, so there is no line-of-sight path to the transmitter. Propagation is therefore mainly by scattering from the surfaces of the buildings and by diffraction over and/or around them. Figure 5.1 illustrates some possible mechanisms by which energy can arrive at a vehicle-borne antenna. In practice energy arrives via several paths simultaneously and a *multipath* situation is said to exist in which the various incoming radio waves arrive from different directions with different time delays. They combine vectorially at the receiver antenna to give a resultant signal which can be large or small depending on the distribution of phases among the component waves.

Moving the receiver by a short distance can change the signal strength by several tens of decibels because the small movement changes the phase relationship between the incoming component waves. Substantial variations therefore occur in the signal amplitude. The signal fluctuations are known as *fading* and the short-term fluctuation caused by the local multipath is known as *fast fading* to distinguish it from the much longer-term variation in mean signal level, known as *slow fading*.

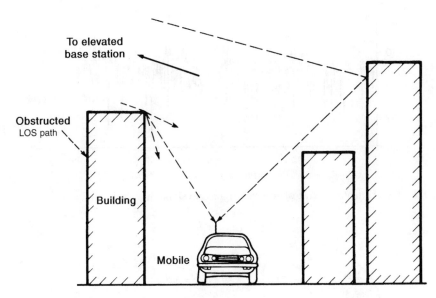

Figure 5.1 Radio propagation in urban areas.

Slow fading was mentioned in Chapter 3 and is caused by movement over distances large enough to produce gross variations in the overall path between the transmitter and receiver. Because the variations are caused by the mobile moving into the shadow of hills or buildings, slow fading is often called *shadowing*. Unfortunately there is no complete physical model for the slow fading, but measurements indicate that the mean path loss closely fits a lognormal distribution with a standard deviation that depends on the frequency and the environment (Chapter 3). For this reason the term *lognormal fading* is also used.

The terms 'fast' and 'slow' are often used rather loosely. The fading is basically a spatial phenomenon, but spatial variations are experienced as temporal variations by a receiver moving through the multipath field. The typical experimental record of received signal envelope as a function of distance shown in Figure 5.2 illustrates this point. The fast fading is observed over distances of about half a wavelength. Fades with a depth less than 20 dB are frequent, with deeper fades in excess of 30 dB being less frequent but not uncommon. The slow variation in mean signal level, indicated in Figure 5.2 by the dotted line, occurs over much larger distances. A receiver moving at 50 kph can pass through several fades in a second, or more seriously perhaps, it is possible for a mobile to stop with the antenna in a fade. Theoretically, communication then becomes very difficult but, in practice, secondary effects often disturb the field pattern, easing the problem significantly.

Whenever relative motion exists between the transmitter and receiver, there is an apparent shift in the frequency of the received signal due to the Doppler effect. We will return to this later; for now it is sufficient to point out that Doppler effects are a manifestation in the frequency domain of the envelope fading in the time domain.

Although physical reasoning suggests the existence of two different fading mechanisms, in practice there is no clear-cut division. Nevertheless, Figure 5.2

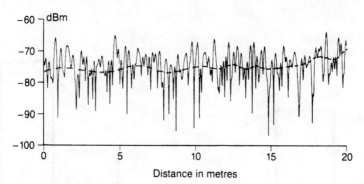

Figure 5.2   Experimental record of received signal envelope in an urban area.

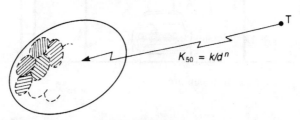

Figure 5.3   Model of mobile radio propagation showing small areas where the mean signal is constant within a larger area over which the mean value varies slowly as the receiver moves.

shows how to draw a distinction between the short-term multipath effects and the longer-term variations of the local mean. Indeed, it is convenient to go further and suggest that in built-up areas the mobile radio signal consists of a local mean value, which is sensibly constant over a small area but varies slowly as the receiver moves; superimposed on this is the short-term rapid fading. In this chapter we concentrate principally on the short-term effects for narrowband channels; in other words, we consider the signal statistics within one of the small shaded areas in Figure 5.3, assuming the mean value to be constant. In this context, 'narrowband' should be taken to mean that the spectrum of the transmitted signal is narrow enough to ensure that all frequency components are affected in a similar way. The fading is said to be flat, implying no frequency-selective behaviour.

5.2   THE NATURE OF MULTIPATH PROPAGATION

A multipath propagation medium contains several different paths by which energy travels from the transmitter to the receiver. If we begin with the case of a stationary receiver then we can imagine a *static multipath* situation in which a narrowband signal, e.g. an unmodulated carrier, is transmitted and several versions arrive sequentially at the receiver. The effect of the differential time delays will be to introduce relative phase shifts between the component waves, and superposition of the different components then leads to either constructive or destructive addition (at any given location) depending upon the relative phases. Figure 5.4 illustrates the two extreme possibilities. The resultant signal arising from propagation via paths A and B will be large because of constructive addition, whereas the resultant signal from paths A and C will be very small.

If we now turn to the case when either the transmitter or the receiver is in motion, we have a *dynamic multipath* situation in which there is a continuous change in the electrical length of every propagation path and thus the relative phase shifts between them change as a function of spatial location. Figure 5.5 shows how the received amplitude (envelope) of the signal varies in the simple case when there are two incoming paths with a relative phase that varies with location. At some positions

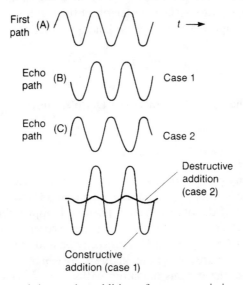

Figure 5.4 Constructive and destructive addition of two transmission paths.

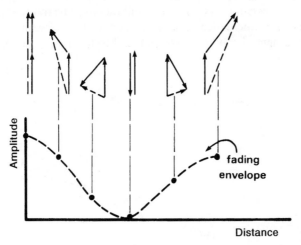

Figure 5.5 How the envelope fades as two incoming signals combine with different phases.

there is constructive addition, at others there is almost complete cancellation. In practice there are several different paths which combine in different ways depending on location, and this leads to the more complicated signal envelope function in Figure 5.2. The space-selective fading which exists as a result of multipath propagation is experienced as time-selective fading by a mobile receiver which travels through the field.

The time variations, or dynamic changes in the propagation path lengths, can be related directly to the motion of the receiver and indirectly to the Doppler effects that arise. The rate of change of phase, due to motion, is apparent as a Doppler frequency shift in each propagation path and to illustrate this we consider a mobile moving with velocity v along the path AA′ in Figure 5.6 and receiving a wave from a scatterer S. The incremental distance d is given by $d = v \Delta t$ and the geometry shows that the incremental change in the path length of the wave is $\Delta l = d \cos \alpha$, where α is the spatial angle in Figure 5.6. The phase change is therefore

$$\Delta \phi = -\frac{2\pi}{\lambda} \Delta l = -\frac{2\pi v \Delta t}{\lambda} \cos \alpha$$

and the apparent change in frequency (the Doppler shift) is

$$f = -\frac{1}{2\pi} \frac{\Delta \phi}{\Delta t} = \frac{v}{\lambda} \cos \alpha \tag{5.1}$$

It is clear that in any particular case the change in path length will depend on the spatial angle between the wave and the direction of motion. Generally, waves arriving from ahead of the mobile have a positive Doppler shift, i.e. an increase in frequency, whereas the reverse is the case for waves arriving from behind the mobile. Waves arriving from directly ahead of, or directly behind the vehicle are subjected to the maximum rate of change of phase, giving $f_m = \pm v/\lambda$.

In a practical case the various incoming paths will be such that their individual phases, as experienced by a moving receiver, will change continuously and randomly. The resultant signal envelope and RF phase will therefore be random variables and it remains to devise a mathematical model to describe the relevant statistics. Such a model must be mathematically tractable and lead to results which are in accordance

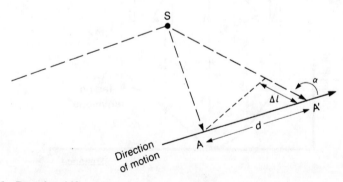

Figure 5.6 Doppler shift.

with the observed signal properties. For convenience we will only consider the case of a moving receiver.

5.3 SHORT-TERM FADING

Several multipath models have been suggested to explain the observed statistical characteristics of the electromagnetic fields and the associated signal envelope and phase. The earliest of these was due to Ossanna [1], who attempted an explanation based on the interference of waves incident and reflected from the flat sides of randomly located buildings. Although Ossanna's model predicted power spectra that were in good agreement with measurements in suburban areas, it assumes the existence of a direct path between transmitter and receiver and is limited to a restricted range of reflection angles. It is therefore rather inflexible and inappropriate for urban areas where the direct path is almost always blocked by buildings or other obstacles.

A model based on scattering is more appropriate in general, one of the most widely quoted being that due to Clarke [2]. It was developed from a suggestion by Gilbert [3] and assumes that the field incident on the mobile antenna is composed of a number of horizontally travelling plane waves of random phase; these plane waves are vertically polarised with spatial angles of arrival and phase angles which are random and statistically independent. Furthermore, the phase angles are assumed to have a uniform probability density function (PDF) in the interval $(0, 2\pi)$. This is reasonable at VHF and above, where the wavelength is short enough to ensure that small changes in path length result in significant changes in the RF phase. The PDF for the spatial arrival angle of the plane waves was specified a priori by Clarke in terms of an omnidirectional scattering model in which all angles are equally likely, so that $p_\alpha(\alpha) = 1/2\pi$. A model such as this, based on scattered waves, allows the establishment of several important relationships describing the received signal, e.g. the first- and second-order statistics of the signal envelope and the nature of the frequency spectrum. Several approaches are possible, a particularly elegant one being due to Gans [4].

The principal constraint on the model treated by Clarke and Gans is its restriction to the case when the incoming waves are travelling horizontally, i.e. it is a two-dimensional model. In practice, diffraction and scattering from oblique surfaces create waves that do not travel horizontally. It is clear, however, that those waves which make a major contribution to the received signal do indeed travel in an approximately horizontal direction, because the two-dimensional model successfully explains almost all the observed properties of the signal envelope and phase. Nevertheless, there are differences between what is observed and what is predicted, in particular the observed envelope spectrum shows differences at low frequencies and around $2f_{\rm m}$.

An extended model due to Aulin [5] attempts to overcome this difficulty by generalising Clarke's model so that the vertically polarised waves do not necessarily travel horizontally, i.e. it is three-dimensional. This is the generic model we will use in this chapter. It is necessarily more complicated than its predecessors and sometimes produces rather different results. The detailed mathematical analysis is available in the original references or in textbooks [6,7]. In this chapter we concentrate on indicating the methods of analysis, the physical interpretation of the results, and ways in which the information can be used by radio system designers.

5.3.1 The scattering model

At every receiving point we assume the signal to be the resultant of N plane waves. A typical component wave is shown in Figure 5.7, which illustrates the frame of reference. The nth incoming wave has an amplitude C_n, a phase ϕ_n with respect to an arbitrary reference, and spatial angles of arrival α_n and β_n. The parameters C_n, ϕ_n, α_n and β_n are all random and statistically independent. The mean square value of the amplitude C is given by

$$E\{C_n^2\} = \frac{E_0}{N} \tag{5.2}$$

where E_0 is a positive constant.

The generalisation in this approach occurs through the introduction of the angle β_n, which in Clarke's model is always zero. The phase angles ϕ_n are assumed to be uniformly distributed in the range $(0, 2\pi)$ but the probability density functions of the spatial angles α_n and β_n are not generally specified. At any receiving point (x_0, y_0, z_0) the resulting field can be expressed as

$$E(t) = \sum_{n=1}^{N} E_n(t) \tag{5.3}$$

where, if an unmodulated carrier is transmitted from the base station,

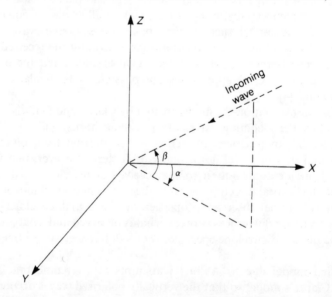

Figure 5.7 Spatial frame of reference: α is in the horizontal plane (XY plane), β is in the vertical plane.

$$E_n(t) = C_n \cos\left(\omega_0 t - \frac{2\pi}{\lambda}(x_0 \cos\alpha_n \cos\beta_n + y_0 \sin\alpha_n \cos\beta_n + z_0 \sin\beta_n) + \phi_n\right)$$

$$(5.4)$$

If we now assume that the receiving point (the mobile) moves with a velocity v in the xy plane in a direction making an angle γ to the x-axis then, after unit time, the coordinates of the receiving point can be written $(v \cos\gamma, v \sin\gamma, z_0)$. The received field can now be expressed as

$$E(t) = I(t) \cos\omega_c t - Q(t) \sin\omega_c t \qquad (5.5)$$

where $I(t)$ and $Q(t)$ are the in-phase and quadrature components that would be detected by a suitable receiver, i.e.

$$I(t) = \sum_{n=1}^{N} C_n \cos(\omega_n t + \theta_n)$$

$$(5.6)$$

$$Q(t) = \sum_{n=1}^{N} C_n \sin(\omega_n t + \theta_n)$$

and

$$\omega_n = \frac{2\pi v}{\lambda} \cos(\gamma - \alpha_n) \cos\beta_n$$

$$(5.7)$$

$$\theta_n = \frac{2\pi z_0}{\lambda} \sin\beta_n + \phi_n$$

In these equations, $\omega_n \ (= 2\pi f_n)$ represents the Doppler shift experienced by the nth component wave. Equations (5.3) to (5.7) reduce to the two-dimensional Clarke model if all waves are confined to the xy plane (i.e. if β is always zero).

If N is sufficiently large (theoretically infinite but in practice greater than 6 [8]) then by the central limit theorem the quadrature components $I(t)$ and $Q(t)$ are independent Gaussian processes which are completely characterised by their mean value and autocorrelation function. Because the mean values of $I(t)$ and $Q(t)$ are both zero, it follows that $E\{E(t)\}$ is also zero. Further, $I(t)$ and $Q(t)$ have equal variance σ^2 equal to the mean square value (the mean power). Thus the PDF of I and Q can be written as

$$p_x(x) = \frac{1}{\sigma\sqrt{2\pi}} \exp\left(-\frac{x^2}{2\sigma^2}\right) \qquad (5.8)$$

where $x = I(t)$ or $Q(t)$ and $\sigma^2 = E\{C_n^2\} = E_0/N$. We will return later to the significance of the autocorrelation function.

5.4 ANGLE OF ARRIVAL AND SIGNAL SPECTRA

If either the transmitter or receiver is in motion, the components of the received signal will experience a Doppler shift, the frequency change being related to the spatial angles of arrival α_n and β_n, and the direction and speed of motion. In terms of the frame of reference shown in Figure 5.7, the nth component wave has a frequency change given by eqn. (5.7) as

$$f_n = \frac{\omega_n}{2\pi} = \frac{v}{\lambda}\cos(\gamma - \alpha_n)\cos\beta_n \tag{5.9}$$

It is apparent that all frequency components in a transmitted signal will be subjected to this Doppler shift. However, if the signal bandwidth is fairly narrow it is safe to assume they will all be affected in the same way. We can therefore take the carrier component as an example and determine the spread in frequency caused by the Doppler shift on component waves that arrive from different spatial directions. The receiver must have a bandwidth sufficient to accommodate the total Doppler spectrum.

The RF spectrum of the received signal can be obtained as the Fourier transform of the temporal autocorrelation function expressed in terms of a time delay τ as

$$E\{E(t)E(t+\tau)\} = E\{I(t)I(t+\tau)\}\cos\omega_c\tau - E\{I(t)Q(t+\tau)\}\sin\omega_c\tau$$
$$= a(\tau)\cos\omega_c\tau - c(\tau)\sin\omega_c\tau \tag{5.10}$$

The correlation properties are therefore expressed by $a(\tau)$ and $c(\tau)$, which Aulin [5] has shown to be

$$a(\tau) = \frac{E_0}{2}E\{\cos\omega\tau\}$$
$$c(\tau) = \frac{E_0}{2}E\{\sin\omega\tau\} \tag{5.11}$$

To proceed further we need to make some assumptions about the PDFs of α and β. Aulin followed Clarke in assuming that waves arrive from all angles in the azimuth (xy) plane with equal probability, i.e.

$$p_\alpha(\alpha) = \frac{1}{2\pi} \tag{5.12}$$

With this assumption, $a(\tau)$ is given by

$$a(\tau) = \frac{E_0}{2}\int_{-\pi}^{+\pi} J_0(2\pi f_m\tau\cos\beta)p_\beta(\beta)\mathrm{d}\beta \tag{5.13}$$

where $J_0(.)$ is the zero-order Bessel function of the first kind and $c(\tau) = 0$.

In general, the power spectrum is given by the Fourier transform of eqn. (5.13); for the particular case of Clarke's two-dimensional model $p_\beta(\beta) = \delta(\beta)$ and in this case eqn. (5.13) becomes

$$a_0(\tau) = \frac{E_0}{2}J_0(2\pi f_m\tau) \tag{5.14}$$

Taking the Fourier transform, the power spectrum of $I(t)$ and $Q(t)$ is given by

$$A_0(f) = F[a_0(\tau)] = \begin{cases} \dfrac{E_0}{4\pi f_m} \left(\dfrac{1}{\sqrt{1 - (f/f_m)^2}} \right) & |f| \leqslant f_m \\ 0 & \text{elsewhere} \end{cases} \tag{5.15}$$

This spectrum is strictly band-limited within the maximum Doppler shift $f_m = \pm v/\lambda$ but the power spectral density becomes infinite at $(f_c \pm f_m)$.

Returning to eqn. (5.13), in order to find a solution in the more general case we must assume a PDF for β. Aulin wrote

$$p(\beta) = \begin{cases} \dfrac{\cos \beta}{2 \sin \beta_m} & |\beta| \leqslant |\beta_m| \leqslant \dfrac{\pi}{2} \\ 0 & \text{elsewhere} \end{cases} \tag{5.16}$$

This is plotted in Figure 5.8(a) and was claimed to be realistic for small β_m. There are sharp discontinuities at $\pm \beta_m$, however, and although it has the advantage of providing analytic solutions, it does not seem to be realistic, except at very small values of β_m (a few degrees). Nevertheless, Aulin used this equation to obtain the RF spectrum as

$$A_1(f) = F[a(\tau)]$$

$$= \begin{cases} 0 & |f| > f_m \\ \dfrac{E_0}{4 \sin \beta_m} \left(\dfrac{1}{f_m} \right) & f_m \cos \beta_m \leqslant |f| \leqslant f_m \\ \dfrac{1}{f_m} \left(\dfrac{\pi}{2} - \arcsin \dfrac{2 \cos^2 \beta_m - 1 - (f/f_m)^2}{1 - (f/f_m)^2} \right) & |f| < f_m \cos \beta_m \end{cases} \tag{5.17}$$

Although Aulin's point that all incoming waves do not travel horizontally is valid, it is equally true that Clarke's two-dimensional model predicts power spectra that have the same general shape as the observed spectra. It is therefore clear that the majority of incoming waves do indeed travel in a nearly horizontal direction and therefore a realistic PDF for β is one that has a mean value of $0°$, is heavily biased towards small angles, does not extend to infinity and has no discontinuities. The PDF shown in Figure 5.8(b) meets all these requirements and can be represented by

$$p_\beta(\beta) = \begin{cases} \dfrac{\pi}{4|\beta_m|} \cos \left(\dfrac{\pi}{2} \dfrac{\beta}{\beta_m} \right) & |\beta| \leqslant |\beta_m| \leqslant \dfrac{\pi}{2} \\ 0 & \text{elsewhere} \end{cases} \tag{5.18}$$

This PDF is limited to $\pm \beta_m$, which depends on the local surroundings. It was originally intended to be relevant for land mobile paths, but with suitable parameters it could also be useful in the satellite mobile scenario.

Using (5.18) in eqn. (5.13) allows us to evaluate the RF power spectrum $A_2(f)$ using standard numerical techniques. Figure 5.9 shows the form of the power spectrum obtained using eqns (5.13) and (5.18), together with the spectrum $A_1(f)$ given by eqn. (5.17) and $A_0(f)$ given by eqn. (5.15). All the spectra are strictly

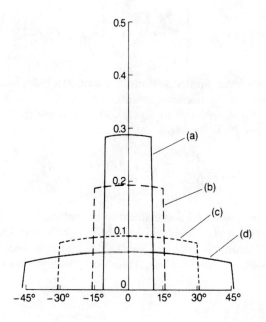

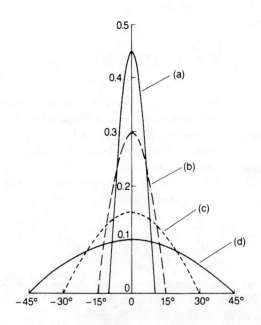

Figure 5.8 Probability density functions for β, the arrival angle in the vertical plane: (top) proposed by Aulin, (bottom) as expressed by equation (5.18). In each case the values of β_m are (a) 10°, (b) 15°, (c) 30°, (d) 45°.

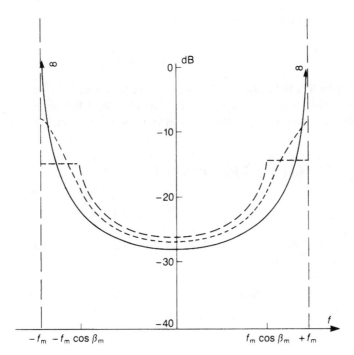

Figure 5.9 Form of the RF power spectrum using different scattering models and $\beta_m = 45°$: (——) Clarke's model, $A_0(f)$; (– – –) Aulin's model, $A_1(f)$; (- - - -) equation (5.18), $A_2(f)$.

band-limited to $|f| < f_m$ but in addition, the power spectral density in the first two cases is always finite. The spectrum given by eqn. (5.17) is actually constant for $f_m \cos \beta_m < |f| < f_m$ but the spectrum obtained from eqn. (5.18) does not have this unrealistic flatness. In contrast, $A_0(f)$ is infinite at $|f| = f_m$. There is a much increased low-frequency content even when β_m is small.

We conclude therefore that the RF signal spectrum is strictly band-limited to a range $\pm f_m$ around the carrier frequency. However, within those limits the power spectral density depends on the PDFs associated with the spatial angles of arrival α and β. The limits of the Doppler spectrum can be quite high; for example, in a vehicle moving at 30 m/s ($\sim$ 70 mph) receiving a signal at 900 MHz the maximum Doppler shift is 90 Hz. Frequency shifts of this magnitude can cause interference with the message information. Hand-portable transceivers carried by pedestrians experience negligible Doppler shift.

5.5 THE RECEIVED SIGNAL ENVELOPE

Practical radio receivers do not normally have the ability to detect the components $I(t)$ and $Q(t)$, they respond to the envelope and/or phase of the complex signal $E(t)$. The envelope $r(t)$ of the complex signal $E(t)$ is given by

$$r(t) = [I^2(t) + Q^2(t)]^{1/2}$$

and it is well known [9] that the PDF of $r(t)$ is given by

$$p_r(r) = \frac{r}{\sigma^2} \exp\left(-\frac{r^2}{2\sigma^2}\right) \tag{5.19}$$

in which σ^2, which is the same as $a(0)$, is the mean power and $r^2/2$ is the short-term signal power. This is the Rayleigh density function, and the probability that the envelope does not exceed a specified value R is given by the cumulative distribution function

$$\text{prob}[r \leqslant R] = P_r(R) = \int_0^R p_r(r)\, dr$$

$$= 1 - \exp\left(-\frac{R^2}{2\sigma^2}\right) \tag{5.20}$$

Several other statistical parameters of the envelope can be expressed in terms of the single constant σ. The mean value (or expectation) of the envelope $E[r]$ is given by

$$r_{\text{mean}} = E\{r\} = \int_0^\infty r p_r(r)\, dr$$

$$= \sigma\sqrt{\frac{\pi}{2}} = 1.2533\sigma \tag{5.21}$$

The mean square value is

$$E\{r^2\} = \int_0^\infty r^2 p_r(r)\, dr = 2\sigma^2 \tag{5.22}$$

The variance is given by

$$\sigma_r^2 = E\{r^2\} - E\{r\}^2$$

$$= 2\sigma^2 - \frac{\sigma^2 \pi}{2}$$

$$= \sigma^2\left(\frac{4-\pi}{2}\right) = 0.4292\sigma^2 \tag{5.23}$$

Finally, the median value r_M, defined as that for which $P_r(r_M) = 0.5$, is obtained from eqn. (5.20) as

$$1 - \exp\left(-\frac{r_M^2}{2\sigma^2}\right) = 0.5$$

hence

$$r_M = \sqrt{2\sigma^2 \ln 2} = 1.1774\sigma \tag{5.24}$$

Figure 5.10 shows the PDF of the Rayleigh function with these points identified.

It is often convenient to express eqns (5.19) and (5.20) in terms of the mean, mean square or median rather than in terms of σ. This is because it is useful to have a measure of the envelope behaviour relative to these parameters. To avoid

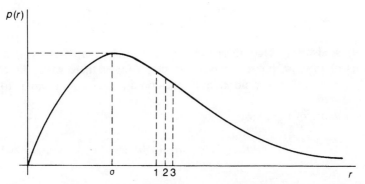

Figure 5.10 PDF of the Rayleigh distribution: $1 = $ median (50%) value, 1.1774σ; $2 = $ mean value, 1.2533σ; $3 = $ RMS value, 1.41σ.

cumbersome nomenclature we write $E\{r\} = \bar{r}$ and $E\{r^2\} = \overline{r^2}$, and in these terms, simple manipulation yields the following results. In terms of the mean square value,

$$p_r(r) = \frac{2r}{\overline{r^2}} \exp\left(-\frac{r^2}{\overline{r^2}}\right)$$

$$P_r(R) = 1 - \exp\left(-\frac{R^2}{\overline{r^2}}\right)$$

$$(5.25)$$

In terms of the mean,

$$p_r(r) = \frac{\pi r}{2\bar{r}^2} \exp\left(-\frac{\pi r^2}{4\bar{r}^2}\right)$$

$$P_r(R) = 1 - \exp\left(-\frac{\pi R^2}{4\bar{r}^2}\right)$$

$$(5.26)$$

In terms of the median,

$$p_r(r) = \frac{2r \ln 2}{r_{\mathrm{M}}^2} \exp\left(-\frac{r^2 \ln 2}{2r_{\mathrm{M}}^2}\right)$$

$$P_r(R) = 1 - 2^{-(R/r_{\mathrm{M}})^2}$$

$$(5.27)$$

Relationships involving the Rayleigh distribution in decibels can be found in Appendix B.

5.6 THE RECEIVED SIGNAL PHASE

The received signal phase $\theta(t)$ is given is terms of $I(t)$ and $Q(t)$ by

$$\theta(t) = \tan^{-1}\left(\frac{Q(t)}{I(t)}\right) \tag{5.28}$$

The argument [9] leading to the conclusion that the envelope is Rayleigh distributed also shows that the phase is uniformly distributed in the interval $(0, 2\pi)$, i.e.

$$p_\theta(\theta) = \frac{1}{2\pi} \tag{5.29}$$

This result is also expected intuitively; in a signal composed of a number of components of random phase it would be surprising if there were any bias in the phase of the resultant. It is random and takes on all values in the range $(0, 2\pi)$ with equal probability.

The mean value of the phase is

$$E\{\theta\} = \int_0^{2\pi} \theta p_\theta(\theta) \, d\theta = \pi \tag{5.30}$$

The mean square value is

$$E\{\theta^2\} = \int_0^{2\pi} \theta^2 p_\theta(\theta) \, d\theta = \frac{4\pi^2}{3} \tag{5.31}$$

and hence the variance is

$$\sigma_\theta^2 = E\{\theta^2\} - E\{\theta\}^2 = \frac{\pi^2}{3} \tag{5.32}$$

We will return later to a consideration of changes in the signal phase.

5.7 BASEBAND POWER SPECTRUM

In Section 5.4 we used the autocorrelation function of the received signal in order to obtain the RF spectrum. We saw that the spectrum was strictly band-limited to $f_c > f_m$ but that the shape of the spectrum within those limits was determined by other factors, in particular the assumed PDFs for the spatial angles α and β.

We can now consider the autocorrelation function of the envelope $r(t)$ and use it to obtain the baseband power spectrum. The mean of the envelope is given by eqn. (5.21) as

$$E\{r(t)\} = \sigma\sqrt{\frac{\pi}{2}} = \sqrt{\frac{\pi}{2} a(0)}$$

and the autocorrelation function is

$$\rho_r(\tau) = E\{r(t)r(t+\tau)\} \tag{5.33}$$

It can be shown [10, Ch. 8] that for a narrowband Gaussian process the envelope autocorrelation can be expressed as

$$\rho_r(\tau) = \frac{\pi}{2} a(0) F\left[-\tfrac{1}{2}, -\tfrac{1}{2}; 1, \left(\frac{a(\tau)}{a(0)} \right)^2 \right] \tag{5.34}$$

where $F[\cdot]$ is the hypergeometric function and $a(\tau)$ is as defined by eqn. (5.13).

The Fourier transform of eqn. (5.34) cannot be carried out exactly, but the hypergeometric function can be expanded in polynomial form and then approximated by neglecting terms beyond the second order. The approximation then becomes

$$\rho_r(\tau) = \frac{\pi}{2}a(0)\left[1 + \frac{1}{4}\left(\frac{a(\tau)}{a(0)}\right)^2\right] \tag{5.35}$$

The justification for taking only the first two terms is that at $\tau = 0$ the value obtained for $\rho_r(\tau)$ is $1.963\sigma^2$, which is only 1.8% different from the true value of $2\sigma^2$ [6]. Since we are principally interested in the continuous spectral content of the envelope, not in the carrier component, we can use the autocovariance function (in which the mean value is removed), thus

$$r_r(\tau) = E\{r(t)r(t + \tau)\} - E\{r(t)\}E\{r(t + \tau)\} \tag{5.36}$$

For a stationary process, $E\{r(t)\} = E\{r(t + \tau)\}$, so

$$r_r(\tau) = \frac{\pi}{2}a(0)\left[1 - \frac{1}{4}\left(\frac{a(\tau)}{a(0)}\right)^2\right] - \frac{\pi}{2}a(0)$$

$$= \frac{\pi}{8a(0)}a^2(\tau) \tag{5.37}$$

It is shown in Appendix A that in noisy fading channels the carrier-to-noise ratio (CNR) is proportional to r^2, so the autocovariance of the squared envelope is also of interest. It has been shown [5] that

$$E[r^2(t)r^2(t + \tau)] = 4[a^2(0) + a^2(\tau)]$$

and we know, from eqn. (5.22) that $E\{r^2(t)\} = 2a(0)$, thus

$$r_{r^2}(\tau) = 4[a^2(0) - a^2(\tau)] - 4a^2(0) = 4a^2(\tau) \tag{5.38}$$

The power spectrum of $r(t)$ and $r^2(t)$ can therefore be written as

$$S(f) = F\{Ca^2(\tau)\}$$
$$= CA(f) * A(f) \tag{5.39}$$

In this expression $A(f)$ can be either $A_1(f)$ as given by eqn. (5.17) or $A_2(f)$ obtained from eqns (5.13) and (5.18). If $A_2(f)$ is used then

$$C = \frac{\pi}{8a(0)} \quad \text{or} \quad 4$$

as appropriate; see equations (5.37) and (5.38).

The convolution represented by eqn. (5.39) can be evaluated exactly for the RF spectrum represented by eqn. (5.15), in which case

$$S_0(f) = CA_0(f) * A_0(f)$$

$$= C\left(\frac{E_0}{4\pi}\right)^2 \frac{1}{f_m} K\left\{\left[1 - \left(\frac{f}{2f_m}\right)^2\right]^{1/2}\right\} \tag{5.40}$$

where $K(\cdot)$ is the complete elliptic integral of the first kind; as $f \to 0$, $S_0(f) \to \infty$.

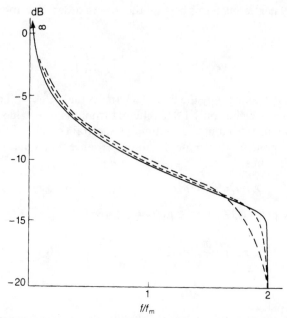

Figure 5.11 Form of the baseband (envelope) power spectrum using different scattering models and $\beta_m = 45°$: (——) Clarke's model, $S_0(f)$; (– – –) Aulin's model, $S_1(f)$; (- - - -) equation (5.18), $S_2(f)$.

Again, in the more general case, eqn. (5.39) can only be evaluated if $p_\beta(\beta)$ is known. The expressions for $p_\beta(\beta)$ given by eqns. (5.16) and (5.18) allow numerical evaluation of baseband spectra $S_1(f)$ and $S_2(f)$ (in the former case, via $A_1(f)$ as given by eqn. (5.17)). A comparison between $S_0(f)$, $S_1(f)$ and $S_2(f)$ is presented in Figure 5.11, which uses a logarithmic scale. Although $S_0(f) \to \infty$ at $f = 0$, $S_1(f)$ and $S_2(f)$ are always finite.

5.8 LCR AND AFD

Figure 5.2 shows that the signal envelope is subject to rapid fading. As the mobile moves, the fading rate will vary, hence the rate of change of envelope amplitude will also vary. Both the two-dimensional and three-dimensional models lead to the conclusion that the Rayleigh PDF describes the first-order statistics of the envelope over distances short enough for the mean level to be regarded as constant. First-order statistics are those for which time (or distance) is not a factor, and the Rayleigh distribution therefore gives information such as the overall percentage of time, or the overall percentage of locations, for which the envelope lies below a specified value. There is no indication of how this time is made up.

We have already commented, in connection with Figure 5.2, that deep fades occur only rarely whereas shallow fades are much more frequent. System engineers are interested in a quantitative description of the rate at which fades of any depth occur and the average duration of a fade below any given depth. This provides a valuable aid in selecting transmission bit rates, word lengths and coding schemes in digital radio systems and allows an assessment of system performance. The required information is provided in terms of *level crossing rate* and *average fade duration* below a specified level. The manner in which these two parameters are derived is illustrated in Figure 5.12.

The level crossing rate (LCR) at any specified level is defined as the expected rate at which the envelope crosses that level in a positive-going (or negative-going) direction. In order to find this expected rate, we need to know the joint probability density function $p(R, \dot{r})$ at the specified level R and the slope of the curve $\dot{r}(= dr/dt)$. In terms of this joint PDF, and remembering that we are interested only in positive-going crossings, the LCR N_R is given by [6, Ch. 1]:

$$N_R = \int_0^\infty \dot{r} p(R, \dot{r}) \, d\dot{r} \tag{5.41}$$

The joint PDF $p(R, \dot{r})$ is

$$p(R, \dot{r}) = \int_{-\infty}^{+\infty} \int_0^{2\pi} p(R, \dot{r}, \theta, \dot{\theta}) \, d\theta \, d\dot{\theta} \tag{5.42}$$

Rice [9] gives an appropriate expression for $p(R, \dot{r}, \theta, \dot{\theta})$ which can be substituted into eqn. (5.42) to show that

$$p(R, \dot{r}) = p_r(R) p_r(\dot{r})$$

from which it follows that R and $\dot{r}$ are independent and hence uncorrelated. The expected (average) crossing rate at a level R is then given by

$$N_R = \sqrt{\frac{\pi}{\sigma^2}} R f_m \exp\left(-\frac{r^2}{2\sigma^2}\right) \tag{5.43}$$

From eqn. (5.22) we know that $2\sigma^2$ is the mean square value and hence $\sqrt{2}\sigma$ is the RMS value. Equation (5.43) can therefore be expressed as

$$N_R = \sqrt{2\pi} f_m \rho \exp(-\rho^2) \tag{5.44}$$

where

$$\rho = \frac{R}{\sqrt{2}\sigma} = \frac{R}{R_{RMS}}$$

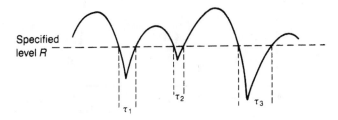

Figure 5.12 LCR and AFD: LCR = average number of positive-going crossings per second, AFD = average of τ_1, τ_2, τ_3, . . ., τ_n.

Equation (5.44) gives the value of N_R in terms of the average number of crossings per second. It is therefore a function of the mobile speed, and this is apparent from the appearance of f_m in the equation. Dividing by f_m produces the number of level crossings per wavelength and this is plotted in Figure 5.13. There are few crossings at high and low levels; the maximum rate occurs when $R = \sigma$, i.e. at a level 3 dB below the RMS level.

It is sometimes convenient to express the LCR in terms of the median value r_M, rather than in terms of the RMS value. Using eqns (5.24) and (5.43) the normalised average number of level crossings per wavelength is then

$$\frac{N_R}{f_m} = \sqrt{2\pi \ln 2} \left(\frac{R}{r_M}\right) 2^{-(R/r_M)^2} \tag{5.45}$$

This expression is independent of both carrier frequency and mobile velocity.

The average duration τ, below any specified level R, is also illustrated in Figure 5.12 and the average fade duration (AFD) is the average period of a fade below that level. The overall fraction of time for which the signal is below a level R is $P_r(R)$, as given by eqn. (5.20), so the AFD is

$$E\{\tau_R\} = \frac{P_r(R)}{N_R} \tag{5.46}$$

Substituting for N_R from eqn. (5.43) gives

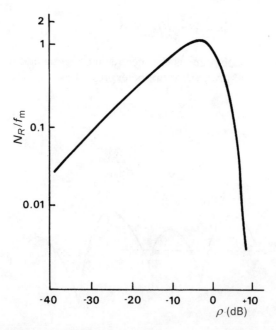

Figure 5.13 Normalised level crossing rate for a vertical monopole under conditions of isotropic scattering.

$$E\{\tau_R\} = \sqrt{\frac{\sigma^2}{\pi}} \frac{\exp(R^2/2\sigma^2) - 1}{R f_m} \tag{5.47}$$

Alternatively, multiplying by f_m enables us to express this in spatial terms, i.e. the average duration in wavelengths is

$$L_R = \sqrt{\frac{\sigma^2}{\pi}} \frac{\exp(R^2/2\sigma^2) - 1}{R} \tag{5.48}$$

Again, this can be expressed in terms of the RMS value as

$$L_R = \frac{\exp(\rho^2) - 1}{\rho \sqrt{2\pi}} \tag{5.49}$$

or, in terms of the median value, as

$$L_R = \frac{1}{\sqrt{2\pi} \ln 2} \frac{2^{(R/r_M)^2} - 1}{R/r_M} \tag{5.50}$$

Normalised AFD is plotted in Figure 5.14 as a function of ρ.

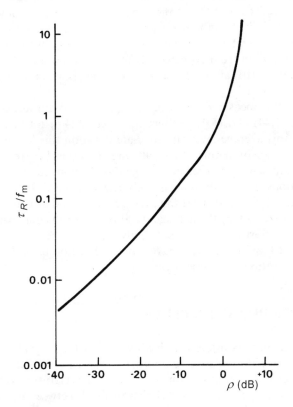

Figure 5.14 Normalised average duration of fades for a vertical monopole under conditions of isotropic scattering.

Table 5.1   Average fade length and crossing rate for fades measured with respect to median value

Fade depth	Average fade length (wavelengths)	Average crossing rate (wavelengths^{-1})
0	0.479	1.043
−10	0.108	0.615
−20	0.033	0.207
−30	0.010	0.066

Table 5.1 gives the AFD and average LCR for various fade depths with respect to the median level and indicates how often a Rayleigh fading signal needs to be sampled in order to ensure that an 'average duration' fade below any specified level will be detected. For example, in order to detect about 50% of the fades 30 dB below the median level, the signal must be sampled every 0.01λ. At 900 MHz this is 0.33 cm.

In practice the median signal level is a very useful measure. Sampling of the signal in order to estimate its parameters will be discussed in Chapter 8 but it is immediately obvious that if a record of signal strength is obtained by sampling the signal envelope at regular intervals of distance or time, then the median value is that exceeded (or not exceeded) by 50% of the samples. This is very easily determined. Furthermore, it is a relatively unbiased estimator since it is influenced only by the number of samples that lie above or below a given level, and not by the actual value of those samples. We note from Appendix B that the mean and RMS values are respectively 0.54 and 1.59 dB above the median, so conversion of the values given in Table 5.1 is straightforward.

In practice [11] the measured average fade rates and durations are closely predicted by eqns. (5.44) and (5.47). Often, however, it is of interest to know the distribution about this average level and for fade duration this has been measured using a Rayleigh fading simulator. The results are shown in Figure 5.15. For fade depths 10 dB or more below the median, all the distributions have identical shapes and for long durations the distributions quickly reach an asymptotic slope of (fade duration)$^{-3}$. In general, fades of twice the average duration occur once in every ten and fades of six or seven times the average duration occur once in every thousand. Very deep fades are short and infrequent. Only 0.2 fades per wavelength have a depth exceeding 20 dB and these fades have a mean duration of 0.03λ. Only 1% of such fades have a duration exceeding 0.1λ.

5.9   THE PDF OF PHASE DIFFERENCE

It is not very meaningful to consider the absolute phase of the signal at any point; in any case it is only the phase relative to another signal, or a reference, that can be measured. It is possible, however, to think in terms of the relative phase between the signals at a given receiving point at two different times, or between the signals at two spatially separated locations at the same time. Both these quantities are meaningful in a study of radio systems.

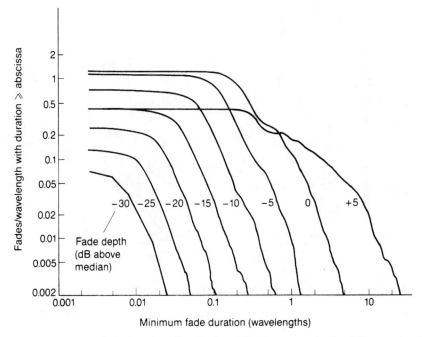

Figure 5.15 Measured fade duration distribution. The data was obtained from a simulator with a Rayleigh amplitude distribution and a parabolic Doppler spectrum.

Unless the value of β_m in eqns. (5.16) and (5.18) is quite large, there is little to choose between the two- and three-dimensional models as far as the PDF of phase difference is concerned [5]. If we consider the phase difference between the signals at a given receiving point as a function of time delay τ, then the PDF of the phase difference can be expressed as [6, Ch. 1]:

$$p(\Delta\theta) = \frac{1 - \rho^2(\tau)}{4\pi^2} \left(\frac{\sqrt{1 - x^2} + x(\pi - \cos^{-1} x)}{(1 - x^2)^{3/2}} \right) \qquad (5.51)$$

where

$$\rho(\tau) = \frac{a(\tau)}{a(0)} \quad \text{and} \quad x = \rho(\tau) \cos \Delta\theta$$

Assuming that $p_\alpha(\alpha) = 1/2\pi$, we can determine the phase difference between the signals at two spatially separated points through the time–distance transformation $l = vt$, and Figure 5.16 shows curves of $p(\Delta\theta)$ for the two-dimensional model for various separation distances.

Two limiting cases are of interest, namely $l \to 0$ (coincident points) and $l \to \infty$. When $l \to 0$, $p(\Delta\theta)$ is zero everywhere except at $\Delta\theta = 0$, where it is a δ-function. When $l \to \infty$, $\Delta\theta$ is uniformly distributed with $p(\Delta\theta) = 1/2\pi$, as would be expected from the convolution of two independent random variables both uniformly distributed in the interval $(0, 2\pi)$. $\Delta\theta$ is also uniformly distributed at all separations for which $J_0(\beta l) = 0$, indicating that at spatial separations for which the envelope is

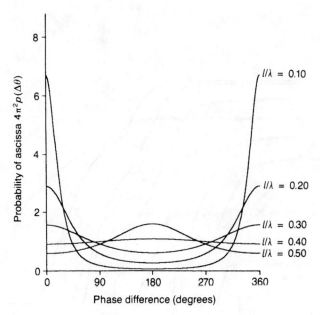

Figure 5.16 The PDF of phase difference $\Delta\theta$ between points spatially separated by a distance l.

uncorrelated then the phase difference is also uncorrelated. This is to be expected since at these separations the electric field signals are uncorrelated.

5.10 RANDOM FM

Since the phase θ varies with location, movement of the mobile will produce a random change of θ with time, equivalent to a random phase modulation. This is usually called random FM because the time derivative of θ causes frequency modulation which is detected by any phase-sensitive detector, e.g. FM discriminator, and appears as noise to the receiver. In simple mathematical terms,

$$\dot{\theta} = \frac{d\theta}{dt} = \frac{d}{dt}\left(\tan^{-1}\frac{Q(t)}{I(t)}\right)$$

The PDF of the random FM can be obtained by appropriate integration of the joint PDF of r, $\dot{r}$, θ and $\dot{\theta}$ (5.42) to give

$$p(\dot{\theta}) = \int_0^{2\pi} \int_{-\infty}^{+\infty} \int_0^{\infty} p(r, \dot{r}, \theta, \dot{\theta}) \, dr \, d\dot{r} \, d\theta \tag{5.52}$$

This has been evaluated in terms of the maximum Doppler shift as

$$p(\dot{\theta}) = \frac{1}{\omega_m \sqrt{2}}\left(1 + 2\left(\frac{\dot{\theta}}{\omega_m}\right)^2\right)^{-3/2} \tag{5.53}$$

The cumulative distribution function is given by

$$P(\dot{\Theta}) = \int_{-\infty}^{\dot{\Theta}} p(\dot{\theta})\, d\dot{\theta}$$

$$= \tfrac{1}{2}\left(1 + \sqrt{2}\,\frac{\dot{\Theta}}{\omega_m}\left(1 + \frac{2\dot{\Theta}^2}{\omega_m^2}\right)^{-1/2}\right) \tag{5.54}$$

Both these functions are shown in Figure 5.17. Although, in Figure 5.17(a) the highest probabilities occur for small values of $\dot{\theta}$, large excursions can also occur.

The spectrum of the random FM can be found from the Fourier transform of the autocorrelation of $\dot{\theta}$, and is given by [5]:

$$E\{\dot{\theta}(t)\dot{\theta}(t+\tau)\} = \frac{1}{2}\left[\left(\frac{\dot{a}(\tau)}{a(\tau)}\right)^2 - \frac{\ddot{a}(\tau)}{a(\tau)}\right]\ln\left[1 - \left(\frac{a(\tau)}{a(0)}\right)\right] \tag{5.55}$$

where, on the right-hand side, a dot denotes differentiation with respect to τ. For the two-dimensional model [6], this becomes

$$\frac{\omega_m^2}{2J_0(\omega_m\tau)}\left[\frac{J_0(\omega_m\tau)J_1(\omega_m\tau)}{\omega_m\tau} - J_0^2(\omega_m\tau) - J_1^2(\omega_m\tau)\right]\ln\left[1 - J_0^2(\omega_m\tau)\right] \tag{5.56}$$

The random FM spectrum can be obtained as the Fourier transform of this expression, and although the evaluation is rather involved, it can be carried out by

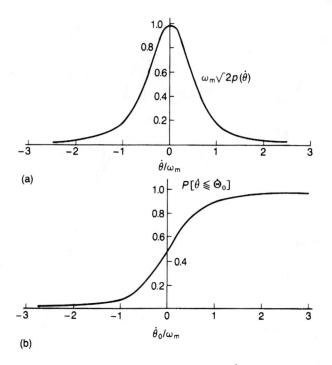

(a)

(b)

Figure 5.17 Probability functions for the random FM $\dot{\theta}$ of the received electric field: (a) probability density function, (b) cumulative distribution.

separating the range of integration into different parts and using appropriate approximations for the Bessel and logarithmic functions. The problem has been studied in some detail by Davis [12] and the power spectrum, plotted on normalised scales, is shown in Figure 5.18. We note that, in contrast to the strictly band-limited power spectrum of the signal envelope (the Doppler spectrum), there is a finite probability of finding the frequency of the random FM at any value. Nevertheless, the energy is largely confined to $2f_m$, from where it falls off as $1/f$ and is insignificant beyond $5f_m$. The majority of energy is therefore confined to the audio band; the larger excursions, being associated with the deep fades, occur only rarely.

The PDF of the difference in random FM between two spatially separated points is of interest in the context of diversity systems, but is not easily obtained. It involves complicated integrals and a computer simulation has been used to produce some results. The PDF can be evaluated, however, when there is either zero or infinite separation between the points. For the case of zero separation a δ-function of unity area at $\Delta\theta = 0$ is obtained. For infinite separation the two values of random FM are independent and the convolution of two equal distributions $p(\theta)$ gives the probability density function [6, Ch. 6] as

$$p(\Delta\theta) = \frac{1}{\omega_m\sqrt{2}} \frac{(1-M)^{5/2}}{4M} \left(K(k) + \frac{2M-1}{1-M} E(k) \right) \tag{5.57}$$

where $K(k)$ and $E(k)$ are complete elliptic integrals of the first and second kind, respectively, and

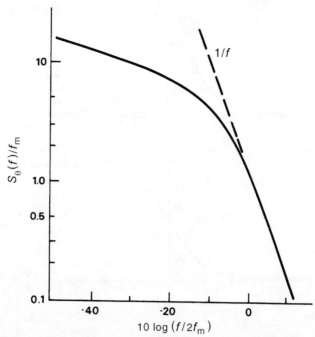

Figure 5.18 Power spectrum of random FM plotted as relative power on a normalised frequency scale.

$$k = \sqrt{M} = \frac{|\Delta\dot{\theta}|}{\omega_m}\left(2 + \left(\frac{\Delta\dot{\theta}}{\omega_m}\right)^2\right)^{1/2}$$

In practice the value of $p(\Delta\theta)$ converges very rapidly to the limiting case of $l = \infty$.

5.11 RICIAN FADING

The discussion up to now has focused on the case where the component waves in the composite signal received at the mobile are of equal (or approximately equal) amplitude. This has led to the conclusion that the envelope is Rayleigh distributed and has enabled us to derive various properties of the envelope and phase.

The assumption of similar-amplitude waves holds in a wide variety of scenarios because in general the mobile has no line-of-sight path to the transmitter and there is no dominant incoming wave. However, there are situations, e.g. in microcells or picocells within a cellular radio system, where there may be a line-of-sight path or at least a dominant specular component. We may then expect the statistics to differ from those already described. The problem is analogous to a sinusoidal wave plus random noise, and this has been extensively treated by Rice [13]. It is intuitively to be expected that there will be fewer deep fades and that the specular component will be a major feature of the spectrum.

The joint PDF of the envelope and phase of a signal with a dominant component r_s is given by

$$p(r, \theta) = \frac{r}{2\pi\sigma^2} \exp\left(-\frac{r^2 + r_s^2 - 2rr_s \cos\theta}{2\sigma^2}\right) \tag{5.58}$$

The envelope PDF can be found by integrating over θ and is given by

$$p_r(r) = \frac{r}{\sigma^2} \exp\left(-\frac{r^2 + r_s^2}{2\sigma^2}\right) I_0\left(\frac{rr_s}{\sigma^2}\right) \tag{5.59}$$

where $I_0(\cdot)$ is the modified Bessel function of the first kind and zero order. This is known as the Rician distribution; it reduces to the Rayleigh distribution in the special case of $r_s = 0$.

In the literature, the Rician distribution is often described in terms of a parameter K defined as

$$K \text{ (dB)} = 10 \log\left(\frac{r_s^2}{2\sigma^2}\right) \tag{5.60}$$

which, in the present context, can be interpreted as the ratio of the power in the steady (dominant) signal to that in the multipath (random) components. Equation (5.59) can be written in terms of K (dB) as

$$p_r(r) = \frac{2r10^{K/10}}{r_s^2} \exp\left(-\frac{10^{K/10}}{r_s^2}(r^2 + r_s^2)\right) I_0\left(\frac{2r10^{K/10}}{r_s}\right) \tag{5.61}$$

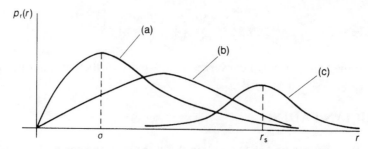

Figure 5.19   Rician probability density function: (a) $K \to 0$, (b) $K \simeq 1$, (c) $K \gg 1$.

The PDF of the envelope, $p_r(r)$, is shown in Figure 5.19 for various values of K. If $K \to 0$ then the PDF tends to a Rayleigh distribution; if $K \gg 1$ then the PDF becomes Gaussian with a mean value r_s.

The PDF of the phase is given by

$$p_\theta(\theta) = \frac{1}{2\pi} \exp\left(-\frac{r_s^2}{2\sigma^2}\right)\left[1 + \sqrt{\frac{\pi}{2}}\frac{r_s \cos\theta}{\sigma} \exp\left(\frac{r^2 \cos^2\theta}{2\sigma^2}\right)\right]\left[1 + \mathrm{erf}\frac{r_s \cos\theta}{\sigma\sqrt{2}}\right]$$

(5.62)

It is clear that the phase will be uniformly distributed in the range $(-\pi, \pi)$, i.e. $p_\theta(\theta) = 1/2\pi$ if $r_s \to 0$. If $K \gg 1$ then the phase will tend to that of the dominant component.

The effect of a dominant component on the RF and baseband spectra is easily envisaged. A horizontally propagating dominant component arriving at an angle α_0 with respect to the direction of vehicle motion experiences a Doppler shift of $f_m \cos\alpha_0$. The resultant RF spectrum therefore contains an additional component (δ-function) at this frequency as shown in Figure 5.20(a) and this leads to two components at $f_m(1 \pm \cos\alpha_0)$ in the baseband spectrum (Figure 5.20(b)). The upper limit of the spectrum remains at $2f_m$.

5.12   SPATIAL CORRELATION OF FIELD COMPONENTS

In mobile radio systems, especially at VHF and above, the effects of fading can be combatted using diversity techniques either at the base station or the mobile. Diversity reception is treated in Chapter 10 and works on the principle that if two or more independent samples (versions) of a random process are obtained, these samples will fade in an uncorrelated manner. It follows that the probability of all the samples being simultaneously below a given level is very much less than the probability of a single sample being below that level; a signal composed of a suitable combination of the various samples will therefore have much less severe fading properties than any individual sample alone.

Space diversity, in which two or more physically separated antennas are used, has received much attention in the literature [6,7,14] but frequency, polarisation and time diversity are also possibilities. Time diversity is attractive in digital communication

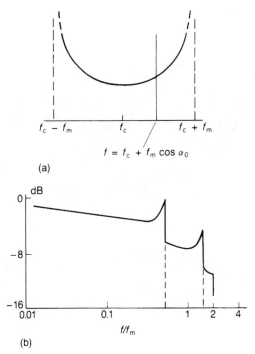

(a)

(b)

Figure 5.20 Spectra in the presence of a dominant component: (a) RF spectrum, (b) envelope spectrum (logarithmic frequency scale).

systems where storage is available at both ends of the radio link [15] and is easy to implement since only one antenna is needed. The question for space diversity is how far apart the antennas need to be, in order to obtain uncorrelated signal envelopes. The question for time diversity is how much time delay will produce the same result.

Information about the required temporal separation can be found from the autocorrelation function of the field components. Strictly, for space diversity we need to find the cross-correlation between the signals at two spatially separated points, but the assumption that $p_\alpha(\alpha) = 1/2\pi$ at the mobile means that the actual direction of motion in the xy plane is irrelevant and the cross-correlation between the signals received from horizontally separated antennas at any spatial separation can be directly related to a point on the autocorrelation curve through the time–distance transformation $l = vt$. This equivalence between the correlation in time and distance is very important.

Expressions for $a(\tau)$ and $a_0(\tau)$ are given in eqns (5.13) and (5.14) and these represent the quantities needed for this purpose. Calculations using $p_\beta(\beta)$ as given by eqns. (5.16) and (5.18) show that $a(\tau)$ and $a_0(\tau)$ are only slightly different for small values of β_m, and Figure 5.21 shows $a_0(\tau)$ plotted against normalised spatial separation $f_m\tau$ ($= l/\lambda$). The correlation reaches zero at $l = 0.38\lambda$ and thereafter is always less than 0.3. The inference is that in this context the simpler two-dimensional model is quite satisfactory for almost all practical situations. For the envelope autocorrelation we can use eqns (5.13) and (5.37) to write

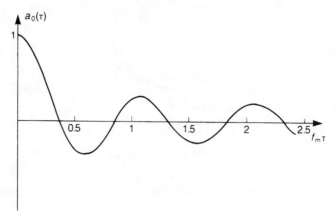

Figure 5.21   Normalised $a_0(\tau)$ as a function of $f_m\tau\ (= l/\lambda)$.

$$r_r(\tau) = \frac{\pi}{8a(0)}\left(\frac{E_0^2}{4}\right)\left(\int_{-\pi}^{+\pi} J_0(2\pi f_m\tau\ \cos\ \beta)p_\beta(\beta)\,d\beta\right)^2 \qquad (5.63)$$

which can be evaluated numerically. If $p_\beta(\beta) = \delta(\beta)$ we have the simpler situation in which

$$
\begin{aligned}
r_r(\tau) &= \frac{\pi}{8a(0)}\left(\frac{E_0^2}{4}\right)J_0^2(2\pi f_m\tau)\\
&= \frac{\pi}{8}a(0)J_0^2(2\pi f_m\tau)
\end{aligned}
\qquad (5.64)
$$

It is interesting to note that the autocorrelation of the field component is proportional to $J_0(\cdot)$ whereas the autocorrelation of the envelope is proportional to $J_0^2(\cdot)$. Again, the simpler model is usually adequate and shows that, at the mobile, sufficient decorrelation can be achieved using a spatial separation of less than $\lambda/2$.

This discussion assumes that the correlation between the envelopes of signals received from spatially separated antennas is the same as the correlation between the envelopes of the appropriate field components at the points where the antennas are located. This is not strictly true, but is a reasonable approximation when the antennas are far enough apart for mutual impedance effects to be negligible. The assumption breaks down, however, at subwavelength separations and the implications of this will be discussed further in Chapter 10.

5.12.1   Cross-correlation

Returning to the more general case of space diversity, vertical separation is also of practical interest. This case explicitly shows the limitations of Clarke's two-dimensional model, which assumes that $\beta = 0$ and leads to the implicit conclusion that the cross-correlation in the vertical plane is unity. Experimental results reported in the literature [16,17], however, show clearly that measured values of cross-correlation for vertically separated antennas at base stations and mobiles can be considerably less than unity. The generic three-dimensional model can cope with this case and a general analysis has been undertaken [18].

Figure 5.22 shows the geometry for mobile reception in which a receiver, moving in the horizontal xy plane receives a number of component waves from scatterers located in three dimensions around it. The angle of arrival of a wave from the ith scatterer, ψ_i, is the angle formed between the line joining the scatterer to the mobile and the direction of motion. It is made up of two components: the vertical angle of arrival β and the horizontal angle of arrival α.

To assess the feasibility of space diversity at the mobile, we need to determine the cross-correlation between the signals received on two antennas separated by a distance d. The general case, when the antennas are separated in both the vertical and horizontal directions, is shown in Figure 5.23. It can be shown that

$$\cos \theta = \cos \beta \cos \gamma \cos \alpha + \sin \beta \sin \gamma \qquad (5.65)$$

where θ_i is the spatial angle formed by the line joining the ith scatterer to the mobile and the line joining the two receiving antennas.

Generally, the cross-correlation between the signals on the two antennas is given by

$$\rho_{cr}(d) = \left| \int_\theta \exp\left(j2\pi \frac{d}{\lambda} \cos \theta \right) p(\theta)\, d\theta \right| \qquad (5.66)$$

In the specific case of vertical separation (for which $\gamma = \pi/2$) and using eqn. (5.18) for $p_\beta(\beta)$, this becomes

$$\rho_{cr}(d)\Big|_{vert} = \left| \frac{\pi}{4|\beta_m|} \int_{-\beta_m}^{+\beta_m} \exp\left(j2\pi \frac{d}{\lambda} \sin \beta \right) \cos\left(\frac{\pi}{2} \frac{\beta}{\beta_m} \right) d\beta \right| \qquad (5.67)$$

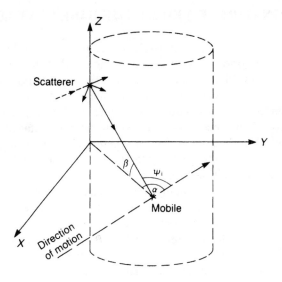

Figure 5.22 Geometry for reception at the mobile, in which waves arrive from a number of scatterers located on a cylinder.

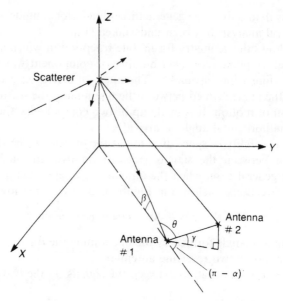

Figure 5.23 Geometry for space diversity reception at the mobile.

which has been numerically evaluated for various values of β_m and is plotted in Figure 5.24. It can be seen that the separation required for a given correlation coefficient decreases rapidly as a function of β_m. Measured values of cross-correlation reported by Feeney [16] and Yamada [17] are also shown in Figure 5.24, and using these values it is possible to estimate that values of β_m in the range 10–20° are reasonable.

5.13 THE SIGNAL RECEIVED AT THE BASE STATION

Up to now we have concentrated exclusively on the properties of the signal received at the mobile. Two assumptions have been made. Firstly, that there are a large number of incoming waves, none of which dominates, and this leads to the conclusion that the received signal has a Rayleigh-distributed envelope and a uniformly distributed phase. The second, independent assumption relates to the spatial angle of arrival of the incoming waves. It has been assumed that in the horizontal plane the angle α is uniformly distributed in the interval $(0, 2\pi)$, whereas in the vertical plane the angle β can be represented by the PDF given by (5.16) or (5.18). These assumptions lead to the spectral and autocorrelation properties described in Section 5.4.

The first assumption holds equally well at base station sites; this is because reciprocity applies and the propagation paths that exist, carry energy equally well in either the base–mobile direction or the mobile–base direction. In this context 'reciprocity' means that if the transmitter and receiver are interchanged, the path loss remains the same over each individual path, hence the overall path loss is also unaltered. Base station sites, however, are deliberately chosen to be well clear of local obstructions in order to give the best coverage of the intended service area, and the scattering objects which produce the multipath effects are located principally in a

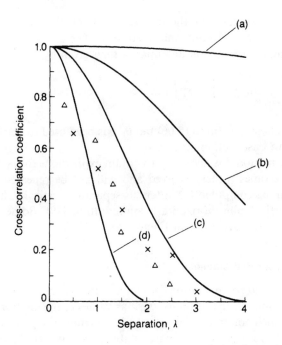

Figure 5.24 Cross-correlation between the envelopes of signals received on two vertically spaced vehicle antennas as a function of their spacing: (a) $\beta_m = 1°$, (b) $\beta_m = 5°$, (c) $\beta_m = 10°$, (d) $\beta_m = 20°$; Δ, $\times$ are experimental points.

small area surrounding the mobile. The assumption of isotropic scattering, i.e. uniformly distributed arrival angle in the azimuth plane at the receiver, is very unlikely to hold at the base station and therefore the signal properties that depend on this assumption are likely to change.

Since it is the mobile that moves, the temporal autocovariance and the received signal spectrum at the base station are the same as those at the mobile when transmission is in the opposite direction; the energy propagates via exactly the same set of scatterers which have the same location with respect to the moving terminal. Gans [4] stated that as far as path loss is concerned, transmission from the base station and reception at the mobile is the same as transmission from the mobile and reception at the base station. Thus he correctly concluded that the temporal autocovariance and power spectral density are identical for base station and mobile reception. However, two antennas located at the base station 'view' the scattering volume around the mobile from only slightly different angles and it is therefore to be expected that the spatial separation required to obtain a given cross-correlation between the signal envelopes will be much greater than the corresponding distance at the mobile. Moreover, it may depend on the orientation of the antennas with respect to a line joining the base and mobile stations.

An estimate of the spatial separation required between base station antennas can be made using the three-dimensional model described earlier in the chapter. The estimate assumes there is no line-of-sight propagation path and that all the waves received at the base station come from scatterers surrounding the mobile unit (as in

Figure 5.22). It also assumes that the distance between the mobile and the base station is much greater than the antenna separation. The cross-correlation between two spatially separated base station antennas is then given by

$$\rho_{cr}(d) = \left| \int_{\theta_b} \exp\left(j2\pi \frac{d}{\lambda} \cos \theta_b \right) p(\theta_b) \, d\theta_b \right| \tag{5.68}$$

This equation is identical in form to (5.66) but θ_b has been used to emphasise that we are now dealing with base station reception.

To evaluate this equation it is necessary to obtain expressions for $p(\theta_b)$ and $\cos \theta_b$. The mathematics becomes rather involved and will not be repeated here, but the cross-correlation can be computed for antennas separated vertically, horizontally or in a composite (vertical plus horizontal) configuration [19]. Some results which illustrate the major effects are given in Figures 5.25 to 5.29.

5.13.1 Vertically separated antennas

For vertically separated base station antennas, Figure 5.25 shows the cross-correlation as a function of separation for various values of the angle σ which defines the direction of vehicle motion. For a vehicle moving along a route which is circumferential with respect to the base station ($\sigma = 0$), the cross-correlation decreases more slowly with vertical separation than when the vehicle moves along a radial route ($\sigma = \pi/2$). Although the difference is fairly small, these results show the same trend as the experimental measurements reported by Feeney [16]. The value

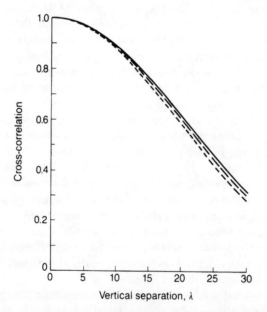

Figure 5.25 Cross-correlation between the envelopes of signals received on two vertically separated base station antennas as a function of the angle σ: (——) $\sigma = 0$ or π; (– – –) $\sigma = \pm 45°$, $\pm 135°$; (- - - -) $\sigma = \pm 90°$.

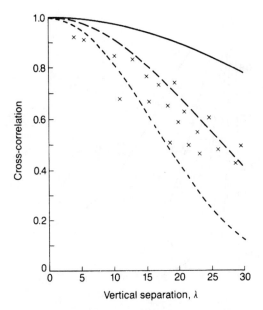

Figure 5.26 Cross-correlation between the envelopes of signals received on two vertically separated base station antennas as a function of the assumed radius of the scattering cylinder: (——) $d_s = 30$ m, (– – –) $d_s = 60$ m, (- - - -) $d_s = 90$ m; $\times$ are experimental points.

$\sigma = \pi/4$ can be taken as a typical or representative value. Using this value, Figure 5.26 shows the effect of different assumptions about the radius of the cylinder containing the scatterers, these having been calculated on the basis that the base–mobile range is approximately 1.2 km.

An obvious, but very important effect is apparent. When the mobile is in a heavily built-up urban area where the buildings (scatterers) are in close proximity to it, the vertical separation between base station antennas required to produce a given cross-correlation is much larger than when the mobile is in suburban or rural areas where the scatterers are further away. Figure 5.26 shows that a cross-correlation < 0.8 is obtainable from antennas vertically separated by 11λ when the effective radius of the scattering cylinder is 90 m but the required separation rises to about 28λ for an effective radius of 30 m. A comparison of these theoretical results with measured cross-correlation values [16], also shown in Figure 5.26, allows an estimation of the effective radius in the area where the experiments were conducted. In this case the estimate for the urban and suburban areas of a large city is 60–70 m. The effective radius will depend heavily on the degree of urbanisation, however, and its value is expected to vary considerably from one area to another.

5.13.2 Horizontally separated antennas

For horizontally separated base station antennas the factor that most obviously affects the value of cross-correlation is α_b, the spatial angle of arrival in the horizontal plane with respect to the line joining the two antennas. Results for various values of α_b are shown in Figure 5.27; the cross-correlation for $\alpha_b = 0$ is unity. As α_b

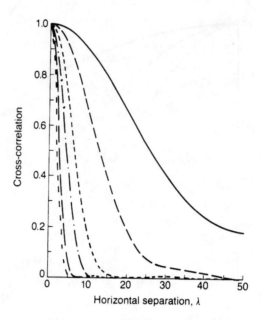

Figure 5.27 Cross-correlation between the envelopes of signals received on two horizontally separated base station antennas as a function of the angle α_b: (——) $\alpha_b = 5°$, (– – –) $\alpha_b = 10°$, (- - - - -) $\alpha_b = 20°$, (–·–·) $\alpha_b = 30°$, (———) $\alpha_b = 60°$, (- - - -) $\alpha_b = 90°$.

increases, the separation needed to obtain a given value of cross-correlation decreases, very rapidly at first and then more slowly. For $\alpha_b > 60°$ the decrease is marginal. Note that even if α_b is only 5°, a cross-correlation of 0.7, which can offer substantial diversity improvement, is obtainable for a separation of 20λ, a distance that is readily obtainable on rooftop sites in the higher UHF band.

   The effect of direction of motion is illustrated in Figure 5.28; this has a greater influence with horizontally spaced antennas than with vertical separation. Once again we can regard the results for $\alpha = \pi/4$ as being representative. The effective radius of the scattering cylinder is also important, as shown by Figure 5.29. Again this shows that a given value of cross-correlation can be obtained with smaller spatial separations when the mobile is in a suburban or rural area where the effective scattering radius is larger. In the case of horizontal separation, the differences are smaller than for vertical separation (compare Figures 5.26 and 5.29). Either of these figures can be used to compare theoretical and measured results and hence to obtain a value of scattering radius representative of the area concerned.

   Generally it can be concluded that low values of cross-correlation are more easily obtained with horizontal rather than vertical separation, although with vertical separation there are no directional effects. It is clear that vertical separation has a noticeable effect when the horizontal separation is small, but its influence decreases and becomes negligible when the horizontal separation is large. This is not surprising; if there is a zero or small separation in one dimension then separation in the other dimension will produce substantial reductions in cross-correlation. However, when the separation in one dimension is sufficient to reduce

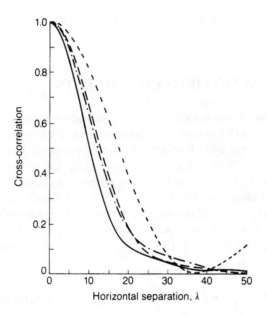

Figure 5.28 Cross-correlation between the envelopes of signals received on two horizontally separated base station antennas as a function of the direction of motion of the mobile: (——) σ = 0°, (- - - -) σ = 45°, (— — — —) σ = 90°, (—·—·) σ = 135°.

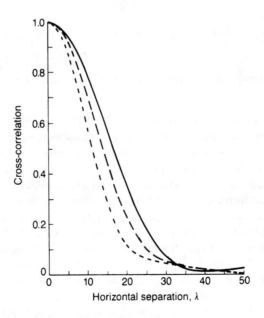

Figure 5.29 Cross-correlation between the envelopes of signals received on two horizontally separated base station antennas as a function of the assumed radius of the scattering cylinder: (——) $d_s = 30$ m, (— — —) $d_s = 60$ m, (- - - -) $d_s = 90$ m.

the cross-correlation to a low value, separation in the other dimension will not improve matters noticeably.

5.14   THE MAGNETIC FIELD COMPONENTS

The previous sections of this chapter have concentrated on the properties of the signal as detected by a vertical monopole or dipole antenna. By implication therefore, we have been concerned only with the electric field component of the vertically polarised electromagnetic waves. Sometimes it is of interest to know the properties of the associated magnetic field; for example, if loop antennas are used or in an assessment of field component diversity. In this section we will briefly survey some of the properties of the magnetic field component, using the two-dimensional field model.

If the various incoming multipath waves are such that the electric field components are all aligned along the vertical axis (i.e. the z-axis) then eqn. (5.4) reduces to

$$E_n(t) = C_n \cos\left(\omega_0 t - \frac{2\pi}{\lambda}(x_0 \cos \alpha_n + y_0 \sin \alpha_n) + \phi_n\right)$$

The magnetic field components all lie in the horizontal (i.e. xy) plane but are randomly oriented because the waves arrive at the receiving point from different directions. It is therefore convenient to resolve these components along the x and y axes such that

$$H_x(t) = -\frac{C_n}{\eta} \sum_{n=1}^{N} \cos\left(\omega_0 t - \frac{2\pi}{\lambda} \sin \alpha_n (x_0 \cos \alpha_n + y_0 \sin \alpha_n) + \phi_n\right) \qquad (5.69)$$

$$H_y(t) = \frac{C_n}{\eta} \sum_{n=1}^{N} \cos\left(\omega_0 t - \frac{2\pi}{\lambda} \cos \alpha_n (x_0 \cos \alpha_n + y_0 \sin \alpha_n) + \phi_n\right) \qquad (5.70)$$

where η is the intrinsic wave impedance ($= 120\pi$).

Two points are worth making. Firstly, it is obvious that because waves arrive at the receiving point from a variety of directions and because magnetic field sensors (e.g. loop antennas) have directional properties, two orthogonal sensors (one aligned along the x-axis and the other along the y-axis) are necessary to detect the field. Secondly, in contrast to the case of a single plane wave, no simple relationship exists between the magnitudes of the electric and magnetic fields at a given point in a multipath field. Equations (5.69) and (5.70) show that H_x and H_y are the sum of a number of components and that the spatial angle α is a factor in determining the resultant amplitude. H_x and H_y can be large or small depending on the relationships that exist, at any receiving point, between the various phases ϕ_n and the various spatial angles α_n. Indeed, it can be shown [6, Ch. 1] that at a given receiving point all three field components are mutually uncorrelated, and this is a very significant result.

As an illustration we can compare the spectra and spatial autocorrelation functions of the different field components. For the two-dimensional model, the RF power spectrum of the electric field is given by eqn. (5.15). A similar analysis, using equations (5.69) and (5.70), leads to expressions for $A_0(f)$ for the magnetic field components H_x and H_y:

$$A_0(f) = \begin{cases} \dfrac{E_0}{4\pi f_{\mathrm{m}}} \sqrt{1 - (f/f_{\mathrm{m}})^2} & \text{for } H_x \\[3mm] \dfrac{E_0}{4\pi f_{\mathrm{m}}} \dfrac{(f/f_{\mathrm{m}})^2}{\sqrt{1 - (f/f_{\mathrm{m}})^2}} & \text{for } H_y \end{cases} \qquad (5.71)$$

The form of these spectra is shown in Figure 5.30 with the spectrum of E_z included for comparison. Analysis, following the method used in Section 5.7, can be used to obtain baseband spectra.

The autocorrelation and cross-correlation functions of the field components are also of interest in the context of diversity systems. For the electric field component, eqn. (5.64) gives the form of the relationship, the normalised value being $J_0^2(2\pi f_{\mathrm{m}}\tau)$. In a similar manner, the normalised covariance functions for the envelopes of the magnetic field components are

$$r_H(\tau) = \begin{cases} [J_0(2\pi f_{\mathrm{m}}\tau) + J_2(2\pi f_{\mathrm{m}}\tau)]^2 & \text{for } H_x \\[2mm] [J_0(2\pi f_{\mathrm{m}}\tau) - J_2(2\pi f_{\mathrm{m}}\tau)]^2 & \text{for } H_y \end{cases} \qquad (5.72)$$

The cross-covariance functions for the envelopes of E_z and H_x, and H_x and H_y are zero for any values of τ, provided $p_\alpha(\alpha)$ is an even function [6]. It only remains, therefore, to evaluate the function for E_z and H_y, which can be shown to be $J_1^2(2\pi f_{\mathrm{m}}\tau)$. These functions are plotted in Figure 5.31.

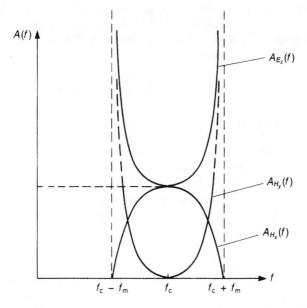

Figure 5.30 RF power spectra of the three field components, assuming uniformly distributed spatial arrival angles.

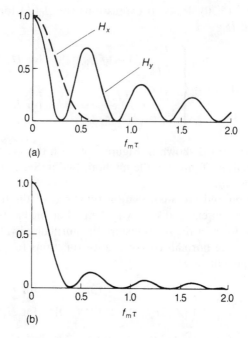

(a)

(b)

Figure 5.31   (a) Normalised covariance functions for signal envelopes; (b) cross-correlation function for E_z and H_y.

The significance of spatial autocovariance functions has already been discussed in connection with space diversity systems and it has been shown that effective diversity systems can be implemented at the mobile end of the link, using antennas less than $\lambda/2$ apart. It is clear from Figure 5.31 that similar considerations apply if the antennas sense the magnetic field rather than the electric field. However, the fact that all three field components are uncorrelated at zero spacing led to an interest in 'field diversity' [20] in which the electric and magnetic fields are sensed by collocated antennas (a monopole and two mutually orthogonal small loops).

5.15   SIGNAL VARIABILITY

Prediction of the received mobile radio signal strength is a two-stage process involving an estimation of both the median received signal within a relatively small area, and the signal variability about that median level. Chapters 3 and 4 were concerned with the problem of median signal strength prediction in small areas. For convenience these areas often coincide with standard community maps and are typically 500 m × 500 m. The reasons for choosing areas with dimensions of this order have already been discussed.

We turn now to the question of signal variability and address the problem of quantifying the extent to which the signal fluctuates within the area under consideration. Once again there are two contributing factors. Firstly there is the variation in the median signal itself as the mobile moves from place to place. This is caused by large-scale variations in the terrain profile along the path to the transmitter and by changes in the nature of the local topography. It is the slow, or lognormal, fading that has been mentioned earlier. Superimposed on this slow fading is the rapid and severe variation in the received signal strength (the fast or Rayleigh fading) caused by multipath propagation in the immediate vicinity of the receiver which has been discussed in this chapter.

A quantitative measure of the signal variability is essential for several reasons. It is only then, for example, that we can estimate the percentage of any given area that has an adequate signal strength, or the likelihood of interference from a distant transmitter. An estimate of the variability is no less important than a prediction of the median signal strength itself.

5.15.1 Statistics of the fast fading

The scattering model [2,5] is based on the assumption that the received signal consists of a large number of randomly phased components and leads to the conclusion that the probability density function of the signal envelope follows a Rayleigh distribution.

This scattering model describes the local, i.e. small area, statistics of the signal envelope in terms of only one parameter σ, the modal value. The mean of the distribution is $\sigma\sqrt{\pi/2}$ and this is the local mean of the signal envelope. As we have seen, factors such as range, the path profile to the transmitter, the type and density of buildings near the receiver, and the width and orientation of the street, combine to influence the value of this mean. It is only over distances sufficiently small to ensure these factors are sensibly constant that the process can be considered statistically stationary.

To test whether the data collected in any small area fit the stationarity hypothesis, the cumulative distribution is often plotted on Rayleigh-scaled graph paper, i.e. graph paper on which Rayleigh-distributed data would appear as a straight line (Appendix A). Departures from Rayleigh are easily apparent but since the scale is highly non-linear, the tails of the distribution are overemphasised and a quantitative judgement is difficult. An alternative, avoiding this drawback, is to plot $P(R)$, the cumulative distribution of the measured data, against $P_r(R)$, the cumulative distribution of a Rayleigh process having the same RMS value. Departures from a straight line of unity slope indicate differences in the statistical distributions of $P(R)$ and $P_r(R)$ with no bias towards any particular range of probabilities. Figure 5.32(a) shows the cumulative distribution of about 6000 data samples collected in London [21] over a distance of 100λ at 168 MHz plotted on Rayleigh paper, and Figure 5.32(b) shows the same data plotted against $P_r(R)$. Comparison of the two figures shows that the data actually departs more seriously from Rayleigh near the middle of the distribution and not at the lower tail, as suggested by Figure 5.32(a).

A statistically valid test of this model would require a large grouping of data to be plotted, but in doing this it is almost certain that the 'small locality' assumption would be violated. Indeed, whenever a reasonably large quantity of data collected over some distance is considered, large departures from Rayleigh always appear. This is accounted for by suggesting that the process is non-stationary, i.e. over each small area the process is Rayleigh but the mean value varies from place to place. To test whether the underlying process is fundamentally Rayleigh, some way must be found to handle the problem of non-stationarity.

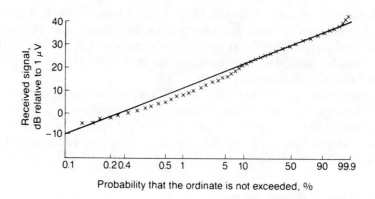

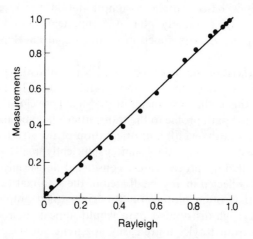

Figure 5.32   (a) Cumulative distribution of 6072 samples obtained in London at 168 MHz over a distance of 100λ; cumulative distribution of the samples in part (a) plotted against the theoretical Rayleigh CDF.

Clarke [2] suggested the technique of normalising the data by using its running mean. He took a set of experimental results and divided each data point by a local mean obtained from averaging the 200 points symmetrically adjacent to it; the resulting normalised random variable was then treated in exactly the same way as the original random variable. The argument was that if a Rayleigh process is normalised to its RMS value, the resultant is another Rayleigh process with an RMS value of unity. If the local Rayleigh process is represented by eqn. (5.19) then the RMS value is $\sigma\sqrt{2}$. If we now normalise to this value, the new variable is $r_n = r/\sigma\sqrt{2}$, and since $p_n(r_n)\,dr_n = p_r(r)\,dr$, the new probability density function will be

$$p_n(r_n) = 2r_n\,\exp(-r_n^2)$$

which is a Rayleigh process with $\sigma^2 = 0.5$ and an RMS value of unity. Normalisation of a Rayleigh process therefore does not change the distribution, it only changes the RMS value.

The problem of determining a distance suitable for normalising the data has been approached both experimentally and theoretically and will be discussed more fully in Chapter 8. For now it is sufficient to say that a few tens of wavelengths is usually considered appropriate and to confirm that in practice normalisation using the running mean technique almost invariably causes the resulting data to display a close approximation to a Rayleigh distribution.

5.15.2   Statistics of the local mean

Measurements reported by Reudink [22], Black and Reudink [23] and Okumura *et al.* [24] are often quoted to suggest that the local mean of signals received at a given range and frequency, and in similar environmental areas, follows a lognormal distribution. These researchers found that when fast fading was averaged out, the variations in the local mean were very closely lognormal. It was experimental evidence such as this which gave rise to the suggestion, mentioned earlier, that a three-stage model might be appropriate to describe urban propagation: an inverse nth power law with range from the transmitter to the area where the receiver is located (with n often very close to 4), lognormal variations of the local mean within that area, and superimposed fast fading which follows a Rayleigh distribution.

Okumura's measurements in Tokyo showed that when the median signal strength was computed over 20 m sectors and the standard deviation determined over areas of diameter 1 to 1.5 km, the values all lay in the range 3–7 dB. They increased slightly with frequency but appeared insensitive to range. It was pointed out that if the size of the area under consideration was increased then the values of standard deviation were likely to be larger and a curve was drawn suggesting that in suburban areas or in rolling hilly terrain, typical values of standard deviation were 7 dB at 200 MHz, rising to 10 dB at 3000 MHz. In urban areas, values were about 2 dB lower.

Analysis of data collected in London [21], in which the mean was computed over 40 m sectors, produced similar results. The cumulative distribution of the local means measured at 2 km range is shown in Figure 5.33(a), the standard deviation being 5 dB at 168 MHz, 5.65 dB at 455 MHz and 6.4 dB at 900 MHz. The values at 9 km range, obtained from Figure 5.33(b), are 4.4 dB and 5.2 dB at 168 MHz and 455 MHz respectively, showing that in a flat city like London the spread at UHF is greater than at VHF. There are two possible explanations for the value being lower at 9 km range than at 2 km. Firstly, the standard deviation might be range dependent, but Okumura's findings did not indicate any range dependence. A more likely explanation is that the value is influenced by the degree of urbanisation. At 2 km, the chosen test area was dominated by high-rise buildings; at 9 km the test area was mainly residential with houses and gardens being the main features. Reudink [22], faced by apparently contradictory results from tests at 800 MHz and 11.2 GHz in Philadelphia and New York, where the standard deviation decreased with range in Philadelphia but increased in New York, also suggested that the influence of the local environment was much stronger than the influence of range from the transmitter.

5.15.3   Large area statistics

We have seen in the previous section that over a relatively small distance (a few tens of wavelengths) the signal is well described by Rayleigh statistics; the local

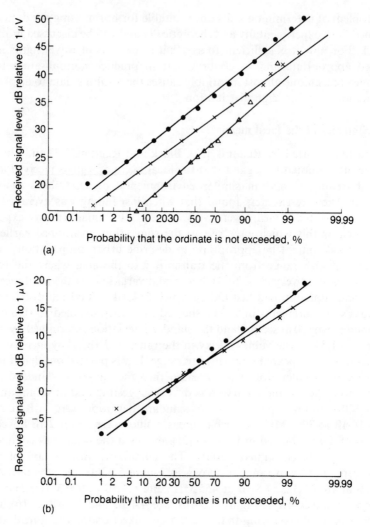

Figure 5.33 Cumulative distributions of the local mean of the received signal for (a) 2 km range and (b) 9 km range: ($\times$) 168 MHz, ($\bullet$) 455 MHz, ($\triangle$) 900 MHz.

mean over a somewhat larger area (with homogeneous environmental character-istics) is lognormally distributed. It is of interest therefore to examine the overall distribution of the received signal in these larger areas. We might reasonably expect it to be a mixture of Rayleigh and lognormal, but it is worthwhile examining other statistical distributions such as the Nakagami and Weibull, both of which contain the Rayleigh distribution as a special case.

The Nakagami-*m* distribution [25] may be represented by the formulation

$$p_m(x) = \frac{2}{\Gamma(m)} \left(\frac{m}{\Omega}\right)^m x^{2m-1} \exp\left(-\frac{m}{\Omega}x^2\right) \begin{array}{l} x \geqslant 0 \\ m \geqslant \frac{1}{2} \end{array} \qquad (5.73)$$

where m and Ω are parameters (Ω being the mean square value) and $\Gamma(\cdot)$ is the gamma function. This reduces to the one-sided Gaussian distribution for $m = 0.5$ and for $m = 1$ it becomes

$$p_m(x) = \frac{2x}{\Omega} \exp\left(-\frac{x^2}{\Omega}\right) \tag{5.74}$$

which is the Rayleigh distribution.

The Weibull distribution can be expressed as

$$p_w(x) = \begin{cases} \alpha w x^{w-1} \exp(-\alpha x^w) & x > 0 \\ 0 & x < 0 \end{cases} \tag{5.75}$$

where $w > 0$ and $\alpha > 0$. For $w = 2$ this becomes the Rayleigh distribution,

$$p_w(x) = 2\alpha x \exp(-\alpha x^2) \tag{5.76}$$

Suzuki [26], Hansen and Meno [27] and Lorenz [28,29] all suggested that the statistics of the mobile radio signal can be represented by a mixture of the Rayleigh and lognormal distributions in the form of a Rayleigh distribution with a lognormally varying mean. Suzuki suggested the formulation

$$p_s(x) = \int_0^\infty \frac{x}{\sigma^2} \exp\left(-\frac{x^2}{2\sigma^2}\right) \frac{M}{\sigma\alpha\sqrt{2\pi}} \exp\left(-\frac{\log(\sigma/\sigma_0)}{2\alpha^2}\right) d\sigma \tag{5.77}$$

where σ is the mode (i.e. most probable value) of the Rayleigh distribution, α is the shape parameter of the lognormal distribution and $M = \log e = 0.434$. The mean square and mean values are

$$E\{x^2\} = 2\sigma_0^2 \exp\left(\frac{2\lambda^2}{M^2}\right) \tag{5.78}$$

$$E\{x\} = \sqrt{\frac{\pi}{2}}\sigma_0 \exp\left(\frac{\lambda^2}{2M^2}\right) \tag{5.79}$$

Equation (5.77) is the integral of the Rayleigh distribution over all possible values of σ, weighted by the PDF of σ, and this attempts to provide a transition from local to global statistics. However, although Suzuki compared the fit of four distributions (Rice, Nakagami, Rayleigh and lognormal) to experimental data, he did not use this mixture distribution, probably because the PDF exists in integral form and presents computational difficulties. It was left to Lorenz [28] to evaluate the expression and it was he who termed this mixture the *Suzuki distribution*. Lorenz introduced the following variable and parameters:

$$F = 20 \log x$$
$$F_{OR} = 20 \log(\sqrt{2}\sigma)$$
$$F_{OS} = 20 \log[\sqrt{2}\sigma_0 \exp(2\lambda^2/M^2)]$$

which represent the signal strength (dB), the RMS value of the Rayleigh distribution (dB) and the RMS value of the Suzuki distribution (dB) respectively. In terms of the parameters, $s = 20\lambda$ and $M_1 = 20M$, the Suzuki distribution as formulated by Lorenz is

$$p_s(F) = \int_{-\infty}^{+\infty} \frac{2}{M_1} \exp\left\{ \frac{2}{M_1}(F - F_{OR}) - \exp\left(\frac{2}{M_1}(F - F_{OR}) \right) \right\}$$

$$\times \frac{1}{s\sqrt{2\pi}} \exp\left\{ -\frac{(F_{OR} - F_{OS} + s^2/M)^2}{2s^2} \right\} dF_{OR} \tag{5.80}$$

To decide which statistical model best fits measured data, it is necessary to obtain appropriate parameters for the distribution under consideration. Suzuki and Lorenz suggested the use of a moment method which equates the theoretical means and variances of the distribution functions to the sample mean and variance of the experimental data. The functional forms of these moments are not simple, but numerical evaluation is possible and the results given by Lorenz [28] are as follows.

For the Nakagami-m distribution:

$$\hat{m} = \frac{4.4}{\sqrt{\hat{u}_2}} + \frac{17.4}{\hat{u}_2 1.29} \quad \text{(a good approximation)}$$

$$\hat{F}_{ON} = \hat{F} - 4.343\{\psi(\hat{m}) - \ln \hat{m}\} \tag{5.81}$$

For the Weibull distribution:

$$\hat{w} = \frac{11.14}{\sqrt{\hat{u}_2}}$$

$$\hat{F}_{OW} = \hat{F} + 4.343 \ln(\Gamma(1 + 2/w)) \tag{5.82}$$

For the Suzuki distribution:

$$\hat{s} = \sqrt{u_2 - 31.025}$$

$$\hat{F}_{OS} = \hat{F} + 2.51 - 0.11\hat{s}^2 \tag{5.83}$$

where hatted variables (e.g. $\hat{m}$) are computed from experimental data

u_2 is the second central moment of the distribution
$\Gamma(\cdot)$ is the gamma function
$\psi(\cdot)$ is the digamma function

Several researchers [26,29,30] have compared the cumulative distribution of the experimental data with the theoretical distributions and have come to the conclusion that the Suzuki distribution provides the best fit to the experimental data, particularly in built-up areas.

For a more detailed comparison between the three distributions it is important to choose a suitable criterion as the basis for comparison. To establish such a criterion it is helpful to remember that interest in the signal statistics stems mainly from the fact that without a reliable statistical model, prediction of the median signal strength is of little help either to the system engineer or to the frequency management bodies. What is really needed is an accurate prediction of values near the tails of the distribution, vital for coverage estimation and frequency reuse planning. Thus the most suitable model is one which, given the median value, will predict with least error the values between say 1% and 20% at one end and between 80% and 99% at the other.

For this purpose, Parsons and Ibrahim [30] computed the quantities $(F_{10} - F_{50})$ and $(F_{90} - F_{50})$ for each theoretical model and plotted them as a function of the appropriate parameter m, w or s. The experimentally measured values of these quantities were then determined from an available set of data and the appropriate parameter $\hat{m}$, $\hat{w}$ or $\hat{s}$ was computed using equations (5.81) to (5.83) on the assumption that the experimental data was drawn from a distribution of the type being considered. The appropriate value was used to plot the experimental points and Figure 5.34 shows data at 900 MHz in comparison with theoretical predictions for the Suzuki distribution. The fit is very good.

Calculations of $\hat{s}$ for various data sets in London [30] show that the value in decibels is normally distributed. The mean value $\hat{s}$ and the standard deviation $\hat{\sigma}_s$ are compared in Table 5.2 with values given by Lorenz for data measured in the Rhine valley.

Notice that $\hat{s}$ is approximately 4 dB and this is also apparent from Figure 5.34. Thus it is possible to use any of the techniques described previously, not only for median signal strength prediction but to estimate values near the tails of the distribution. Figure 5.35 shows various quantiles of the Suzuki distribution in relation to the median so that for a given value of s, the difference between the median and values near the tails of the distribution can be found. The dotted line in Figure 5.35 shows that for $s = 4$ dB the values of F_{90} and F_{10} can be estimated, given the median value F_{50}, by subtracting 9.5 dB and adding 7.5 dB respectively. Values corresponding to other values of s can also be obtained from that diagram.

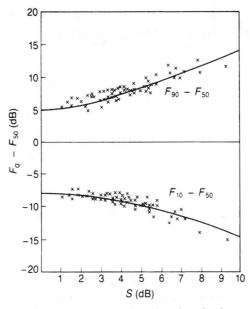

Figure 5.34 Experimental data at 900 MHz plotted against the theoretical Suzuki distribution.

Table 5.2 Comparison of the average value of s and the value of σ_s in the Rhine valley and London[a]

	s (dB)	σ_s (dB)
Rhine valley measurements (450 MHz)		
Forests	3.7	1.7
Small city	3.9	1.9
Medium-sized city	3.3	1.5
London measurements (large city)		
168 MHz	4.1	1.4
455 MHz	3.7	1.0
900 MHz	3.3	1.6

[a]Values computed directly from the dB record.

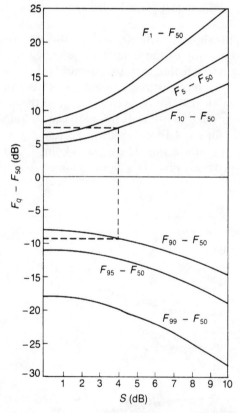

Figure 5.35 Quantiles of the Suzuki distribution in relation to the median value.

The lognormal approximation

We have already established that the mobile radio signal is composed of two fading components, fast fading caused by local multipath propagation and slow fading due to shadowing. The envelope of the received signal can be expressed as

$$z(t) = x(t)y(t) \tag{5.84}$$

where $x(t)$ is the fast fading envelope which closely follows a Rayleigh distribution and $y(t)$ is the slow fading component which is lognormally distributed with a standard deviation in the range 4–12 dB. The envelope $z(t)$ has a Suzuki distribution as described previously.

The Suzuki distribution, as described by eqns (5.77) and (5.80), is rather complicated and not easy to handle mathematically. It would be very useful to have an approximation, and it has been suggested [31] that under certain conditions $z(t)$ can be approximated by a lognormal distribution. This is very convenient since such a distribution can be completely specified in terms of two parameters, the mean and standard deviation. The original paper appears to give incorrect values for these parameters but correct values can be obtained as follows.

We can express eqn. (5.84) in dB units by writing

$$20 \log_{10} z(t) = 20 \log_{10} x(t) + 20 \log_{10} y(t)$$

i.e.

$$z_{\mathrm{dB}}(t) \quad = \quad x_{\mathrm{dB}}(t) \quad + \quad y_{\mathrm{dB}}(t) \tag{5.85}$$

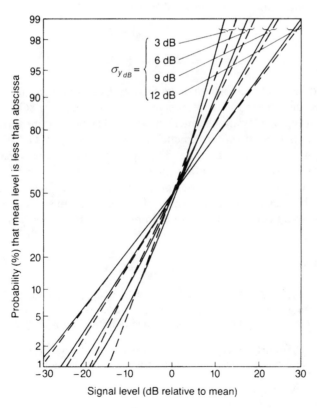

Figure 5.36 Approximation for z_{dB} using various values of $\sigma_{y_{\mathrm{dB}}}$ (the approximation is reasonable if $\sigma \geqslant 6$ dB): (– – –) theoretical.

Table 5.3 Mean and standard deviation of the resultant approximated lognormal distribution compared with the theoretical values

$\sigma_{y_{dB}}$	$E\{z_{dB}\}$		$\sigma_{z_{dB}}$	
	From expt	From (5.86)	From expt	From (5.87)
3	-2.47	-2.5	6.33	6.33
6	-2.49	-2.5	8.19	8.19
9	-2.50	-2.5	10.59	10.58
12	-2.52	-2.5	13.24	13.23

The mean and standard deviation of $x_{dB}(t)$ are given in Appendix B as

$$E\{x_{dB}\} = -2.5 \text{ dB}$$

assuming $x(t)$ has a zero mean value, and

$$\sigma_{x_{dB}} = 5.57 \text{ dB}$$

Now, $x_{dB}(t)$ and $y_{dB}(t)$ are independent random processes and this enables us to write the mean and standard deviation of z as [32]:

$$E\{z_{dB}\} = E\{y_{dB}\} - 2.5 \tag{5.86}$$

$$\sigma_{z_{dB}} = \sqrt{\sigma_{y_{dB}}^2 + (5.57)^2} \tag{5.87}$$

A verification designed to test whether the Suzuki distribution can be approximated by a lognormal distribution with parameters given by eqns. (5.86) and (5.87) has been obtained using a software channel simulator. The results are shown in Figure 5.36. Here $z_{dB}(t)$ is plotted on 'arithmetic' graph paper scaled such that a normal distribution appears as a straight line. The approximation is reasonable if y_{dB} is greater than about 6 dB, which is often the case in practice.

Values of the mean and standard deviation of z_{dB}, obtained from the simulation, are given in Table 5.3 for selected values of $\sigma_{y_{dB}}$, where they are compared with values obtained using eqns (5.86) and (5.87) under the assumption that $E\{y_{dB}\} = 0$.

REFERENCES

1. Ossanna J.F. (1964) A model for mobile radio fading due to building reflections: theoretical and experimental fading waveform power spectra. *Bell Syst. Tech. J.*, **43**, 2935–71.
2. Clarke R.H. (1968) A statistical theory of mobile radio reception. *Bell Syst. Tech. J.*, **47**, 957–1000.
3. Gilbert E.N. (1965) Energy reception for mobile radio. *Bell Syst. Tech. J.*, **44**, 1779–803.
4. Gans M.J. (1972) A power-spectral theory of propagation in the mobile radio environment. *IEEE Trans.*, **VT21**(1), 27–38.
5. Aulin T. (1979) A modified model for the fading signal at a mobile radio channel. *IEEE Trans*, **VT28**(3), 182–203.
6. Jakes W.C. (ed.) (1974) *Microwave Mobile Communications*. John Wiley, New York.
7. Lee W. C.-Y. (1982) *Mobile Communications Engineering*. McGraw-Hill, New York.

8. Bennett W.R. (1948) Distribution of the sum of randomly phased components. *Quart. Appl. Math.*, **5**, 385–93.
9. Rice S.O. (1944) Mathematical analysis of random noise. *Bell Syst. Tech. J.*, **23**, 292–332.
10. Davenport W.B. and Root W.L. (1958) *An Introduction to the Theory of Random Signals and Noise.* McGraw-Hill, New York.
11. Bodtmann W.F. and Arnold H.W. (1982) Fade duration statistics of a Rayleigh-distributed wave. *IEEE Trans.*, **COM30**(3), 549–53.
12. Davis B.R. (1971) FM noise with fading channels and diversity. *IEEE Trans.*, **COM19**(6), 1189–200.
13. Rice S.O. (1948) Statistical properties of a sine wave plus random noise. *Bell Syst. Tech. J.*, **27**(1), 109–57.
14. Parsons J.D. and Gardiner J.G. (1989) *Mobile Communications Systems.* Blackie, Glasgow.
15. Adachi F., Feeney M.T. and Parsons J.D. (1988) Level crossing rate and average fade duration for time diversity reception in Rayleigh fading conditions. *Proc. IEE. Part F*, **135**(4), 501–6.
16. Feeney M.T. (1989) The complex narrowband UHF mobile radio channel. PhD thesis, University of Liverpool.
17. Yamada Y., Ebine Y. and Nakajima N. (1987) Base station/vehicular antenna design techniques employed in high-capacity land mobile communications systems. *Rev. Elec. Commun. Lab.*, Nippon Telegraph and Telephone Public Corporation (NTT), **35**, 115–21.
18. Parsons J.D. and Turkmani A.M.D. (1991) Characterisation of mobile radio signals: model description. *Proc. IEE Part I*, **138**(6), 549–56.
19. Turkmani A.M.D. and Parsons J.D. (1991) Characterisation of mobile radio signals: base station cross-correlation. *Proc. IEE Part I*, **138**(6), 557–65.
20. Lee W. C.-Y. (1967) Theoretical and experimental study of the properties of the signal from an energy-density mobile-radio antenna. *IEEE Trans.*, **VT16**(1), 25–32.
21. Ibrahim M.F. and Parsons J.D. (1983) Signal strength prediction in built-up areas. Part 1: median signal strength. *Proc. IEE Part F*, **130**(5), 377–84.
22. Reudink D.O. (1972) Comparison of radio transmission at X-band frequencies in suburban and urban areas. *IEEE Trans.*, **AP20**, 470–3.
23. Black D.M. and Reudink D.O. (1972) Some characteristics of mobile radio propagation at 836 MHz in the Philadelphia area. *IEEE Trans.*, **VT21**, 45–51.
24. Okumura Y., Ohmori E., Kawano T. and Fukuda K. (1968) Field strength and its variability in VHF and UHF land mobile service. *Rev. Elec. Commun. Lab.*, **16**, 825–73.
25. Nakagami M. (1960) The *m*-distribution. A general formula of intensity distribution of rapid fading. In *Statistical Methods in Radio Wave Propagation* (ed. W.C. Hoffman). Pergamon, Oxford.
26. Suzuki H. (1977) A statistical model for urban radio propagation. *IEEE Trans.*, **COM25**, 673–80.
27. Hansen F. and Meno F.I. (1977) Mobile radio fading – Rayleigh and lognormal superimposed. *IEEE Trans.*, **VT26**, 332–5.
28. Lorenz R.W. *Theoretical distribution functions of multipath fading processes in mobile radio and determination of their parameters by measurements.* Technischer Bericht 455, TBr 66, Forschungsinstitut der Deutschen Bundespost (in German).
29. Lorenz R.W. (1980) Field strength prediction method for a mobile telephone system using a topographical data bank. *IEE Conference Publication 188*, pp. 6–11.
30. Parsons J.D. and Ibrahim M.F. (1983) Signal strength prediction in built-up areas. Part 2: signal variability. *Proc IEE Part F*, **130**(5), 385–91.
31. Muammar R. and Gupta S.C. (1982) Cochannel interference in high capacity mobile radio systems. *IEEE Trans.*, **COM30**(8), 1973–8.
32. Stremler F.G. (1982) *Introduction to Communication Systems.* Addison-Wesley, New York.

Chapter 6

Wideband Channel Characterisation

6.1 INTRODUCTION

The mobile radio propagation environment clearly places fundamental limitations on the performance of radio communication systems. Signals arrive at a receiver via a scattering mechanism, and the existence of multiple propagation paths (multipath) with different time delays, attenuations and phases gives rise to a highly complex, time-varying transmission channel.

In order for systems engineers to determine optimum methods of mitigating the impairments caused by multipath propagation, it is essential that the transmission channel be properly characterised. Previously we dealt with narrowband characterisation which is appropriate for transmissions where the inverse of the signal bandwidth is very much greater than the spread in propagation path delays. For narrowband transmissions in a mobile environment, the multipath then results in rapid fading of the received signal envelope and an associated Doppler spread is apparent in the received spectrum. The signal statistics appropriate to narrowband transmissions are usually determined from measurements carried out at a single frequency and, as we have seen, the Rayleigh distribution is usually a good approximation for the envelope statistics.

However, the distribution departs significantly from Rayleigh when a strong direct path is present, and in that case the envelope statistics are better described by a Rician distribution. Even so, it is the areas of low signal strength, where multipath dominates, that are of the greatest importance in determining the limits on mobile radio system performance, and departures from the Rayleigh model in areas of high signal strength do not detract from its usefulness for most system analysis applications.

The slower variation in the average signal strength caused by gross changes along the propagation path and in the local environment can be described by lognormal statistics. A description of the channel in terms of Rayleigh-distributed fast fading and lognormally distributed shadow fading is usually adequate for the evaluation of narrowband systems [1].

6.2 FREQUENCY-SELECTIVE FADING

The earlier discussion concentrated in general on describing the envelope and phase variations of the signal received at a mobile when an unmodulated carrier is radiated by the base station transmitter. The question now arises as to the adequacy of this channel description when real signals, which occupy a finite bandwidth, are radiated. It is clear that in practice we need to consider the effects of multipath propagation on these signals and to illustrate the point we consider the case of two frequency components within the message bandwidth. If these frequencies are close together then the different propagation paths within the multipath medium have approximately the same electrical length for both components and their amplitude and phase variations will be very similar. In other words, although there will be fading due to multipath, the two frequency components will behave in a very similar way.

More generally, provided the message bandwidth is sufficiently small, all frequency components within it behave similarly and *flat fading* is said to exist. As the frequency separation increases, however, the behaviour at one frequency tends to become uncorrelated with that at the other, because the phase shifts along the various paths are different at the two frequencies. The extent of the decorrelation depends on the spread of time delays since the phase shifts arise from the excess path lengths. For large delay spreads, the phases of the incoming components can vary over several radians even if the frequency separation is quite small.

Signals which occupy a bandwidth greater than the bandwidth over which spectral components are affected in a similar way will become distorted since the amplitudes and phases of the various spectral components in the received version of the signal are not the same as they were in the transmitted version. This phenomenon is known as *frequency-selective fading* and appears as a variation in received signal strength as a function of frequency. In analogue FM systems the frequency selectivity limits the maximum usable frequency deviation for a given amount of signal distortion. The bandwidth over which the spectral components are affected in a similar way is known as the *coherence*, or *correlation bandwidth*.

The fact that the lengths of the individual propagation paths vary with time due to motion of the vehicle provides a method of gaining further insight into the propagation mechanism, since the changing time of arrival suggests the possibility of associating each delayed version of the transmitted signal with a physical propagation path. Indeed, if a number of distinct physical scatterers are involved, it may be possible to associate each scatterer with an individual propagation path. However, it is not possible to distinguish between different paths merely by considering the difference between the time of arrival; the spatial direction of arrival also has to be taken into account. If we consider only single-scattered paths then all scatterers associated with a certain path length can be located on an ellipse with the transmitter and receiver at its foci. Each time delay between transmitter and receiver defines a confocal ellipse as shown in Figure 6.1. If we consider scatterers located at A, B and C, then we can distinguish between paths TAR and TBR, which have the same angle of arrival, by their different time delays; and we can distinguish between TAR and TCR, which have the same time delay, by their different angles of arrival.

The angles of arrival can be determined by Doppler shift. Whenever the receiver or transmitter is in motion the received RF signal experiences a Doppler shift; the

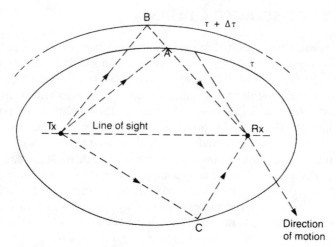

Figure 6.1 Geometry for single scattering.

frequency shift being related to the cosine of the spatial angle between the direction of arrival of the wave and the direction of motion of the vehicle. So if we transmitted a short RF pulse and measured its time of arrival and its Doppler shift at the receiver, we could identify the length of the propagation path and the angle of arrival. Of course, there is left/right ambiguity inherent in the Doppler shift measurement but this could be resolved, if necessary, by the use of directional antennas.

An important and instructive feature of Figure 6.1 is that, for a particular receiver location, a suitably scaled diagram with several confocal ellipses can be produced in the form of a map overlay. Coordinated use of this overlay, together with experimental results for the location in question, allows the identification of significant single scatterers or scattering areas, and gives an indication of the extent of the contribution from multiple scattering.

These time-delayed echoes can overlap as Figure 6.2 shows causing errors in digital systems due to intersymbol interference. In this case, increasing the signal-to-noise ratio will not cause a reduction in error rate, so the delay spread sets the lower bound on error performance for a specified data rate. This limit is often termed the irreducible bit error rate (IBER), although in practice the performance can be further improved by the use of channel equalisation techniques (Chapter 10).

The characterisation of mobile radio channels in terms of their effect on narrowband signals is well developed (Chapter 5) and is important in connection with PMR systems and first-generation analogue cellular radio systems (AMPS, TACS, etc.). However, second-generation digital systems such as GSM are inherently wideband in nature and third-generation (UMTS) systems will be even more so. The signal parameters identified in Chapter 5 remain relevant, but for these newer systems the effects of frequency-selective fading are equally important.

6.2.1 Channel characterisation

In general, characterisation of mobile radio propagation channels can be developed from the general description of linear time-variant channels [2]. The behaviour of the

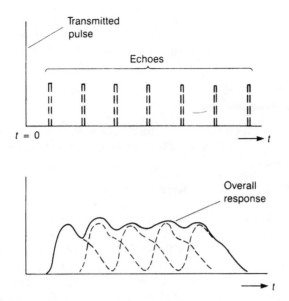

Figure 6.2 The receiver responses to echoes of a transmitted pulse can overlap to produce intersymbol interference.

channel can then be described in terms of system functions which give an insight into the physical mechanisms which dominate the channel behaviour – an important consideration for engineers.

The first general analytical treatment along these lines was that of Zadeh [3], who dealt with time-variant linear filters. Kailath [4] subsequently produced further work, with an emphasis on channel characterisation. He dealt with some canonical sampling models of time-variant channels and some theorems on measurability. In a seminal paper, Bello [2] further developed the work of his predecessors and presented it in a form which readily showed the compactness, symmetry and application of the characterisation approach to general and restricted classes of radio channels. He developed some symmetrical relationships between system functions in the time and frequency domains employing duality and Fourier transformations and subsequently generalised the concept of time–frequency duality [5] and channel measurability [6].

Bajwa [7] took a more restricted approach by concentrating primarily on the characterisation of practical channels. Alternative descriptions for restricted classes of channels have also been presented [8] and will be discussed later. Although it is clear from earlier chapters that mobile radio channels have characteristics that vary randomly with the location of the mobile, it is easier and more convenient to introduce the concepts by assuming, in the first instance, that the channel is deterministic.

6.3 CHARACTERISATION OF DETERMINISTIC CHANNELS

The radio propagation channel may be envisaged as a system element which transforms input signals into output signals. It is therefore analogous to a linear filter but since the channel behaviour is generally time-variant, we must also allow the transmission characteristics of the equivalent filter to be time-varying. The inputs and outputs of a linear filter can be described in both the time and frequency domains, and this leads to four possible transmission functions that can be used to describe the channel.

6.3.1   The time domain function

In the discussion that follows, it is convenient to represent real bandpass signals by their complex envelopes. Conventionally the relationship between real and complex signals is expressed as

$$x(t) = \text{Re}[z(t)\exp\{j2\pi f_c t\}] \tag{6.1}$$

where, Re[·] is the real part of a complex function, $z(t)$ is the complex envelope of $x(t)$ and f_c is a nominal carrier frequency.

The time domain description of a linear system is specified by the time impulse response of the system. Application of the superposition principle then expresses the system output, for a known input signal, in the time domain. Since the channel is time-variant, the impulse response is also a time-varying function. If the complex envelope of the time-variant impulse response of the channel equivalent filter is given by $h(t, \tau)$, where τ is a delay variable, then the complex envelope of the filter output, $w(t)$, is related to the complex envelope of the input, $z(t)$, by the convolution relationship

$$w(t) = \int_{-\infty}^{+\infty} z(t-\tau)h(t, \tau)\,d\tau \tag{6.2}$$

Equation (6.2) provides a physical representation of the channel as a continuum of non-moving, scintillating scatterers, with each elemental scatterer having a gain fluctuation $h(t, \tau)\,dt$ and providing delays in the range $(\tau, \tau + d\tau)$. Physically, $h(t, \tau)$ can be interpreted as the channel response at time t to an impulse input τ seconds in the past. Since a physical channel cannot have an output before the input has arrived, $h(t, \tau)$ must be subject to the constraint that it vanishes for $\tau < 0$. Therefore, for a physically realisable channel, observed over a finite period T, the limits of integration in eqn. (6.2) become $(0, T)$. However, for simplicity, the limits will remain written as $(-\infty, \infty)$ with the constraint that the integrand becomes zero outside the range $(0, T)$, thereby ensuring physical realisability. In his discussion of channel characterisation using system functions, Bello [2] termed the time-variant impulse response $h(t, \tau)$, the *input delay-spread function*.

Writing the convolution relationship of eqn. (6.2) as a summation,

$$w(t) = \Delta\tau \sum_{m=1}^{n} z(t - m\Delta\tau)h(t, m\Delta\tau) \tag{6.3}$$

enables us to envisage a physical representation (Figure 6.3) in the form of a densely tapped delay line composed of differential delay elements and modulators [3,4,8]. Note that eqn. (6.3) leads to a model where the input is first delayed and then multiplied by the differential scattering gain.

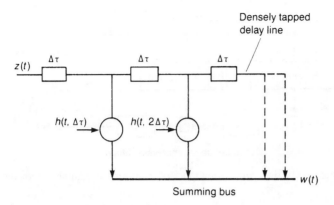

Figure 6.3 Tapped delay line model of a multipath channel (time domain representation).

6.3.2 The frequency domain function

A general channel characterisation is also possible in terms of frequency variables through the use of a function which is the dual of the time-variant impulse response. This dual channel function $H(f, v)$ relates the channel output spectrum to the channel input spectrum in an identical manner to the way in which $h(t, \tau)$ relates the input/output time functions. This dual characterisation involves representing the output spectrum $W(f)$ as a superposition of elemental Doppler-shifted and filtered replicas of the input spectrum $Z(f)$. The transmission characteristics are then described in terms of frequency and frequency-shift variables by the input/output relationship

$$W(f) = \int_{-\infty}^{+\infty} Z(f - v)H(f - v, v)\, dv \tag{6.4}$$

Although the input delay-spread function $h(t, \tau)$ provides an insight into the contributions from scatterers having different path lengths, i.e. multipath, it does not provide an explicit illustration of the time-varying behaviour of the channel. Such an illustration is possible, however, through a characterisation in terms of $H(f, v)$, where the frequency-shift variable v can be visualised as the Doppler shift experienced in such channels. Again, writing eqn. (6.4) as a summation,

$$W(f) = \Delta v \sum_{m=1}^{n} Z(f - m\Delta v)H(f - m\Delta v, m\Delta v) \tag{6.5}$$

allows a further physical model of the channel to be envisaged in the form of a dense frequency conversion chain, analogous to the tapped delay line model used to represent eqn. (6.3). Figure 6.4 represents eqn. (6.5) through the use of a bank of filters having transfer functions $H(f, v)\,\Delta v$, followed by Doppler-shifting frequency converters producing Doppler shifts in the range $(v, v + \Delta v)$ hertz. Bello [2] referred to $H(f, v)$ as the *output Doppler-spread function*.

6.3.3 The time-variant transfer function

Sections 6.3.1 and 6.3.2 have shown that characterisation of a time-variant channel in terms of the input delay-spread function $h(t, \tau)$ relates the output time function to

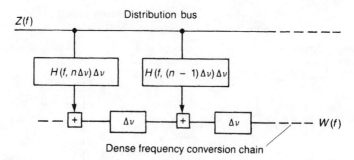

Figure 6.4 Frequency shifting convertor model of a multipath channel (frequency domain representation).

the input time function, whereas a characterisation in terms of the output Doppler-spread function $H(f, v)$ relates the output spectrum to the input spectrum. Another characterisation approach is possible in which the output time function is expressed in terms of the input spectrum to the channel equivalent filter [2]. This function is known as the *time-variant transfer function* $T(f, t)$ and it was first introduced by Zadeh [3]. The input/output relationship in this case is given by

$$w(t) = \int_{-\infty}^{+\infty} Z(f)T(f, t) \exp\{j2\pi ft\} \, df \qquad (6.6)$$

The time-variant transfer function is the Fourier transform of the input delay-spread function with respect to the delay variable, and also the inverse Fourier transform of the output Doppler-spread function with respect to the Doppler-shift variable, i.e.

$$T(f, t) = \int_{-\infty}^{+\infty} h(t, \tau) \exp\{-j2\pi f\tau\} \, d\tau = \int_{-\infty}^{+\infty} H(f, v) \exp\{j2\pi vt\} \, dv \qquad (6.7)$$

$T(f, t)$ can be envisaged as the frequency transmission characteristic of the channel and can be determined by direct measurement of the channel cissoidal response. Each of the system functions provides a description of the channel behaviour as a function of two specific variables, and $T(f, t)$ represents the frequency transfer function of the channel as a function of time.

6.3.4   The delay/Doppler-spread function

Any linear time-variant channel can be represented as a continuum of stationary scintillating scatterers through the use of the input delay-spread function, or as a continuum of filters and hypothetical Doppler-shifting elements through use of the output Doppler-spread function. These two functions therefore provide an explicit description of only one aspect of the channel's dispersive behaviour, either the time delay or the Doppler shift. From the engineer's viewpoint, it would be useful to have a system function that simultaneously provides a description in both time-delay and Doppler-shift domains.

The system functions introduced in the preceding sections were classified according to whether the channel model had its delay operation, or Doppler-shift operation, at the input or output. Since both time delays and Doppler shifts occur in this new characterisation, one of the two operations has to be constrained to the input, and the other to the output. A characterisation which has the time-delay operation at the input and the Doppler-shift operation at the output can be termed a *delay-Doppler domain* characterisation.

The delay-Doppler domain system function is obtained by representing the input delay-spread function $h(t, \tau)$ as the inverse Fourier transform of its spectrum $S(\tau, v)$, i.e.

$$h(t, \tau) = \int_{-\infty}^{+\infty} S(\tau, v) \exp\{j2\pi vt\} \, dv \qquad (6.8)$$

Substitution of eqn. (6.8) in eqn. (6.2) yields

$$w(t) = \int_{-\infty}^{+\infty} \int_{-\infty}^{+\infty} z(t - \tau) S(\tau, v) \exp\{j2\pi vt\} \, dv \, d\tau \qquad (6.9)$$

Equation (6.9) shows that the output is represented as the sum of delayed and then Doppler-shifted signals. Signals corresponding to delays in the range $(\tau, \tau + d\tau)$, and Doppler shifts in the range $(v, v + dv)$ have a differential scattering amplitude $S(\tau, v) \, dv \, d\tau$. The delay Doppler-spread function $S(\tau, v)$ therefore explicitly describes the dispersive behaviour of the channel in terms of both time delays and Doppler shifts and can be physically interpreted in terms of Figure 6.1 [2].

6.3.5 Relationships between system functions

Figure 6.5 shows the interrelationships between the various system functions that can be used to characterise deterministic time-variant linear channels. The lines labelled F and F^{-1} connecting any two system functions indicate that they are related via Fourier or inverse Fourier transforms. Each system function involves two variables, and any two system functions connected by an F or F^{-1} have one common variable. This should be regarded as a fixed parameter when employing the Fourier

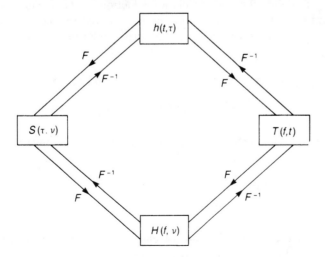

Figure 6.5 Relationships between system functions.

transform relationships involving the other two variables, one of which will be a time variable and the other a frequency variable. To make the F notation unique, the standard convention is applied: a negative exponent (i.e. Fourier transform) is used when transforming from a time variable to a frequency variable; a positive exponent (i.e. inverse Fourier transform) is used when transforming from a frequency variable to a time variable.

6.4 RANDOMLY TIME-VARIANT LINEAR CHANNELS

Having introduced the various channel descriptions and the relationships between them assuming deterministic behaviour, we can now extend the analysis to a discussion of real radio channels, which are randomly time-variant. The system functions then become stochastic processes. In order to describe the statistical characterisation of such a channel exactly, a knowledge of the multidimensional joint probability density functions of all the system functions is required. This is a formidable requirement and although it is necessary for a precise assessment of the channel behaviour, in practice it is unlikely to be achieved. A less accurate but more realistic approach is based on obtaining a statistical characterisation in terms of correlation functions for the various system functions [2,8]. This approach is attractive because it enables the autocorrelation function of the channel output to be determined. Furthermore, if the output is a Gaussian process then a description in terms of the mean and autocorrelation function is statistically complete. In the following discussion it is assumed for convenience and ease of representation that each of the system functions has a zero ensemble average.

6.4.1 Channel correlation functions

In using complex envelopes to represent real bandpass processes, a problem arises when attempting to define the autocorrelation function of the original real process since, in general, two autocorrelation functions are required to specify it uniquely [2]. This can be shown from calculation of the autocorrelation function of the real process $x(t)$, i.e.

$$E[x(t)x(s)] = \tfrac{1}{2}\mathrm{Re}[E[z(t)z^*(s)]\exp\{\mathrm{j}2\pi f_c(s-t)\}]$$
$$+ \tfrac{1}{2}\mathrm{Re}[E[z(t)z(s)]\exp\{\mathrm{j}2\pi f_c(s+t)\}] \tag{6.10}$$

where $E[\cdot]$ is the ensemble average and $z^*(s)$ is the complex conjugate of $z(t)$. Two autocorrelation functions, defined as

$$R_z(t,s) = E[z(t)z^*(s)] \tag{6.11}$$
$$\tilde{R}_z(t,s) = E[z(t)z(s)] \tag{6.12}$$

are therefore required to specify the autocorrelation function of the real process. In practice, fortunately, the narrowband process is such that $\tilde{R}_z(t,s) = 0$ and only the autocorrelation function defined in eqn. (6.11) is required.

The correlation functions that will be used for the four system functions in Section 6.3 can therefore be defined as follows:

$$E[h(t, \tau)h^*(s, \eta)] = R_h(t, s; \tau, \eta) \tag{6.13}$$

$$E[H(f, v)H^*(m, \mu)] = R_H(f, m; v, \mu) \tag{6.14}$$

$$E[T(f, t)T^*(m, s)] = R_T(f, m; t, s) \tag{6.15}$$

$$E[S(\tau, v)S^*(\eta, \mu)] = R_S(\tau, \eta; v, \mu) \tag{6.16}$$

In these equations, τ and η are time-delay variables, v and μ are frequency-shift variables.

Through use of the channel input/output relationships defined in eqns. (6.13) to (6.16), it is possible to determine the relationships between the autocorrelation function of the output and the autocorrelation functions of the system functions. The input delay-spread function will be considered as an example; the input/output correlation function relationships for the other system functions can be derived in a similar manner.

Using eqn. (6.2), the autocorrelation function of the channel output, $R_w(t, s)$, can be expressed as

$$R_w(t, s) = E[w(t)w^*(s)] = E\left[\int_{-\infty}^{+\infty} \int_{-\infty}^{+\infty} z(t - \tau)z^*(s - \eta)h(t, \tau)h^*(s, \eta)\, d\tau\, d\eta\right] \tag{6.17}$$

When the input $z(t)$ is deterministic, this becomes

$$R_w(t, s) = \int_{-\infty}^{+\infty} \int_{-\infty}^{+\infty} z(t - \tau)z^*(s - \eta)E[h(t, \tau)h^*(s, \eta)]\, d\tau\, d\eta \tag{6.18}$$

The term $E[h(t, \tau)h^*(s, \eta)]$ was defined in eqn. (6.13) as the autocorrelation function of the input delay-spread function, $R_h(t, s; \tau, \eta)$. Equation (6.18) therefore reduces to

$$R_w(t, s) = \int_{-\infty}^{+\infty} \int_{-\infty}^{+\infty} z(t - \tau)z^*(s - \eta)R_h(t, s; \tau, \eta)\, d\tau\, d\eta \tag{6.19}$$

This shows that the autocorrelation function of the channel output, $R_w(t, s)$, can be determined provided that the autocorrelation function $R_h(t, s; \tau, \eta)$ of the input delay-spread function $h(t, \tau)$ is known. For physical channels, $R_h(t, s; \tau, \eta)$ can be measured in the form $E[h(t, \tau)h^*(s, \eta)]$ using impulse sounding techniques.

6.4.2 Relationships between the functions

In Section 6.3.5 the relationships between the four system functions were shown in terms of single Fourier transforms. In a similar way, the autocorrelation functions of the system functions are related through double Fourier transforms. These relationships are illustrated in Figure 6.6; the lines marked DF or DF^{-1} indicate a double Fourier transform or double inverse Fourier transform relationship between the connected correlation functions, and the subscripts h, H, T and S indicate the appropriate system function.

Since the channel correlation functions comprise four variables, any two correlation functions linked by DF or DF^{-1} must have two common variables

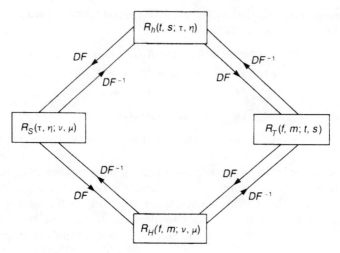

Figure 6.6 Relationships between channel correlation functions.

which should be considered as fixed parameters when employing the double Fourier transform involving the remaining variables. The standard convention outlined in Section 6.3.5 is again used to make the necessary transformations.

6.5 CLASSIFICATION OF PRACTICAL CHANNELS

Up to this point we have dealt with the ways in which deterministic and random time-variant channels can be characterised and represented. These general approaches can now be made more specific by considering practical channels which are subject to certain constraints.

6.5.1 The wide-sense stationary channel

Many physical channels possess fading statistics that can be assumed stationary over short periods of time or over small spatial distances. Although these channels are not necessarily stationary in the strict sense, they can be categorised as stationary in the wide sense (the term 'weakly stationary' is often used). *Wide-sense stationary* (WSS) channels have the property that the channel correlation functions are invariant under a translation in time, i.e. the fading statistics do not change over a short interval of time ξ. This means that the autocorrelation functions for a WSS channel depend on the variables t and s only through $\xi \, (= s - t)$. For a WSS channel the autocorrelation functions of the input delay-spread function $h(t, \tau)$ and the time-variant transfer function $T(f, t)$ become

$$R_h(t, t + \xi; \tau, \eta) = R_h(\xi; \tau, \eta) \qquad (6.20)$$
$$R_T(f, m; t, t + \xi) = R_T(f, m; \xi) \qquad (6.21)$$

It can be demonstrated that WSS channels give rise to uncorrelated Doppler-shift scattering. Using the double Fourier transform relationships in Figure 6.6, the autocorrelation function of the delay Doppler-spread function $S(\tau, v)$, in terms of the autocorrelation function of the input delay-spread function $h(t, \tau)$ is given by

$$R_S(\tau, \eta; v, \mu) = \int_{-\infty}^{+\infty} \int_{-\infty}^{+\infty} R_h(t, s; \tau, \eta) \exp\{j2\pi(vt - \mu s)\} \, dt \, ds \qquad (6.22)$$

Noting that $\xi = s - t$ for a WSS channel, and using eqn. (6.20), this becomes

$$R_S(\tau, \eta; v, \mu) = \int_{-\infty}^{+\infty} \int_{-\infty}^{+\infty} R_h(\xi; \tau, \eta) \exp\{j2\pi(vt - \mu t - \mu \xi)\} \, dt \, d\xi \qquad (6.23)$$

Rearranging,

$$R_S(\tau, \eta; v, \mu) = \int_{-\infty}^{+\infty} \exp\{j2\pi t(v - \mu)\} \, dt \int_{-\infty}^{+\infty} R_h(\xi; \tau, \eta) \exp\{-j2\pi\mu\xi\} \, d\xi \qquad (6.24)$$

The first integral in eqn. (6.24) can be recognised as a unit impulse at $v = \mu$. The second integral can be expressed in terms of the delay-Doppler cross-power spectral density $P_S(\tau, \eta; v)$, noting that $P_S(\tau, \eta; v)$ is the Fourier transform of $R_h(\xi; \tau, \eta)$ with respect to the variable ξ, i.e.

$$P_S(\tau, \eta; v) = \int_{-\infty}^{+\infty} R_h(\xi; \tau, \eta) \exp\{-j2\pi v\xi\} \, d\xi \qquad (6.25)$$

Therefore, eqn. (6.24) reduces to

$$R_S(\tau, \eta; v, \mu) = \delta(v - \mu) P_S(\tau, \eta; v) \qquad (6.26)$$

The singular behaviour of the channel correlation function $R_S(\tau; \eta; v, \mu)$ with respect to the Doppler shift variable suggests the following physical interpretation. In terms of the channel model composed of a number of elemental scatterers each producing delay and Doppler shift, *the contributions from elemental scatterers are uncorrelated if they produce different Doppler shifts.*

In a similar manner, it can be shown that

$$R_H(f, m; v, \mu) = \delta(v - \mu) P_H(f, m; v) \qquad (6.27)$$

where $P_H(f, m; v)$ is the Fourier transform of $R_T(f, m; \xi)$ with respect to the delay variable ξ, i.e.

$$P_H(f, m; v) = \int_{-\infty}^{+\infty} R_T(f, m; \xi) \exp\{-j2\pi v\xi\} \, d\xi \qquad (6.28)$$

In terms of a circuit model representation, the singular behaviour of $R_H(f, m; v, \mu)$ implies that *the transfer functions of the random filters associated with different Doppler shifts are uncorrelated.*

6.5.2 The uncorrelated scattering channel

Several physical channels (e.g. troposcatter and Moon reflection) have been modelled approximately as a continuum of uncorrelated scatterers [2]. An *uncorrelated scattering* (US) channel is defined as a channel in which the contributions from elemental scatterers with different path delays are uncorrelated. So, by analogy with eqn. (6.26), we expect the autocorrelation of the channel functions to be singular in the time-delay variable. The autocorrelation functions may therefore be expressed in terms of delta functions in the time-delay domain as

$$R_h(t, s; \tau, \eta) = \delta(\eta - \tau) P_h(t, s; \tau) \tag{6.29}$$

$$R_S(\tau, \eta; v, \mu) = \delta(\eta - \tau) P_S(\tau; v, \mu) \tag{6.30}$$

where

$$P_h(t, s; \tau) = \int_{-\infty}^{+\infty} R_T(\Omega; t, s) \exp\{j2\pi\tau\Omega\} d\Omega \tag{6.31}$$

$$P_S(\tau; v, \mu) = \int_{-\infty}^{+\infty} R_H(\Omega; v, \mu) \exp\{j2\pi\tau\Omega\} d\Omega \tag{6.32}$$

Equations (6.31) and (6.32) define the delay and delay-Doppler cross-power spectral densities respectively.

Bello [2] showed that US and WSS channels are time–frequency duals. Consequently, the US channel can be regarded as possessing WSS statistics in the frequency variable so the autocorrelation functions depend only on the frequency difference Ω between the variables. For example, as far as f and m are concerned it is Ω ($= m - f$) which is important. The autocorrelation functions of the output Doppler-spread function $H(f, v)$ and the time-variant transfer function $T(f, t)$ therefore become

$$R_H(f, f + \Omega; v, \mu) = R_H(\Omega; v, \mu) \tag{6.33}$$

$$R_T(f, f + \Omega; t, s) = R_T(\Omega; t, s) \tag{6.34}$$

An example serves to verify eqns. (6.29) and (6.30). From Figure 6.6 the relationship between the autocorrelation of the input delay-spread function, $R_h(t, s; \tau, \eta)$, and the autocorrelation of the time-variant transfer function, $R_T(f, m; t, s)$, is

$$R_h(t, s; \tau, \eta) = \int_{-\infty}^{+\infty} \int_{-\infty}^{+\infty} R_T(f, m; t, s) \exp\{-j2\pi(f\tau - m\eta)\} df dm \tag{6.35}$$

Noting that $\Omega = m - f$ for a US channel, and using eqn. (6.34), this becomes

$$R_h(t, s; \tau, \eta) = \int_{-\infty}^{+\infty} \int_{-\infty}^{+\infty} R_T(\Omega; t, s) \exp\{-j2\pi(f\tau - f\eta - \Omega\eta)\} df d\Omega \tag{6.36}$$

Rearranging gives

$$R_h(t, s; \tau, \eta) = \int_{-\infty}^{+\infty} \exp\{j2\pi f(\eta - \tau)\} df \int_{-\infty}^{+\infty} R_T(\Omega; t, s) \exp\{j2\pi\eta\Omega\} d\Omega \tag{6.37}$$

The first integral in this equation can be recognised as a unit impulse at $\eta = \tau$. The second integral can be expressed in terms of the Delay cross-power spectral density $P_h(t, s; \tau)$ by noting that $P_h(t, s; \tau)$ is the Fourier transform of $R_T(\Omega; t, s)$ with respect to the variable Ω, i.e.

$$P_h(t, s; \tau) = \int_{-\infty}^{+\infty} R_T(\Omega; t, s) \exp\{j2\pi\tau\Omega\} d\Omega \tag{6.38}$$

This is identical with eqn. (6.31). Equation (6.37) therefore reduces to

$$R_h(t, s; \tau, \eta) = \delta(\eta - \tau)P_h(t, s; \tau) \tag{6.39}$$

which is identical with eqn. (6.29). Equation (6.30) can be verified in a similar manner.

The singular behaviour of the channel correlation function $R_S(\tau, \eta; v, \mu)$ with respect to the time-delay variable has the following interpretation for a physical channel. In terms of the channel model composed of a number of elemental scatterers, producing delays and Doppler shifts, *the complex amplitudes of the contributions from the elemental scatterers are uncorrelated if the scatterers produce different time delays.*

6.5.3 The WSSUS channel

We can now move on to *wide-sense stationary uncorrelated scattering* (WSSUS) channels, an important class of practical channels which simultaneously exhibit wide-sense stationarity in the time variable and uncorrelated scattering in the time-delay variable. This is the simplest non-degenerate class, displaying uncorrelated dispersiveness in both the time-delay and Doppler-shift domains, that can be described in terms of channel correlation functions [2]. Fortunately, many radio channels can be characterised as WSSUS.

From the earlier sections on WSS and US channels, it can be inferred that the simultaneous constraints placed on a WSSUS channel result in singular behaviour in both the time-delay and Doppler-shift variables. Therefore, the autocorrelation functions of the channel system functions have the form

$$R_h(t, t + \xi; \tau, \eta) = \delta(\eta - \tau)P_h(\xi; \tau) \tag{6.40}$$

$$R_H(f, f + \Omega; v, \mu) = \delta(v - \mu)P_H(\Omega; v) \tag{6.41}$$

$$R_T(f, f + \Omega; t, t + \xi) = R_T(\Omega; \xi) \tag{6.42}$$

$$R_S(\tau, \eta; v, \mu) = \delta(\eta - \tau)\delta(v - \mu)P_S(\tau; v) \tag{6.43}$$

These equations lead to the following physical pictures for the WSSUS channel:

- The autocorrelation function of the input delay-spread function, $R_h(t, t + \xi; \tau, \eta)$, indicates wide-sense stationarity in the time variable and uncorrelated scattering in the time-delay variable. In terms of a differential circuit model in the form of a densely tapped delay line, the channel can be represented as a continuum of uncorrelated, randomly-scintillating scatterers having wide-sense stationary statistics.
- The autocorrelation function of the output Doppler-spread function, $R_H(f, f + \Omega; v, \mu)$, exhibits wide-sense stationarity in the frequency variable and uncorrelated scattering in the Doppler-shift variable. In terms of a circuit model, the channel appears as a continuum of uncorrelated filtering–Doppler shifting elements, with the filter transfer functions having wide-sense stationary statistics in the frequency variable.
- The autocorrelation function of the time-variant transfer function, $R_T(f, f + \Omega; t, t + \xi)$, displays wide-sense stationarity in both time and frequency variables. Previously this has been used to determine the correlation

between two signals which are separated by Ω hertz [1,9,10]. In this context a correlation function of practical interest is the spaced frequency correlation function [11], given by

$$R_T(\Omega; 0) = R(\Omega) \tag{6.44}$$

which represents the correlation between the signal amplitudes at two frequencies separated by Ω hertz. It has been shown that the frequency coherence for the variable Ω can also be determined by pulse sounding techniques [9].

- The autocorrelation function of the delay Doppler-spread function, $R_S(\tau, \eta; v, \mu)$, reveals uncorrelated scattering in both time-delay and Doppler-shift variables. In terms of a differential circuit model, the channel can be depicted as a continuum of non-scintillating, uncorrelated scatterers causing both time delays and Doppler shifts. Such a representation is closer to the phenomenological description of dispersive radio channels [8]. For WSSUS channels the delay-Doppler cross-power spectral density $P_S(\tau; v)$ is identical to the radar target scattering function $\sigma(\tau; v)$; and although $\sigma(\tau; v)$ was originally defined for radar targets [11], it has more general applications and can be incorporated into a study of propagation in mobile radio channels.

The relationships between the channel correlation functions for WSSUS channels are illustrated in Figure 6.7, and take the form of single Fourier transforms.

6.6   CHANNEL CHARACTERISATION USING THE SCATTERING FUNCTION

The statistics presented in Section 6.5.3 showed how a characterisation in terms of the delay/Doppler-spread function explicitly revealed the dispersive behaviour of the channel. It was also stated that the delay/Doppler-spread function $P_S(\tau; v)$ and the target scattering function $\sigma(\tau; v)$ were identical. A physical interpretation of the scattering function can now be obtained through a simple channel model [8].

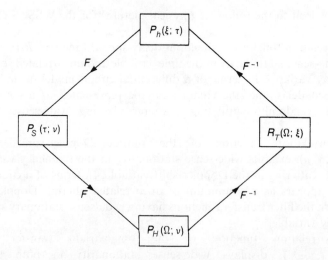

Figure 6.7   Relationships between correlation functions in WSSUS channels.

6.6.1 The point scatterer description

Assuming that propagation through the mobile radio channel takes place purely through single scattering, the channel can be represented as a set of independent scatterers as shown in Figure 6.8. Energy arriving at the receiver from the ith scatterer is related to its scattering cross-section, ρ_i^2, where ρ_i determines the amplitude of the scattered waveform. A propagation time delay, T_i is associated with the ith scatterer, but since the position of the scatterers with respect to the mobile changes due to its motion, the propagation time delay must be a function of time, i.e. $T_i(t)$. Considering $T_i(t)$ as a linear function of time gives

$$T_i(t) = \xi_i + \dot{\xi}_i t \tag{6.45}$$

where $\dot{\xi}_i$ is the rate of change of delay.

The transmitted signal $x(t)$, expressed in terms of its complex envelope $z(t)$, is given by eqn. (6.1), hence the contribution of the ith scatterer $x_i(t)$ to the received waveform $v_i(t)$ is merely a delayed and attenuated replica of the transmitted signal, i.e.

$$v_i(t) = A\rho_i \mathrm{Re}[z(t - \xi_i - \dot{\xi}_i t) \exp\{j\omega_c(t - \xi_i - \dot{\xi}_i t)\}] \tag{6.46}$$

where A is an unimportant constant.

If the variation in $\dot{\xi}_i t$ is small compared to the reciprocal bandwidth of $z(t)$ then its variation in the argument of z may be ignored. Also, provided the signal is narrowband, i.e. the bandwidth of $z(t)$ is much smaller than the carrier frequency, then differences of π/ω_c in the value of ξ_i will not significantly change the value of $z(t - \xi_i)$. However, they will appreciably affect the value of $v_i(t)$ because they alter the exponent by π radians.

Since each value of ξ_i is rarely known exactly, and since small perturbations are important, it seems reasonable to represent each ξ_i as the sum of a gross delay τ_i and a perturbation delay $\delta\tau_i/\omega_c$, i.e.

$$\xi_i = \tau_i + \frac{\delta\tau_i}{\omega_c} \tag{6.47}$$

where $\delta\tau_i$ is specified as being a variable which takes values in the range $(-\pi, \pi)$.

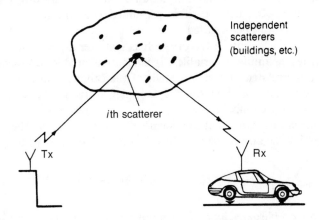

Figure 6.8 Point scatterer representation of a multipath mobile radio channel.

Applying these conditions and using eqn. (6.47), eqn. (6.46) can be rewritten as

$$v_i(t) = A\rho_i \text{Re}[z(t - \tau_i) \exp\{j(\omega_c(t - \tau_i - \dot{\xi}_i t) - \delta\tau_i)\}] \tag{6.48}$$

Rewriting this equation to show the Doppler shift v_i associated with the contribution from the ith scatterer gives

$$v_i(t) = A\rho_i \text{Re}[z(t - \tau_i) \exp\{j(2\pi(f_c - v_i)t - 2\pi f_c\tau_i - \delta\tau_i)\}] \tag{6.49}$$

where

$$2\pi v_i = \omega_c \dot{\xi}_i \tag{6.50}$$

and v_i is in hertz.

Summation of the individual contributions from all the scatterers comprising the channel, produces the total received waveform due to single scattering:

$$v(t) = \sum_i v_i(t) = A \, \text{Re}\left[\sum_i \rho_i z(t - \tau_i) \exp\{j(2\pi(f_c - v_i)t - 2\pi f_c\tau - \delta\tau_i)\}\right] \tag{6.51}$$

Although this expression describes the received waveform, the parameters $\rho_i, \tau_i, \delta\tau_i$ and v_i are all random variables and a statistical description of eqn. (6.51) is therefore required.

6.6.2 Statistical point scatterer model

It would be possible to obtain a statistical description of the channel by considering the joint probability distributions of the variables $\rho_i, \tau_i, \delta\tau_i$ and v_i. But as we shall see, a simpler approach is possible through consideration of the individual probability distributions.

The perturbation term $\delta\tau_i$ corresponds to a maximum phase ambiguity of π radians, i.e. a delay uncertainty of half a wavelength, and since the total delay may be many thousands of wavelengths (particularly at UHF), the percentage uncertainty is sufficiently small that it is reasonable to assume that $\delta\tau_i$ is uniformly distributed over the range $(-\pi, \pi)$. In addition, it is presumed that the $\delta\tau_i$ are statistically independent and therefore uncorrelated.

The cross-sections ρ_i^2 account for factors such as scatterer aspect and consequently they can be regarded as random variables. It is also reasonable to assume that the values of ρ_i^2 are uncorrelated and independent of the other parameters. A uniform distribution is sometimes assumed for the Doppler shifts v_i. These assumptions suffice for an elementary channel model.

The above assumptions imply that the mean value of $v(t)$ is zero. They also imply that, using eqn. (6.51), the autocorrelation function $R_v(t, s)$ of $v(t)$ can be expressed as

$$R_v(t, s) = \text{Re}\left[\sum_i \frac{A^2}{2} E[\rho_i^2 z(t - \tau_i)z^*(s - \tau_i)] \exp\{j2\pi(f_c - v_i)(t - s)\}\right] \tag{6.52}$$

However, since the complex envelope of the input $z(t)$ is deterministic, $z(t - \tau_i)$ is also deterministic and this equation simplifies to

$$R_v(t, s) = \text{Re} \left[\sum_i \frac{A^2}{2} E[\rho_i^2] z(t - \tau_i) z^*(s - \tau_i) \exp\{j2\pi(f_c - v_i)(t - s)\} \right] \qquad (6.53)$$

The dispersive behaviour of the channel is now displayed by the average cross-sections $E[\rho_i^2]$ for each delay τ_i and Doppler shift v_i. However, this correlation function is rather awkward to manipulate and interpret as it stands, and a more suitable form can be derived.

6.6.3 The scattering function

The expression on the right-hand side of eqn. (6.53) depends only on the total average cross-section associated with each pair of values of τ_i and v_i, and is independent of the number of scatterers involved. Hence it is possible to introduce an average scatterer cross-section $\bar\sigma(\tau; v)$ for each pair of τ_i and v_i, i.e.

$$\bar\sigma(\tau; v) = \sum_i E[\rho_i^2] \qquad (6.54)$$

where the summation is for all cross-sections that correspond to delays $\tau_i = \tau$, and Doppler shifts $v_i = v$.

The autocorrelation function $R_v(t, s)$ can now be expressed as

$$R_v(t, s) = \text{Re} \left[\sum_i \frac{A^2}{2} z(t - \tau) z^*(s - \tau) \bar\sigma(\tau; v) \exp\{j2\pi(f_c - v)(t - s)\} \right] \qquad (6.55)$$

where the summation is over all pairs of τ and v for which $\bar\sigma(\tau; v)$ is non-zero.

Because of practical limitations on $z(t)$, there will be times when it is impossible to differentiate between contributions from individual scatterers. In other words, time delays τ which differ by less than the reciprocal bandwidth of $z(t)$, and Doppler-shifts v which differ by less than the reciprocal time duration of $z(t)$ are not resolvable. When this occurs, the contributions merge together to produce an average contribution. The discrete form of $\bar\sigma(\tau; v)$ can therefore be replaced by a continuous function, and the summation can be replaced by an integral. Equation (6.55) then becomes

$$R_v(t, s) = \text{Re} \left[\int_{-\infty}^{+\infty} \int_{-\infty}^{+\infty} \frac{A^2}{2} z(t - \tau) z^*(s - \tau) \bar\sigma(\tau; v) \exp\{j2\pi(f_c - v)(t - s)\} \, d\tau \, dv \right]$$

$$(6.56)$$

In its continuous form, the scattering function $\bar\sigma(\tau; v)$ can be envisaged as a scatterer cross-section density for all values of τ and v. The cross-section corresponding to delays in the range $(\tau, \tau + d\tau)$ and Doppler shifts in the range $(v, v + dv)$ is then given by $\bar\sigma(\tau; v) \, d\tau \, dv$.

The function $\bar\sigma(\tau; v)$ describes the distribution of average cross-section and the total amount of such cross-section; $z(t)$ describes the structure of the transmitted waveform and its energy level. In both cases the former attribute relates to the general structure of the received process and the latter attribute merely determines the average received energy.

It is reasonable to assume that $z(t)$ can be scaled to a unit norm, since this is easily satisfied by redefining A so that

$$\int_{-\infty}^{+\infty} |z(t)|^2 \, \mathrm{d}t = 1 \tag{6.57}$$

We can now introduce the normalised density of cross-section $\sigma(\tau; v)$, defined as

$$\sigma(\tau; v) = \bar{\sigma}(\tau; v) \bigg/ \int_{-\infty}^{+\infty} \int_{-\infty}^{+\infty} \bar{\sigma}(\tau; v) \, \mathrm{d}\tau \, \mathrm{d}v \tag{6.58}$$

$\sigma(\tau; v)$ is more usually called the *channel scattering function* [2,8,12]. It is obvious from eqn. (6.58) that

$$\int_{-\infty}^{+\infty} \int_{-\infty}^{+\infty} \sigma(\tau; v) \, \mathrm{d}\tau \, \mathrm{d}v = 1 \tag{6.59}$$

Also, the average received energy E_R, is defined as follows:

$$E_\mathrm{R} = \int_{-\infty}^{+\infty} E[v(t)]^2 \, \mathrm{d}t = \int_{-\infty}^{+\infty} R_v(t, t) \, \mathrm{d}t = \frac{A^2}{2} \int_{-\infty}^{+\infty} \int_{-\infty}^{+\infty} \bar{\sigma}(\tau; v) \, \mathrm{d}\tau \, \mathrm{d}v \tag{6.60}$$

Using eqns (6.58) and (6.60) in eqn. (6.56) gives

$$R_v(t, s) = \mathrm{Re}\left[E_\mathrm{R} \int_{-\infty}^{+\infty} \int_{-\infty}^{+\infty} z(t - \tau) z^*(s - \tau) \sigma(\tau; v) \exp\{\mathrm{j}2\pi(f_\mathrm{c} - v)(t - s)\} \, \mathrm{d}\tau \, \mathrm{d}v \right] \tag{6.61}$$

Finally, for WSS channels eqn. (6.61) reduces to

$$R_v(t, t + \xi) = \mathrm{Re}\left[E_\mathrm{R} \int_{-\infty}^{+\infty} z(t - \tau) z^*(t + \xi - \tau) \sigma(\tau; v) \exp\{-\mathrm{j}2\pi(f_\mathrm{c} - v)\xi\} \, \mathrm{d}\tau \, \mathrm{d}v \right] \tag{6.62}$$

This is an expression for the autocorrelation function of the wide-sense stationary channel output $R_v(t, t + \xi)$, in terms of the scattering function $\sigma(\tau; v)$.

Alternatively, a similar expression can be obtained using the relationships in Section 6.4.1. The autocorrelation function of the channel output given in eqn. (6.19) can also be expressed as

$$R_w(t, s) = \int_{-\infty}^{+\infty} \int_{-\infty}^{+\infty} z(t - \tau) z^*(s - \eta) R_h(t, s; \tau, \eta) \, \mathrm{d}\tau \, \mathrm{d}\eta \tag{6.63}$$

From Section 6.4.2 the double Fourier transform relationship relating $R_h(t, s; \tau, \eta)$ to $R_S(\tau, \eta; v, \mu)$ is

$$R_h(t, s; \tau, \eta) = \int_{-\infty}^{+\infty} \int_{-\infty}^{+\infty} R_S(\tau, \eta; v, \mu) \exp\{-\mathrm{j}2\pi(vt - \mu s)\} \, \mathrm{d}v \, \mathrm{d}\mu \tag{6.64}$$

Substitution in eqn. (6.63) gives

$$R_w(t, s) = \int_{-\infty}^{+\infty} \int_{-\infty}^{+\infty} z(t - \tau)z^*(s - \eta)$$

$$\times \int_{-\infty}^{+\infty} \int_{-\infty}^{+\infty} R_S(\tau, \eta; v, \mu) \exp\{-j2\pi(vt - \mu s)\}\, dv\, d\mu\, d\tau\, d\eta \quad (6.65)$$

In a WSSUS channel, the function $R_S(\tau, \eta; v, \mu)$ can be replaced by the delay-Doppler cross-power spectral density $P_S(\tau; v)$ (6.43), so $R_w(t, s)$ becomes

$$R_w(t, s) = \int_{-\infty}^{+\infty} \int_{-\infty}^{+\infty} z(t - \tau)z^*(s - \eta)$$

$$\times \int_{-\infty}^{+\infty} \int_{-\infty}^{+\infty} \delta(v - \mu)\delta(\eta - \xi)P_S(\tau; v) \exp\{-j2\pi(vt - \mu s)\}\, dv\, d\mu\, d\tau\, d\eta$$
$$(6.66)$$

Simplifying,

$$R_w(t, s) = \int_{-\infty}^{+\infty} \int_{-\infty}^{+\infty} z(t - \tau)z^*(s - \tau)P_S(\tau; v) \exp\{-j2\pi v(t - s)\}\, dv\, d\tau \quad (6.67)$$

The autocorrelation of the real bandpass signal $R_v(t, s)$ can be obtained from $R_w(t, s)$ by using the equivalence relationship (6.1).

Therefore, $R_v(t, s)$ is given by

$$R_v(t, s) = \mathrm{Re}\left[\int_{-\infty}^{+\infty} \int_{-\infty}^{+\infty} z(t - \tau)z^*(s - \tau)P_S(\tau; v) \right.$$

$$\left. \times \exp\{-j2\pi v(t - s)\} \exp\{j2\pi f_c(t - s)\}\, dv\, d\tau \right] \quad (6.68)$$

Now, noting from Section 6.5.3 that $P_s(\tau; v)$ is equivalent to the scattering function $\sigma(\tau; v)$, for WSS channels this equation becomes

$$R_v(t, t + \xi) = \mathrm{Re}\left[\int_{-\infty}^{+\infty} \int_{-\infty}^{+\infty} z(t - \tau)z^*(t + \xi - \tau)\sigma(\tau; v) \right.$$

$$\left. \times \exp\{-j2\pi(f_c - v)\xi\}\, dv\, d\tau \right] \quad (6.69)$$

Apart from the constant E_R, eqns (6.62) and (6.69) are identical, showing how the scattering function description based on a simple physical model can be derived from the more strict channel characterisation of Section 6.4. We can therefore see how knowledge of $\sigma(\tau; v)$ allows a description of the received process for a given input signal. A statistical channel characterisation in terms of $\sigma(\tau; v)$ provides an insight into the physical mechanisms of multipath propagation and a pictorial representation of the distribution of received energy.

Figure 6.9 shows a typical scattering function which provides a vivid illustration of the relationship between received power, time delay and Doppler shift. Interpretation of the Doppler shift in terms of the spatial arrival angle allows the identification of dominant scatterers and helps to build up a physical picture of the propagation

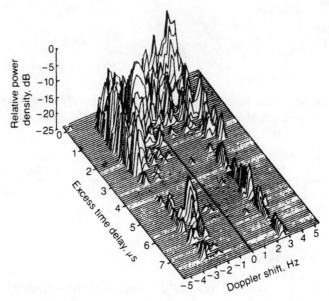

Figure 6.9 Example of a scattering function in a severe multipath area.

mechanism in the area concerned. Experimental techniques for obtaining the scattering function and other channel descriptors will be discussed in Chapter 8. Note that if the statistics of the received process are Gaussian, then a description in terms of $\sigma(\tau; v)$ is statistically complete.

6.7 MOBILE RADIO CHANNEL CHARACTERISATION

The theoretical analysis in this chapter has so far been concerned with time-variant radio channels in general and then with practical radio channels which, being subject to some constraints, allowed a less general approach. We can now be even more specific by considering mobile radio channels.

Practically all mobile radio communication channels can be characterised as linear in their effect on message signals transmitted through them. It therefore seems reasonable to consider mobile radio channels as special cases of random time-variant linear filters.

Mobile radio links often require communication between fixed base stations and mobile transceivers. We know from the discussion in earlier chapters that in these circumstances the channel is non-stationary. However, characterisation of mobile radio channels proves extremely difficult unless stationarity can be assumed over short intervals of time. In order to obtain a fairly complete statistical description of the channel, a two-stage characterisation has been proposed.

Firstly, the channel is characterised over a period of time, or a geographical area, which is small in comparison to the period of the slow channel variations, so that the mean received signal strength appears virtually constant. It is further assumed that over this small interval or small area the prominent features of the environment remain unchanged, i.e. the significant scattering centres do not change. The large-scale behaviour of the channel is then obtained by examining the behaviour of the small-scale statistics over larger areas. This two-stage model was first proposed by Bello [2], and was subsequently used by Cox [9] and Bajwa [7]. This class of channel has been called *quasi-wide-sense stationary* (QWSS).

A further simplification in the characterisation of mobile radio channels can be effected by assuming that contributions from scatterers with different path delays are uncorrelated. The channel can then be described in terms of WSSUS statistics and can be depicted as a continuum of uncorrelated scatterers, in both time delays and Doppler shifts, having elemental cross-section $\sigma(\tau; v)\,d\tau\,dv$, the channel being specified here in terms of the scattering function $\sigma(\tau; v)$. Useful characterisation of mobile radio channel behaviour can be provided by application of the various correlation functions and their interrelationships.

6.7.1 Small-scale channel characterisation

The small-scale channel transmission characteristics of primary interest are the input delay-spread function $h(t, \tau)$ and the time-variant transfer function $T(f, t)$. In Section 6.4.1 we saw how the autocorrelation function of the complex envelope of the received signal $R_w(t, s)$, could be obtained from the autocorrelation function of the input delay-spread function, $R_h(t, s; \tau, \eta)$. As shown in Section 6.5.3, certain simplifications in the characterisation arise when the channel can be classed as WSSUS. A sufficient characterisation of the WSSUS channel, in terms of its dispersive behaviour, is possible through knowledge of $R_h(t, s; \tau, \eta)$ and this can be illustrated for both the time and frequency domains.

Time domain description

The time domain description of the channel is obtained by expressing the autocorrelation function of the channel output, $R_w(t, s)$, in terms of the autocorrelation function of the input delay-spread function, $R_h(t, s; \tau, \eta)$ as in eqn. (6.19).

Furthermore, using eqn. (6.40) for a WSSUS channel, we can write

$$R_w(t, t + \xi) = \int_{-\infty}^{+\infty} \int_{-\infty}^{+\infty} z(t - \tau)z^*(t + \xi - \eta)\delta(\eta - \tau)P_h(\xi; \tau)\,d\tau\,d\eta \qquad (6.70)$$

For the case $\xi = 0$, i.e. when the time separation of the observation is zero, $P_h(\xi; \tau)$ becomes

$$P_h(\tau; \tau) = P_h(\tau) \qquad (6.71)$$

that is, the cross-power spectral density $P_h(\xi; \tau)$ becomes a simple delay-power spectral density $P_h(\tau)$. Thus, eqn. (6.70) simplifies to

$$R_w(t, t) = \int_{-\infty}^{+\infty} |z(t - \tau)|^2 P_h(\tau)\,d\tau \qquad (6.72)$$

and if $|z(t)|^2$ is an impulse function this becomes

$$R_w(t, t) = P_h(t) \qquad (6.73)$$

For WSSUS channels we therefore have the important result that the auto-correlation function of the channel output is described by the profile of the time distribution of received power, the so-called *power-delay profile*. Equation (6.73) is valid on condition that $|z(t)|^2$ appears to be impulsive with respect to $P_h(t)$; this is true provided the time duration of $z(t)$ is much smaller than the spread of multipath delays within the channel. More precisely, eqn. (6.73) holds provided the Fourier transform of $|z(t)|^2$ is constant over the frequency interval where the Fourier transform of $P_h(t)$ is non-zero [6]. For convenience, $P_h(t)$ normally has its time origin redefined so as to position the earliest received echo at $t = 0$, and the function is then defined in terms of the excess time-delay variable τ, i.e.

$$P_h(\tau) = P_h(t - t_0) \tag{6.74}$$

where t_0 is the time delay for the shortest echo path. Provided the received signal has Gaussian statistics, the channel behaviour will be completely described by $P_h(\tau)$. A knowledge of $P_h(\tau)$ will typically specify some gross features of the channel; these are obtained by regarding $P_h(\tau)$ as a statistical distribution of echo strengths. A typical power-delay profile is shown in Figure 6.10. *It can be regarded as the scattering function averaged over all Doppler shifts.* Two statistical moments of $P_h(\tau)$ of practical interest are the average delay D and the delay spread S. The average delay is the first central moment of $P_h(\tau)$, and the delay spread is the square root of the second central moment. These are expressed as

$$D = \frac{\int_0^\infty \tau P_h(\tau)\,\mathrm{d}\tau}{\int_0^\infty P_h(\tau)\,\mathrm{d}\tau} \tag{6.75}$$

$$S = \sqrt{\frac{\int_0^\infty (\tau - D)^2 P_h(\tau)\,\mathrm{d}\tau}{\int_0^\infty P_h(\tau)\,\mathrm{d}\tau}} \tag{6.76}$$

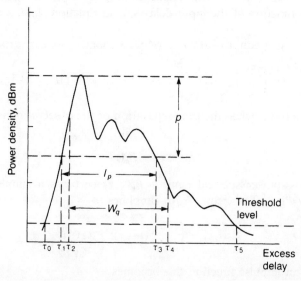

Figure 6.10 A power-delay profile illustrating the measurement of delay window W_q and delay interval I_p.

Although these parameters are estimates, they constitute relevant design parameters for WSSUS channels. The average delay is related to ranging error in phase ranging systems, and the delay spread places limits on communication system performance [14], as outlined in Section 6.2.

It has been recognised [15] that these two parameters are not sufficient to describe all the important characteristics of the channel, and two further time domain parameters have been recommended for use. They can be computed, as can D and S, either from a single power-delay profile or from profiles averaged over a distance of a few wavelengths. The delay window W_q is the duration of the middle portion of the delay profile that contains $q\%$ of the total energy in that profile. As illustrated in Figure 6.10, it is

$$W_q = (\tau_4 - \tau_2)_q \tag{6.77}$$

The boundaries τ_4 and τ_2 are defined by

$$\int_{\tau_2}^{\tau_4} P_h(\tau)\, d\tau = q \int_{\tau_0}^{\tau_5} P_h(\tau)\, d\tau = q P_{\text{tot}} \tag{6.78}$$

and the energy outside the window is split into two equal parts.

The delay interval I_p is the difference in time delay between the points where the power-delay profile first crosses a point p dB below its maximum value and the point where it falls below that threshold for the last time. It is also illustrated in Figure 6.10 and can be expressed as

$$I_p = (\tau_3 - \tau_1)_p \tag{6.79}$$

Frequency domain description

The frequency-selective behaviour of the mobile channel is readily obtained by observing the correlation between two signals, at different frequencies, at the receiver. The existence of different time delays for the constituent propagation paths causes the statistical properties of two different radio frequencies to become essentially independent if their separation is sufficiently large. The maximum frequency difference for which the signals are still strongly correlated is called the *coherence bandwidth* of the channel. The coherence bandwidth is a useful parameter in assessing the performance and limitations of various modulation and diversity reception techniques.

In Section 6.3.3 it was shown how the time-variant transfer function $T(f, t)$ characterises a channel in response to a cissoidal time function. Random time-variant channels necessitate a characterisation in terms of the autocorrelation function of the time-variant transfer function, $R_T(f, m; t, s)$, as detailed in Section 6.4.1. For WSSUS channels the autocorrelation function reduces to

$$R_T(f, f + \Omega; t, t + \xi) = R_T(\Omega; \xi) \tag{6.80}$$

and $R_T(\Omega; \xi)$ has been called the *time–frequency correlation function* [2].

The interrelationships between autocorrelation functions for WSSUS channels (Section 6.5.4) can be used to show that $R_T(\Omega; \xi)$ is related to $P_h(\xi; \tau)$ via a Fourier transform, i.e.

$$R_T(\Omega; \xi) = \int_{-\infty}^{+\infty} P_h(\xi; \tau) \exp\{-j2\pi\Omega\tau\} \, d\tau \tag{6.81}$$

This emphasises that a separate measurement of $R_T(\Omega; \xi)$ is not required in order to provide a frequency domain description of the channel.

When $\xi = 0$, i.e. the time separation of the observation is zero,

$$R_T(\Omega; 0) = R_T(\Omega) \tag{6.82}$$

$$P_h(0; \tau) = P_h(\tau) \tag{6.83}$$

and

$$R_T(\Omega) = \int_{-\infty}^{+\infty} P_h(\tau) \exp\{-j2\pi\Omega\tau\} \, d\tau \tag{6.84}$$

$R_T(\Omega)$ is known as the *frequency correlation function* [2], and the *coherence bandwidth* B_c, is the smallest value of Ω for which $R_T(\Omega)$ equals some suitable correlation coefficient, e.g. 0.5 or 0.9. A typical example is shown in Figure 6.11.

The interrelationships between the channel correlation functions for WSSUS channels were shown in Section 6.5.4, and it can be seen that the scattering function $\sigma(\tau; v)$, the delay-power spectral density $P_h(\tau)$, and the frequency correlation function $R_T(\Omega)$ are simply related through Fourier transforms. Therefore, alternative channel descriptions can be easily obtained from practical channel measurements performed in either the time or frequency domain.

6.7.2 Large-scale channel characterisation

For small spatial distances, of the order of a few wavelengths, the dispersive behaviour of the channel can be modelled as quasi-wide-sense stationary in the time domain. However, over larger distances the changes in terrain and local environment give rise to temporal non-stationarity in the statistical characterisation of the multipath. Therefore, while it is not possible to apply the small-scale statistics directly to the characterisation of multipath over areas where non-local scattering can be observed, it is possible to use the scattering function over contiguous wide-

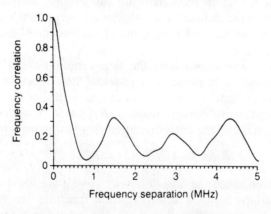

Figure 6.11 Typical frequency correlation function in an urban area.

sense stationary, and spatially homogeneous, sections in order to investigate the scattering behaviour for larger areas.

Direct data reduction of the statistical moments of the small-scale characteristics provides distributions of parameters such as the delay spread and coherence bandwidth over the larger area. This method has proved popular [2,5,9], since it gives rise to parameters that are useful for systems designers.

Although this method yields useful design parameters, they represent 'static' measures of the channel performance. A more powerful characterisation would facilitate the production of an accurate channel simulation. In attempting to achieve this aim, several studies [5,13,16,17] have concentrated on fitting global probability distributions to the echo amplitudes, path delays, carrier phases and Doppler shifts. While representative simulators have been developed [16,17], the basis for each model has been that the time delays conform to a modified Poisson sequence. This restricts the models to computer simulations, since the time delays must be generated from random numbers. This drawback can be overcome and a simple channel simulation based upon a tapped delay line model will be described in Chapter 8.

REFERENCES

1. Jakes W.C. (ed.) (1974) *Microwave Mobile Communications.* John Wiley, New York.
2. Bello P.A. (1963) Characterization of randomly time-variant linear channels. *IEEE Trans.*, **CS11**, 360–93.
3. Zadeh L.A. (1950) Frequency analysis of variable networks. *Proc. IRE*, **38**, 291–9.
4. Kailath T. (1959) *Sampling models for linear time-variant filters.* Report 352, MIT Research Lab of Electronics, Cambridge MA.
5. Bello P.A. (1964) Time-frequency duality. *IEEE Trans.*, **IT10**(1), 18–33.
6. Bello P.A. (1969) Measurement of random time-variant linear channels. *IEEE Trans.*, **IT15**(4), 469–75.
7. Bajwa A.S. (1979) Wideband characterisation of UHF mobile radio propagation in urban and suburban areas. PhD thesis, Department of Electronic and Electrical Engineering, University of Birmingham.
8. Kennedy R.S. (1969) *Fading Dispersive Communication Channels.* John Wiley, New York.
9. Cox D.C. and Leck R.P. (1975) Correlation bandwidth and delay spread multipath propagation statistics for 910 MHz urban mobile radio channels. *IEEE Trans.*, **COM23**(11), 1271–80.
10. Gans M.J. (1972) A power-spectral theory of propagation in the mobile-radio environment. *IEEE Trans.*, **VT21**(1), 27–38.
11. Price R. and Green P.E. (1960) *Signal processing in radar astronomy.* Report 234, MIT Lincoln Lab, Lexington MA.
12. Schwartz M., Bennett W.R. and Stein S. (1966) *Communication Systems and Techniques.* McGraw-Hill, New York.
13. Turin G.L., Clapp F.D., Johnston T.L., Fine S.B. and Lavry D. (1972) A statistical model of urban multipath propagation. *IEEE Trans.*, **VT21**(1), 1–9.
14. Lorenz, R.W. (1986) *Impact of frequency-selective fading on binary and quadrature phase modulation in mobile radio communication demonstrated by computer simulations using the WSSUS channel model.* COST207 Technical Document TD(86) No. 1.
15. European Commission (1989) *Digital land mobile radio communications.* COST207 Final Report, European Commission, Brussels.
16. Suzuki H. (1977) A statistical model for urban radio propagation. *IEEE Trans.*, **COM25**(7), 673–80.
17. Hashemi H. (1979) Simulation of the urban radio propagation channel. *IEEE Trans.*, **VT28**(3), 213–25.

Chapter 7

Other Mobile Radio Channels

7.1 INTRODUCTION

A great deal of attention has been given to propagation in built-up areas, in particular to the situation where the mobile is located in the streets, i.e. when it is outside the buildings. It is apparent, however, that other important scenarios exist. For example, hand-portable equipment can be taken inside buildings, and in recent years there has been a substantial increase in the use of this type of equipment. As a result, interest in characterising the radio communication channel between a base station and a mobile located inside a building has become a priority. Propagation totally within buildings is also of interest for applications such as cordless telephones, paging, cordless PABX systems and wireless local area networks. In city areas there are tunnels and underpasses in which radio coverage is needed, and away from cities there are suburban and rural areas where the losses due to buildings are not necessarily the dominant feature.

Before dealing with such channels, it is worth pausing to clarify a few points and to identify the ways in which the characteristics of the various channels differ. We wish to distinguish between differences which are merely those of scale and more fundamental differences of statistical character relating to the signal or the interference. Differences of scale are exemplified by the urban radio channel. This is characterised by Rayleigh plus lognormal fading and is the same whether the mobile is vehicle-borne or hand-portable. The differences are apparent because the fading rate experienced by a moving vehicle is generally much greater than the fading rate experienced by a hand-portable. Although these differences do not represent a fundamental change in the statistical nature of the channel, they may not be trivial as far as system designers are concerned. For vehicles moving at a reasonable speed, it is often adequate to determine the system performance averaged over the (Rayleigh) fading. For a hand-portable it may be more meaningful to determine the maximum error rate over a specified large percentage of locations. Changes of statistical character are exemplified by indoor radio channels where the interference environment differs markedly in magnitude and nature from that outside, and the rural channel where the signal statistics are not well described by the Rayleigh model.

7.2 RADIO PROPAGATION INTO BUILDINGS

During recent years there has been a marked increase in the use of hand-portable equipment, i.e. transceivers carried by the person rather than installed in a vehicle. Such equipment is particularly useful in cellular and personal radio systems and now completely dominates the market. It is essential for radio engineers to plan systems that encompass this need, and a knowledge of the path losses between base stations and transceivers located inside buildings is a vital factor that needs to be evaluated.

The problem of modelling radio wave penetration into buildings differs from the more familiar vehicular case in several respects. In particular:

- The problem is truly three-dimensional because at a fixed distance from the base station the mobile can be at a number of heights depending on the floor of the building where it is located. In an urban environment this may result in there being an LOS path to the upper floors of many buildings, whereas this is a relatively rare occurrence in city streets.
- The local environment within a building consists of a large number of obstructions. These are constructed of a variety of materials, they are in close proximity to the mobile, and their nature and number can change over quite short distances.

There have been several investigations of radio wave penetration into buildings, particularly in the frequency bands used in cellular systems [1–7]. They can be divided into two main categories:

- Those that consider base station antenna heights in the range 3.0–9.0 m and mobiles mainly operating in one- or two-storey suburban houses.
- Those which consider the problem for base station antenna heights similar to those used in cellular systems and mobiles operating in multi-storey office buildings.

Investigations in the first category all originated in connection with the design of a proposed Universal Portable Radio Telephone System [8]. Because such a system would need to cater for large numbers of very low-power portables, it is based on a very small cell size (< 1.0 km radius). Moreover, in such a system it is considered that coverage within multi-storey office buildings will be provided by a number of cells within the building. It is for these reasons that the studies have used low base station antenna heights, base-to-mobile distances less than 1 km, and have concentrated on taking measurements in buildings the size of suburban houses.

In existing cellular systems, base stations for macrocells are typically located on the roof of a tall building which may be 100 m or more above the local terrain, and base-to-mobile distances of 1 km or more are of interest. Consequently, it is difficult to use the results directly in the design of current-generation systems. However, these studies have shown that the signal in small areas within buildings is approximately Rayleigh distributed with the scatter of the medians being approximately lognormally distributed. In other words, the signal statistics within a building can be modelled as superimposed small-scale (Rayleigh) and large-scale (lognormal) processes – the model used for radio propagation outside buildings in urban areas. The variation of signal level with antenna height is consistent with the presence of a reflecting ground plane.

Cox *et al.* investigated the power–range law by fitting results to an equation of the form

$$L_{50} = S + 10n \log_{10} d \qquad (7.1)$$

where S is a constant and d is the distance between transmitter and receiver. The experiments were conducted using a fixed receiver and a hand-held transmitter which was moved around in areas of $4\,ft^2$ $(0.37\,m^2)$ throughout the building. The values of n were found to be 4.5, 3.9, 3.0 and 2.5 for measurements outside the building, on the first floor, on the second floor and in the basement, respectively.

With one exception [6], studies in the second category have been concerned with the statistical characterisation (median or mean, variance and CPD) of the 'building loss', a term first introduced by Rice [9], to denote the difference between the median signal on a given floor of a building and the median signal level outside, in the streets immediately adjacent to the building. However, in reading the literature there is a need for some care; this definition has been interpreted in different ways. There are two obvious possibilities, either to take a number of measurements in the streets that surround the building to produce an average external measurement as suggested by Rice, or alternatively to use the signal level at a point immediately outside the building in line with the centre of the building and the transmitter location [2].

The second method has merit when an LOS path exists between the transmitter and the building concerned, but generally when this is not the case, and energy enters the building via a number of scattered paths, the first method seems more realistic. The method of data analysis also differs, although in almost all investigations the signal has been sampled at fixed intervals of time or distance. In general the different methods of data analysis do not significantly affect the measured value of mean building penetration loss, but calculations of the signal variability can be affected depending upon whether this is described in terms of a standard deviation or as a statistical distribution function.

For these reasons it is sometimes difficult to compare the results from the different investigations. The penetration loss depends on a number of factors, central among them being the carrier frequency, the propagation conditions along the path and the height of the receiver within the building. However, there are several other influencing factors which include the orientation of the building with respect to the base station, the building construction (the construction materials and the number and size of windows) and the internal building layout. Their influence and relative importance will become apparent later. Almost all models for predicting signal strength in buildings have used the technique proposed by Rice, i.e. firstly predict the median signal level in the neighbouring streets using one of the known methods and then add the building penetration loss.

An investigation by Barry and Williamson in New Zealand [10] concentrated originally on buildings where the majority of floors had a line-of-sight path to the base station. By using criteria similar to those for the vehicular environment, i.e. that the best statistical descriptor was one which adequately predicted values near the tails, it was found that the signal on any floor was best fitted by Suzuki statistics and at 900 MHz the standard deviation of the lognormal part of the distribution was 6.7 dB. It was also suggested that mirror-glass windows could introduce an additional loss of the order of 10 dB.

A series of experiments in the UK at frequencies of 441, 896.5 and 1400 MHz [11] produced general conclusions about signal variability similar to those from previous investigations, and they also provided an insight into the effects of transmission conditions and carrier frequency. The transmission conditions appear to have a strong effect on the value of the standard deviation and on the departure of the distribution from lognormal.

Table 7.1 shows the penetration loss for three different frequencies (441, 896.5 and 1400 MHz) for a receiver located in a modern six-storey building. The penetration loss decreases by around 1.5 dB as the frequency is increased from 441 to 896.5 MHz and by a further 4.3 dB when the frequency is raised to 1400 MHz. These results (the decrease in penetration loss at higher frequencies) are consistent with the conclusions drawn by Rice [9] and Mino [12].

A different series of measurements using a number of large buildings has produced ground-floor penetration loss values of 14.2, 13.4 and 12.8 dB at 900, 1800 and 2300 MHz respectively. It can be argued that for system designers, the penetration loss at ground-floor level is the most important because if a system is designed to give adequate service to mobiles at ground-floor level, then service on higher floors within a building will almost certainly be as good if not better.

It is worth re-emphasising that the total loss between the base station and the mobile has been split into two parts: the loss from the base station to points in the streets surrounding the building concerned and the additional penetration loss from the street into the building itself. This has the advantage that established methods can be used to estimate the first component, and the penetration loss then becomes an additional factor. Although the penetration loss, as defined, decreases with frequency in the range considered above, the path loss from the base station to the streets outside will increase. This factor dominates, so the total path loss between transmitter and receiver will always increase as the frequency is raised.

The transmission conditions have a strong influence on the value of the standard deviation and also on the departure of the distribution from lognormal. Figure 7.1 shows that when no LOS path exists, the large-scale signal variations exactly fit a lognormal distribution and that the standard deviation is about 4 dB. In other circumstances where there is an LOS path to the whole building or part of the building, the large-scale signal variations depart somewhat from the lognormal and have a higher standard deviation. For complete LOS the standard deviation is 6–7 dB. These values are very close to those reported by Cox [2].

Two building construction effects have been noted. First, the standard deviation of the large-scale variations is related to the floor area of the building concerned; smaller floor areas lead to lower values of standard deviation and vice versa. Secondly, the penetration loss generally reduces as the receiver is moved higher

Table 7.1 Mean penetration loss on various floors of a six-storey building[a]

Floor level	Penetration loss (dB)		
	441.0 MHz	896.5 MHz	1400.0 MHz
Ground	16.37	11.61	7.56
1	8.11	8.05	4.85
2	12.76	12.50	7.98
3	13.76	11.18	9.11
4	11.09	8.95	6.04
5	5.42	5.98	3.31
6	4.20	2.53	2.54

[a]Figures are relative to the signal measured outside the building in the adjacent streets.

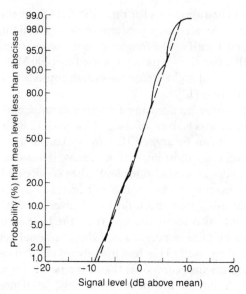

Figure 7.1 Cumulative distribution of the large-scale variations of the signal at 900 MHz within a building when no line-of-sight path exists: (——) measured, (– – –) theoretical lognormal distribution with standard deviation 4 dB.

within a building; indeed there may be an LOS path to the higher floors of a building when no such path exists to the streets outside or to lower floors of the building. Occasionally, however, it has been found that the penetration loss increases at high levels within a building. A result of this kind was reported without discussion by Walker [7], where the penetration loss increased from −1.4 dB at floor 9 to 15.3 dB at floor 12 of the same building. It seems likely that such increases result from the specific propagation conditions existing between the transmitter and receiver locations. Figure 7.2 [11] shows a change of about 2 dB per floor, and this agrees very closely with the findings of other workers [4,7,13].

In summary, when the transmitter is outside, the signal within a building can be characterised as follows:

- The small-scale signal variation is Rayleigh distributed.
- The large-scale signal variation is lognormally distributed with a standard deviation related to the condition of transmission and the area of the floor.
- The building penetration loss, as defined, decreases at higher frequencies.
- When no line-of-sight path exists between the transmitter and the building concerned (i.e. scattering is the predominant mechanism) the standard deviation of the local mean values is approximately 4 dB. When partial or complete line-of-sight conditions exist, the standard deviation rises to 6–9 dB.
- The rate of change of penetration loss with height within the building is about 2 dB per floor.

Finally we comment briefly on the matter of modelling. Most of the outdoor propagation models in Chapter 4 were developed and optimised for macrocells, and without further validation they are not necessarily reliable for microcellular propagation where the antenna height is low. In addition, predicting first the

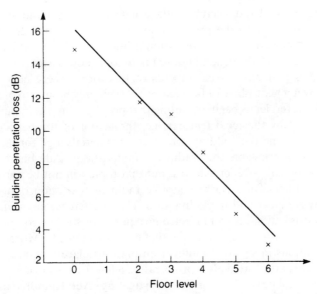

Figure 7.2 Building penetration loss as a function of height within the building: × are experimental points.

average signal level in the streets surrounding a building using a method which has limited accuracy and then adding a building penetration loss, itself subject to statistical variation, inevitably leads to a reduction in accuracy. It seems clear that the prediction of path loss from an external transmitter to a receiver located within a building will be more accurate if it is undertaken directly and not merely as an extension of outdoor modelling. Indeed, Barry and Williamson [14] suggested combining factors associated with propagation into buildings with factors associated with propagation inside buildings to produce a comprehensive model.

Toledo *et al.* [15] undertook a multiple regression analysis of a large database and investigated the relationships between a number of variables. The best results were obtained by including three variables in the regression equations, the distance d between transmitter and receiver, the floor area A_f of the building concerned and a factor S_Q which represents the number of sides of the building which have an LOS path to the receiver. The models at 900 and 1800 MHz respectively are

$$L_{50} = -37.7 + 40 \log_{10} d + 17.6 \log_{10} A_f - 27.5 S_Q$$
$$L_{50} = -27.9 + 40 \log_{10} d + 23.3 \log_{10} A_f - 20.9 S_Q \tag{7.2}$$

The root mean square errors between these equations and the measurements from which they were derived are 2.4 and 2.2 dB respectively, slightly lower than those obtained by Barry and Williamson from their measurements in Auckland [14].

7.3 PROPAGATION INSIDE BUILDINGS

In cordless telephone systems the indoor portion of the subscriber line is replaced by a radio link so that the telephone handset can be carried about freely within a limited area, calls being initiated and received in the usual way. The demand for such systems has prompted research into the propagation characteristics of radio signals where both the transmitter and receiver are within a building. The possibility of cordless telephone exchanges and the general interest in indoor radio systems of various kinds are added factors that have given impetus to this topic. There have been several investigations over a wide range of frequencies; we will only be able to present a rather brief review. However, let us begin by noting that propagation within buildings is very strongly influenced by the local features, i.e. the layout of the particular building under consideration and the building construction materials used for the walls, floors and ceilings. It is conceivable that radio communication inside buildings could be aided by the use of leaky-feeder systems, but that topic will not be considered here.

Indoor radio differs from normal mobile radio in two important respects: the interference environment and the fading rate. The interference environment is often caused by spurious emissions from electronic equipment such as computers, and the level can sometimes be much greater than that measured outside. Moreover, there are substantial variations in signal strength from place to place within a building. The signal can be highly attenuated after propagating a few metres through walls, ceilings and floors or may still be very strong after propagating several hundred metres along a corridor. The signal-to-interference ratio is unpredictable and highly variable.

The slow fading rate makes it inappropriate to calculate system performance by averaging over the fading; it is more appropriate to envisage two possibilities as follows. First if the user of, say, a cordless telephone is moving around slowly during the conversation then the antenna will pass through several fades, albeit rather slowly. This situation can best be described in terms of the percentage of time for which the signal-to-interference ratio falls below an acceptable threshold or, in a digital system, the percentage of time for which the error rate exceeds a given value. However, because of secondary effects (e.g. motion of other people, doors being opened and closed), these probabilities will change slowly with time. Survey papers exist [16,17] which discuss the literature available at the time of writing.

Unsatisfactory performance in wideband systems can also be caused by intersymbol interference due to delay spread, and this limits the data rate. Thus, in narrowband systems, multipath and shadow fading limit the coverage, whereas interference causes major problems even within the intended coverage area. Interference, discussed in Chapter 9, can be natural or man-made noise or it can come from other users in a multi-user system. It limits the number of users that can be accommodated within the coverage area. Techniques such as dynamic channel assignment, power control and diversity [18] can help used to reduce the problems.

7.3.1 Propagation characteristics

Several investigations have been undertaken to determine radio propagation characteristics in houses [3,19–21], office buildings [22–24] and factories [25]. One early investigation, prompted by the proposed introduction of a cordless telephone system in Japan, was concerned with the 250 MHz and 400 MHz bands [19]. As a result of measurements made using a low-power (10 mW) transmitter, it was concluded that the median path loss follows the free space law for very short distances (up to 10 m), it then increases almost in proportion to distance. If the propagation path was blocked by furniture of various kinds, the characteristics were affected in different ways and no general statements were made. The short-term variations in signal about the median value were closely represented by a Rayleigh distribution as a result of scattering from walls, floors, ceilings and furniture.

A law relating path loss to distance from the transmitter can be used to predict signal strength in a building of a given structure, but it is difficult to make general statements. The best approximations to straight-line characteristics are most likely to occur where rooms are of a similar size, uniformly arranged, with walls of uniform attenuation between each room [20]. The exponent n in the power law varies from approximately 2 (free space) along hallways and corridors to nearly 6 over highly cluttered paths.

Motley and Keenan [26] reported the results of experiments in a multi-storey office block at 900 and 1700 MHz. A portable transmitter was moved around selected rooms in the building while a stationary receiver, located near the centre of the office block monitored the received signal levels. The conventional power–distance law was expressed in the form of equation (7.1) as

$$P = P' + kF = S + 10n \log_{10} d$$

where F represents the attenuation provided by each floor of the building and k is the number of floors traversed. When P' was plotted against distance d, on a logarithmic scale, the experimental points lay very close to a straight line. Table 7.2 summarises the values of the measured parameters. Notice that n is similar at both frequencies but F and S are respectively 6 dB and 5 dB greater at 1700 MHz. These results were confirmed by tests in another multi-storey building with metal partitioning. Overall the measured path loss at 1700 MHz was 5.5 dB more than at 900 MHz, which agrees well with theoretical predictions based on reduced effective antenna aperture.

Other workers [27] have obtained a loss of 3–4 dB through a double plasterboard wall and a loss of 7–8 dB through a breeze block or brick wall. These values are less than through a floor, probably because floors often have metal beams and reinforcing meshes which are not present in the walls. It seems that at 1700 MHz there is a greater tendency for RF energy to be channelled via stairwells and lift shafts than at 900 MHz. It has been reported that the losses between floors are influenced by the construction materials used for the external walls, the number and size of windows and the type of glass [28].

The external surroundings also have to be considered since there is evidence [29,30] that energy can propagate outwards from a building, be reflected and scattered from adjacent buildings and re-enter the building at a higher and/or lower level depending upon the location of the antenna and its polar pattern. Experiments have also shown that the attenuation between adjacent floors is greater than the

Table 7.2 Propagation parameters within buildings

	F (dB)	S (dB)	n
Frequency = 900 MHz	10	16	4
Frequency = 1700 MHz	16	21	3.5

incremental attenuation caused by each additional floor and that after five or six floors there is little further attenuation. Several workers [2,31] have published information about signal losses caused by propagation through various building materials over a wide range of frequencies.

It appears that propagation totally within buildings is more dependent on building layout and construction in the 1700 MHz band than it is at 900 MHz. The lower band (860 MHz) is already used for the Digital European Cordless Telephone (DECT) system, which is designed for domestic and business environments. It offers good quality speech and other services for voice and data applications, and it provides local mobility to users of portable equipment in conjunction with an in-building exchange. Although propagation losses increase with frequency, the 1700 MHz band may also be viable for an in-building cordless telephone system where, in any case, the number of base stations is dictated by capacity and performance requirements rather than by the limitations of signal coverage.

Experiments reported by Bultitude [24] give an indication of signal variability within buildings at 900 MHz. Although it might be anticipated that for locations where there is no line-of-sight path, the data would be well represented by a Rayleigh distribution as reported at lower frequencies [19], this did not prove to be the case. Data representing such locations was generally found to be Rician distributed with a specular/random power ratio K of approximately 2 dB. Exceptional locations were found where Rayleigh statistics fitted well. For any fixed location having these Rician statistics there is a 90% probability that the signal is greater than -7 dB but less than $+4$ dB with respect to that determined by losses along the transmitter–receiver path. Temporal variations in the received signal envelope are also apparent as a result of movement of people and equipment. These variations are slow and have characteristics that depend upon the floor plan of the building.

In buildings which are divided into individual rooms, fading is likely to occur in bursts lasting several seconds with a dynamic range of about 30 dB. In open office environments fading is more continuous with a smaller dynamic range, typically 17 dB. These temporal envelope variations are Rician with a value of K between 6 and 12 dB. The value of K is a function of the extent to which motion within the building alters the multipath structure near the receiver location. Terminal motion also causes fading due to movement through the spatially varying field. This is adequately described, as above, by a Rician distribution with $K \approx 2$ dB.

There have been several attempts to model indoor radio propagation using an extension of eqn. (7.1):

$$L_{50} = S + 10n \log_{10} d + X_\sigma \tag{7.3}$$

where X_σ is a lognormal variable (normally in dB) with standard deviation σ. Anderson *et al.* [32] give typical values of n and σ for a variety of buildings over a range of frequencies, n lying in the range 1.6–3.3 and σ being between 3.0 and 14 dB. Seidel [28] also gave values for a variety of situations in different buildings, derived from measurements in a large number of locations. These values were used to model propagation using an equation of the form

$$L_{50} = S + 10n_{SF} \log_{10} d + F \tag{7.4}$$

where n_{SF} represents the value of the exponent for measurements on the same floor. Assuming that a good estimate of n_{SF} exists, the path loss on a different floor can be found by adding an appropriate value of the floor attenuation factor F. Alternatively, in eqn. (7.4) F can be removed by using an exponent n_{MF} which already includes the effect of multiple-floor separation. The propagation equation then becomes

$$L_{50} = S + 10n_{MF} \log_{10} d \qquad (7.5)$$

Devasirvatham [33] found that the in-building path loss could be modelled as the free space loss plus an additional loss that increased exponentially with distance, thus implying that the total loss could be expressed by a modification of eqn. (7.4):

$$L_{50} = S + 10n_{SF} \log_{10} d + \alpha d + F \qquad (7.6)$$

where α is a suitable attenuation constant in decibels per metre (dB/m). This model and others are summarised by Rappaport [34]. Finally, using the basic equation (7.1) as a reference, Toledo and Turkmani [35,36] undertook a multiple regression analysis using a number of other factors, in order to establish those which were most influential. Their final equations for predicting the path loss, at 900 and 1800 MHz respectively, from a transmitter to a given room in a multi-storey building were

$$L_{50} = 18.8 + 39.0 \log_{10} d + 5.6k_f + 13.0S_{win} - 11.0G - 0.024A_f$$
$$L_{50} = 24.5 + 33.8 \log_{10} d + 4.0k_f + 16.6S_{win} - 9.8G - 0.017A_f \qquad (7.7)$$

In these equations k_f is the number of floors separating the transmitter and receiver; S_{win} is a factor representing the amount of energy which leaves and re-enters the building (it takes into account the position of the transmitter relative to the external walls of the building); G represents the observed tendency for the signal to be stronger on the lowest two floor of the building; and A_f is the floor area of the room containing the receiver. S_{win} is given a value between 0 and 1 depending on the relative location of the radio terminals.

For rooms on the same side of the building as the transmitter, $S_{win} = 1$; for rooms on the opposite side $S_{win} = 0.25$; and for those on the two sides perpendicular to the side where the transmitter is located, $S_{win} = 0.5$. For internal rooms with no external windows $S_{win} = 0$. Some judgement is needed to assign values to rooms close to the transmitter, to corridors and to areas separated from the transmitter only by, say, a single wooden door which may or may not be open at any time.

The factor G was set equal to 1 on the lower two floors and it was 0 elsewhere. Although it may be difficult to predict the path loss accurately for receiver locations close to the transmitter, this is of academic interest only since the signal is likely to be high, providing good communication. The best signal coverage of any building is usually achieved by locating the transmitter in a large room as near as possible to the centre of the building [30]

7.3.2 Wideband measurements

In addition to narrowband measurements designed to determine how median signal strength varies with distance and to evaluate signal variability, there have also been several investigations of the wideband characteristics of propagation within buildings.

Measurements of time-delay spread in office buildings and residences have been reported by Devasirvatham [37–39] using equipment operating at 850 MHz with a time-delay resolution capability of 25 ns (i.e. paths differing in length by 7.5 m or more can be resolved). It appears that the detailed shape of the individual power–delay profiles have little impact on the performance of a radio system [40,41], so effort was concentrated on evaluating the average delay and the RMS delay spread.

In general, the delays and delay spreads are smaller than corresponding values measured outside buildings. The averaged time-delay profile in Figure 7.3 represents data collected in a large, six-storey building and has an RMS time-delay spread of 247 ns. Figure 7.4 shows the cumulative distribution of time-delay spread for this office building and a smaller two-level building. A portable communications system would have to work under worst-case delay spread, which for both these office buildings is about 250 ns. Larger delay spreads, in the range 300–420 ns, were measured at residential locations, particularly on inside-to-outside paths, but the limited number of locations that were used makes general conclusions rather difficult to draw. Note, however, that whenever a line-of-sight path exists between transmitter and receiver, the RMS delay spread is significantly reduced, typically to less than 100 ns.

Bultitude *et al.* [42] compared indoor characteristics at 900 MHz and 1.75 GHz using equipment with parameters the same as Devasirvatham's. Measurements were made in a four-storey brick building and in a modern building of reinforced concrete blocks, both in Ottawa, Canada. There were perceivable differences in the measured characteristics, but these seemed to be more a function of the location than a function of the transmission frequency. In one building, RMS delay spreads were slightly greater at 1.75 GHz for over 90% of locations (28 ns compared with 26 ns), whereas in the other building the reverse was true for about 70% of locations. Although the results indicated that coverage would be less uniform in both buildings at 1.75 GHz, they also showed that coverage would be less uniform in one of the buildings than in the other, regardless of the transmission frequency. It seems difficult, on the basis of this work, to conclude anything except that there is little difference between the wideband frequency correlation statistics in the two frequency bands.

A statistical model for indoor multipath propagation has been presented by Salah and Valenzuela [43] based on measurements at 1.5 GHz using 10 ns radar-like pulses in a medium-sized office building. Their results showed that the indoor channel is quasi-static, i.e. it varies very slowly, principally as a result of people moving around. The nature and statistics of the channel impulse response are sensibly independent of the polarisation of the transmitter and receiver provided that no line-of-sight path exists. The maximum delay spread observed was 100–200 ns within rooms, but occasionally values greater than 300 ns were measured in hallways. It is very interesting to note that the measured RMS delay spread within rooms had a median value of 25 ns and a worst-case value of 50 ns, five times smaller than Devasirvatham's results from a much larger building.

A simple statistical model was proposed in which the rays that make up the received signal arrive in clusters. The ray amplitudes are independent Rayleigh random variables with variances that decay exponentially with cluster delay as well as with ray delay within a cluster. The corresponding phase angles are independent random variables uniformly distributed in the range $(0, 2\pi)$. The clusters, and the

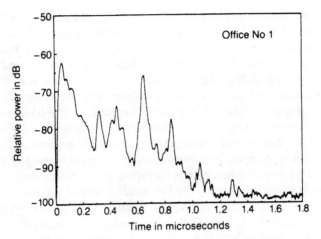

Figure 7.3 Measured time-delay profile within a large six-storey building (after Devasirvatham).

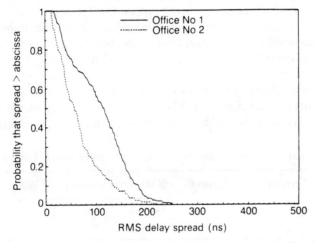

Figure 7.4 Cumulative distribution of time-delay spread within two office buildings.

rays within a cluster, form Poisson arrival processes with different but fixed rates, and the clusters and the rays have exponentially distributed interarrival times. The formation of the clusters is determined by the building structure and the rays within a cluster are formed by multiple reflections from objects in the vicinity of the transmitter and receiver. Both discrete and continuous versions of the model are possible. However, it has been suggested [44] that the discrepancies actually arise as a result of the Poisson arrival assumption and that a modified Poisson process is more representative. Furthermore, the path amplitudes have been shown to follow a lognormal distribution rather than a Rayleigh distribution.

Finally, Rappaport *et al.* [25,45,46], again using similar equipment, have studied multipath propagation in factory buildings at 1300 MHz. Substantial physical differences exist between such buildings and offices or residential houses in respect of construction techniques, contents and placement of walls and partitions. It might be expected, therefore, that propagation characteristics would also be different. In fact, it was found that the path loss exponent *n* was approximately 2.2 and that Rician fading was the norm. The RMS delay spread ranged between 30 and 300 ns, the median values being 96 ns for line-of-sight paths along aisles and 105 ns for obstructed paths across aisles. The worst-case measured value was 300 ns. These values are comparable with those measured in large office buildings [38].

Table 7.3 brings together some of the in-building figures that have been reported. Definitive conclusions are not easy because the propagation conditions are so variable. It seems that where line-of-sight paths exist, the propagation law exponent is usually near 2, indicating that a free space mode is dominant, and this is accompanied by Rician rather than Rayleigh fading. For obstructed paths the exponent rises to 4 or more, and although in many cases the fading is still characterised by Rician statistics, Rayleigh characteristics have also been reported. It is likely that Rician channels will support higher data rates. Wideband measurements have been made at frequencies in the range 850–1750 MHz but there are no obvious effects that can be attributed to changes in the carrier frequency. There is no evidence to suggest that the scattering and reflecting properties of the materials used for construction change appreciably over this frequency range, as the delay spreads do not exhibit any significant statistical difference.

It might be expected that delay spread would decrease with frequency due to increased attenuation by the structural materials, but this is certainly not apparent below 2 GHz. On the other hand, there is some evidence [21,47] that at 60 GHz the propagation mechanism is different since the radio waves are effectively screened by any metal partitions. Although at this frequency there is some leakage through doors and windows, this is insufficient to give room-to-room coupling except where a line-of-sight path exists. At this frequency the transmission, reflection and absorption

Table 7.3 Measured parameters from propagation experiments inside buildings

Investigators	Frequency	Environ-ment	RMS delay spread (ns)		Worst case (ns)	Propagation law exponent *n*
			Median value	Standard deviation		
Bultitude *et al.*	910 MHz 1.75 GHz	Within brick and concrete office buildings	26–30 28–29	8–11 17–22		
Saleh and Valenzuela	1.5 GHz	Within office buildings	25–50		100–200	3–4
Devasirvatham and Murphy	850 MHz 1.7 GHz	Within office buildings	In the range 50–150		400	
Rappaport	1.3 GHz	In factory buildings	96 (LOS) 105 (NLOS)		300	2.2

LOS = line-of-sight.
NLOS = non-line-of-sight.

properties of materials commonly used for building construction vary very widely. However, no wideband measurements have been reported.

7.4   RAY TRACING: A DETERMINISTIC APPROACH

In Chapter 3 we noted that the availability of high-resolution databases makes it more attractive to move towards deterministic propagation methods. We can never hope for 100% accuracy of course because databases are rarely completely up to date, and there are always factors such as moving vehicles, trees in or out of leaf and, inside buildings, changes of furniture location, which introduce uncertainties.

Nevertheless, propagation methods based on ray theory have been the subject of many investigations in recent years. They have been used for both indoor and outdoor environments, and in theory they have enormous potential. If a number of rays can be traced from a given transmitter location to a given receiver location, the electrical lengths of the various ray paths give the amplitudes and phases of the component waves and they can be used to calculate the signal strength. In this context, due account must be taken of changes in amplitude and phase caused by propagation through, or reflection from, obstacles along the ray path. Moreover, the physical lengths of the ray paths allow calculation of the propagation times along those paths, thus permitting evaluation of delay spread and other similar parameters.

The characteristics of the antennas used at both ends of the link can be built into the prediction algorithm, so methods based on ray theory have the potential to provide a complete channel characterisation as far as propagation is concerned. This can be in two or three dimensions depending upon the nature of the available databases. In outdoor environments, sophisticated processing techniques can be used to convert aerial or satellite photographs into 3D databases; in indoor environments, architectural drawings and other layout information can serve the same purpose. However, the extent to which any given ray will penetrate, be reflected from, or be diffracted around a given obstacle depends crucially on the electrical properties of the material or materials from which the obstacle is constructed as well as on its geometrical shape.

The equations in Section 2.3.1 show that the reflection coefficient of a plane surface depends on the polarisation of the incident wave, the angle of incidence and very importantly on the dielectric constant and conductivity of the material. Precise values of conductivities and dielectric constants are needed if accurate predictions are to be obtained. Reflection from a curved surface, surface roughness and diffraction were all discussed in Chapter 2 and have a part to play in prediction methods based on ray theory.

The propagation model normally recognises that when an obstacle exists in the path of a ray, the ray can be specularly reflected, scattered, transmitted (and partially absorbed in the process) or in some cases diffracted around the edge of the obstacle. Specular reflection is characterised by the incident and reflected rays making equal angles with the normal to the surface, transmission obeys Snell's law of refraction, and diffraction effects can be estimated using any of the methods discussed in Chapter 3, e.g. UTD. Scattering is not so easy to deal with and is often neglected on the basis that the vast majority of the energy is contained in the specularly reflected component. Whether this is justified or not, depends on the particular propagation scenario. Reflected and transmitted rays have an inverse square law power dependence (cf. free space propagation) depending on the total distance travelled.

Care is necessary in applying the reflection coefficients given by eqns (2.9) and (2.10); for smooth surfaces, conservation of energy dictates that the transmission coefficient is (1 − reflection coefficient). The proper reflection coefficient must be used depending on the polarisation of the ray relative to the obstacle concerned. For example, in an indoor environment, when a vertically polarised ray launched from a transmitter meets the floor or ceiling, the E-field is normal to the surface and eqn (2.10) applies. On the other hand, the E-field is parallel to walls, so eqn (2.9) should be used. Oblique incidence can be treated by resolving the incident ray into two orthogonal components and proceeding appropriately.

Two basic methods appear in the literature, the ray launching or 'brute force' method [48] and ray tracing [49]. Reciprocity applies as far as each individual propagation path is concerned, but it is customary and more intuitive to trace rays assuming that they start at the transmitter, since the single-transmitter/multiple-receiver scenario is by far the most common. This is particularly relevant in the ray launching method which works as follows.

A software program checks for an LOS between the specified transmitter and receiver locations. Next it launches and traces a ray away from the transmitter in a specified direction and detects whether it intersects an obstruction specified on the database. If it does not, the process stops and a new source ray in a different direction is launched. If an intersection is found, the program determines whether the reflected ray from the intersection point has an unobstructed path to the receiver, and the reflected and transmitted rays are then traced to the receiver or to another obstruction. This recursive process – launching a ray at a given angle and tracing its path – continues for each ray until the ray reaches the receiver, until a specified number of intersections is exceeded, until the ray energy falls below a specified threshold (e.g. rays which pass through obstructions such as walls) or until no further intersections occur. Of course, rays launched in certain directions will never reach the receiver because the geometry is such that no path exists.

To determine all possible rays that propagate between the transmitter and receiver, it is necessary to consider all possible angles of launch from the transmitter and arrival at the receiver. One way of doing this is to consider a large number of rays, each separated from its neighbouring rays by a small but constant angle in 3D space. It appears that an acceptable trade-off between coverage and computation time is attained with an angular separation of about 1° [50]. It is also necessary to decide whether any ray has reached the receiver, by applying a minimum distance test. Since it would be unrealistic to regard the receiving location as being infinitesimally small, an imaginary sphere of small radius is constructed around the receiving point and any ray which intersects this sphere is considered to have been received. The signal strength calculated from the phasor addition of all received rays is considered to be the mean signal over the area defined by the sphere.

Image-based ray tracing differs from ray launching and appears to have some advantages. Instead of using the 'brute force' approach of launching many rays (often up to 40 000) at very similar angles, the technique considers all obstructions as potential reflectors and calculates their effect using the method of images. This is a strictly analytical approach which does not require the use of a receiving sphere, paths are neither duplicated or missed out and in simple environments the computation time is much less because only paths which actually exist between the transmitter and receiver are considered.

A database is used and the transmitter and receiver locations are specified using a 3D coordinate system. As in ray launching methods, the strengths of reflected and transmitted rays are computed using geometrical optics, and diffracted rays are treated using one of the standard techniques; UTD is very popular in current approaches [51]. The existence, or not, of an LOS path is established then virtual source or 'image' data is generated by reflecting the source to the opposite side of all relevant obstacles.

To make this process more manageable, a wall sequence diagram is created. Figure 7.5(b) is a partial wall sequence diagram (which resembles a tree structure) for the simple layout in Figure 7.5(a). There are four obstructions, wall 1 is the first reflector, and paths with up to three reflections are considered. Consecutive reflections from the same wall are not possible and they do not exist in the diagram. Figure 7.5(b) shows there are a total of 13 possible paths with wall 1 as the first reflector (1 single reflection, 3 double reflections and 9 triple reflections). Similar diagrams can be drawn with walls 2, 3 and 4 as the first reflector, so 52 possible paths exist (plus the LOS path) in this simple scenario if up to three reflections are considered. The wall sequence diagram only identifies the possibilities; it does not imply that all these paths actually exist.

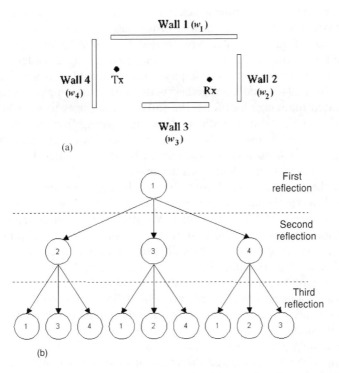

Figure 7.5 (a) Simple indoor propagation scenario; (b) partial wall sequence diagram for part (a).

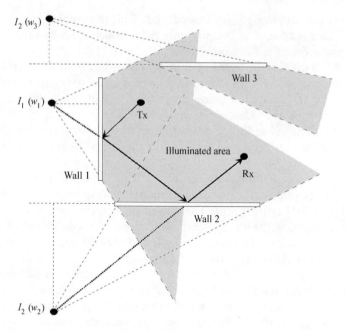

Figure 7.6 The process of image generation.

Figure 7.6 gives a partial example which illustrates the process of image generation in an indoor environment. $I_1(w_1)$ is the first-order image of Tx in wall 1. Two second-order images, i.e. images of $I_1(w_1)$, are created in wall 2 (extended) and wall 3 (extended) and they are designated $I_2(w_2)$ and $I_2(w_3)$. Higher-order images are generated as appropriate. A more complete picture would also show first-order images of Tx in walls 2 and 3, together with appropriate higher-order images. Having calculated the image locations, the software then tests to see whether each image is capable of providing a path. It does this starting from the highest-order images and working back towards the transmitter (this gives it the name 'backward method'). Images which do not provide any paths are eliminated from the stored data before any propagation calculations are made.

To illustrate the conditions that have to be met, Figure 7.7(a) shows a simple two-wall situation. $I_1(w_1)$ is the first-order image of Tx in wall 1 and $I_2(w_2)$ is the second-order image, i.e. image of $I_1(w_1)$, in wall 2. We draw a line joining $I_2(w_2)$ and the receiver to establish the proper refection point on wall 2. Clearly the point P_2 does not coincide with any physical point on wall 2, so the double reflection path Tx–w_1–w_2–Rx does not exist in practice. It is clear from Figure 7.7(a) that a necessary condition for the path to exist is that the reflection point P_2 coincides with a physical location on wall 2 and that this is only possible if Rx lies in the illuminated area defined by $I_2(w_2)$ and wall 2. But this condition, although necessary, is not sufficient.

Figure 7.7(b) shows Rx within the illuminated area, thus ensuring that the necessary reflection point on wall 2 physically exists. In this case the necessary reflection point on wall 1 is outside the physical limits of that wall, so again the path does not actually exist. Now, however, it is easy to see what is necessary. We have established that the reflection point P_2 on wall 2 exists physically, provided

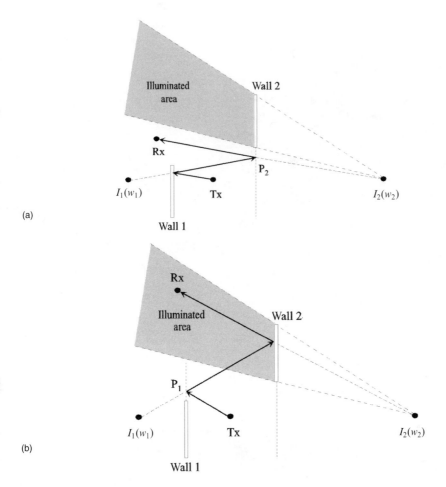

Figure 7.7 (a) The required reflection point P_2 on wall 2 does not exist; (b) the required reflection point on wall 1 does not exist.

Rx lies within the illuminated area shown. It follows that the necessary reflection point P_1 on wall 1 also exists provided P_2 lies within the illuminated area defined by $I_1(w_1)$ and wall 1. Figure 7.7(c) illustrates this.

For the path to exist a necessary and sufficient condition is that P_2 lies on that part of wall 2 which lies within the illuminated area defined by $I_2(w_2)$ and wall 2 and the illuminated area defined by $I_1(w_1)$ and wall 1. This is the shaded area in Figure 7.7(c). It follows that if no part of wall 2 falls within the illuminated area defined by $I_1(w_1)$ and wall 1 then the path being considered does not exist for any position of Rx within the given illuminated area. In general, the above process can be applied recursively, starting from Rx and working back towards Tx, to establish whether each of the necessary reflection points physically exists along any multiple-reflection path.

One further idea is illustrated in Figure 7.7(d). Here both reflection points exist and meet the above criteria, but Rx is on the 'wrong' side of wall 2. This is a

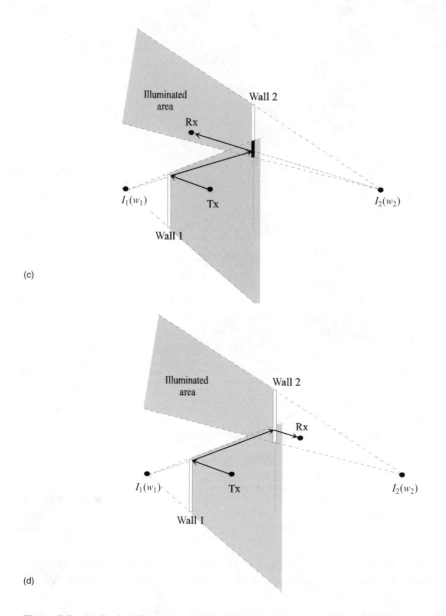

Figure 7.7 (c) Both reflection points exist, so the path is valid; (d) the receiver is not in the illuminated area.

reminder that the images are virtual sources which can be used to compute the reflected paths, but the defined illuminated area only exists on the side of the wall remote from the image (the shaded area). Of course, in Figure 7.7(d) there is a single-reflection path from Tx to Rx via wall 1, and for a different position of Rx within the sector there could be a single reflection from wall 1 and subsequent transmission through wall 2, but the path illustrated does not exist.

As an example we can return to Figure 7.6. In this case walls 1 and 2 meet all the necessary conditions. Wall 3 does not produce a ray path, however, because although the whole of wall 3 lies within the illuminated area of $I_1(w_1)$, the line connecting $I_2(w_3)$ and Rx is not in the illuminated area projected from $I_2(w_3)$.

To improve computational efficiency in practice, prespecified conditions are introduced; for example, no rays which undergo more than n reflections or have a strength more than X dB below that of the strongest path will be considered. As far as the strength condition is concerned, rays can be attenuated by normal spreading loss, by reflection, and by transmission through obstacles. However, as a very simple example we consider a two-path situation comprising a direct ray and a single-reflected ray and we impose the condition that the reflected ray will only be considered if the power it produces at the receiver is greater than -30 dB relative to the power of the direct ray. This gives

$$10 \log_{10} \left(\frac{P_{\text{refl}}}{P_{\text{dir}}} \right) > -30 \tag{7.8}$$

Since $P \propto 1/d^2$, this can be expressed as

$$d_{\text{refl}} < |\rho| \sqrt{1000} \, d_{\text{dir}} \tag{7.9}$$

which means that a reflected ray path with a length less than $|\rho| \sqrt{1000} \times$ (the length of the direct path) will be taken into account in the computation. The above equation can be generalised for any order of reflection as

$$d_{\text{refl}}^{(n)} < |\rho|^n \sqrt{10^t} \, d_{\text{dir}} \tag{7.10}$$

where n is the order of reflection and t is the threshold index evaluated from X dB $= 10^{x/10} = 10^t$. It is a matter of judgment as to what threshold level is appropriate in any given situation. Clearly if it is set too high then many weak paths will be included and computation time will become quite high. On the other hand, if it is set too low, accuracy will suffer.

In summary, the last decade has seen the emergence of ray tracing as an important technique for modelling microcell and indoor picocell propagation. Accuracy depends crucially on the availability of up-to-date high-resolution databases and the availability of computational techniques for extracting relevant information quickly and in an appropriate form. The electrical properties of natural and man-made materials that are used to construct walls, doors, windows, etc., also need to be known with some accuracy. There have been investigations of the accuracy and sensitivity of such methods [52–54] and comparisons with measurements [55]. Where LOS paths exist, they are dominant and only strong low-order reflected paths need to be considered. If there is no LOS path then it is necessary to consider multi-reflected and diffracted rays.

Diffraction is a very important mechanism in some cases since despite the complexity that it adds to the models, without it they would often fail completely in non-LOS areas. Good accuracy has been reported with up to 7 orders of reflection and 2 orders of diffraction, although if computation time is important, 5 reflections with 1 diffraction appears a reasonable compromise for a coverage study. Default values of 10^{-3} S/m for conductivity and 5 for relative permittivity gave reasonable results (RMS prediction error $\sim 4\,\mathrm{dB}$) in a typical suburban area.

7.5 RADIO PROPAGATION IN TUNNELS

There have been some investigations of radio propagation in tunnels at frequencies of interest for mobile communications. In the VHF band the attenuation is very high [56] and it is only the use of highly directional antennas that makes communication possible within tunnels over distances exceeding a few tens of metres. It is well known that a car radio tuned to a normal FM broadcast station loses signal very rapidly when the vehicle enters a tunnel. At higher frequencies there is some improvement, although severe problems remain.

Propagation in tunnels is exemplified by an experiment conducted by Reudink [57] in New York. He reports work undertaken in the Lincoln Tunnel that connects Manhattan to New Jersey under the Hudson River. The tunnel has a rectangular cross section of dimensions approximately 4 m × 7.5 m and is about 2425 m in length. Propagation tests were made at seven frequencies between 153 MHz and 11.2 GHz using transmitters located within the tunnel, about 300 m from the entrance. Figure 7.8 shows some of the results plotted on a logarithmic scale. Attenuation is very high at VHF but decreases as the frequency is increased. Signal attenuation that follows a simple d^n law appears as a straight line with a slope that depends on the value of n. Figure 7.8 shows that n is approximately 4 at 900 MHz, reducing to 2 at 2400 MHz. Above this frequency the loss is less than the free space path loss, indicating that some kind of guiding mechanism exists. At frequencies above 2.4 GHz the attenuation is quite low, making it much more feasible to design a working system. Theories of radio propagation in tunnels and similar structures have been published [58,59].

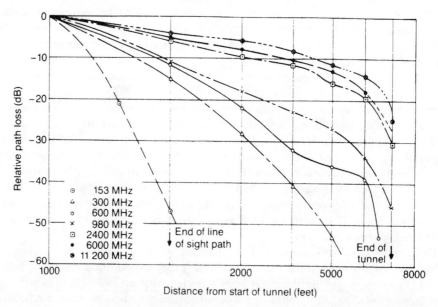

Figure 7.8 Path loss within a tunnel at several different transmission frequencies (after Jakes).

In modern cities it is not uncommon to find an underpass where major roads cross each other. It has been reported [60, Ch. 2] that at 900 MHz a 10–15 dB drop in signal level can be expected in these circumstances and radio communication systems can be severely affected. In general, at frequencies used for mobile radio systems, propagation problems in tunnels and underpasses are very severe and reliable communication cannot be guaranteed. The best solution may involve the use of leaky feeders or rebroadcasting the signal into the tunnel (maybe from both ends) using highly directional antennas.

7.6 PROPAGATION IN RURAL AREAS

7.6.1 Introduction

Multipath fading models of the type described by Aulin and Clarke have proved adequate for the urban mobile radio channel. They predict that the statistical distribution of the field strength values follows a Rayleigh distribution, reaching this conclusion by arguments based on superimposing a large number of components having similar magnitudes scattered from different reflecting and diffracting obstacles in the immediate vicinity of the mobile. In practice this situation does not always exist; we have seen (Section 7.3.1) that the fading characteristics within buildings sometimes have Rician statistics.

In a rural area the number of scatterers may be quite small and the magnitudes of the individual scattered components can vary, with line-of-sight paths being common. The effect of these conditions is to cause the fast fading signal statistics to be non-Rayleigh, and if system planners were to use the Rayleigh model then they would overestimate the severity of the signal fading. The resultant design would therefore be based on pessimistic modelling, so the transmitter power would be unnecessarily high, possibly leading to interference problems.

Some general observations may be made from Figure 7.9, a recording of the fast fading signal envelope measured at 900 MHz along a rural route. This represents

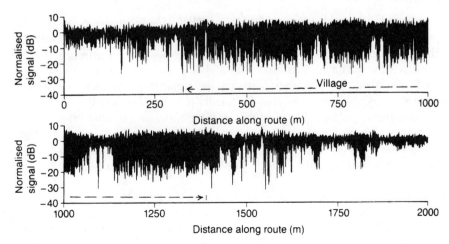

Figure 7.9 Normalised recording of the signal received at a mobile moving along a rural route and passing through a village.

data collected as the mobile travelled along a major road in the direction of the transmitter and passed through a village.

There is an obvious change in the statistics of the received signal for the relatively 'open aspect' areas either side of the village. Referring to the arguments above, a strong direct component may be received in these locations, and this will cause the envelope statistics to differ from those in the surrounding areas. It would be misleading simply to use the complete 2 km section of data shown here to estimate the characteristics of the fast fading, since this route covers different types of terrain, in each of which there may be a different signal distribution. A detailed investigation is necessary to analyse the variations of the signal statistics along routes such as this.

Data extracted from Figure 7.9 reveals that there are large sections of rural routes where the signal statistics do not conform to the Rayleigh distribution. A distinct feature is that the topography of the area immediately surrounding the mobile has a prominent influence on the signal variability. In many relatively open areas, between towns and villages, the Rayleigh model does not appear to be a good approximation to the received signal statistics, whereas in built-up areas the statistics tend to conform more readily to Rayleigh. This is a significant conclusion as it means that, in order to determine the effectiveness of the Rayleigh model within rural areas, the efficiency of the model should be investigated within discrete terrain environments. Thus, small-scale signal variations have to be analysed within typical 'small-area' locations, such as towns, villages, rural lanes and woodland.

There is a noticeable correlation between the mean signal values and the signal variability. For example, in locations where there is an increase in the mean signal strength, there is a corresponding decrease in the standard deviation of the fast fading envelope, and the measured CDF does not conform to Rayleigh statistics. This is reasonable, because whenever a dominant signal component is received, there is a reduction in the signal fading, accompanied by an increase in the local mean signal value.

7.6.2 Signal statistics

Data collected in rural areas at 900 MHz [61] has been analysed in 200–300 m sections to produce the cumulative distribution functions in Figure 7.10. These distributions are plotted on Rayleigh-scaled paper with the theoretical Rayleigh distribution shown as a broken line. Although some of the measured distributions are well described by Rayleigh statistics, a significant number are not; this characteristic has also been observed by Suzuki [62] and Davis [63].

This discovery prompted a further investigation to establish the statistical distribution which best described the signal envelope measured in rural areas. In addition to the Rayleigh distribution, the Rice, Nakagami and Weibull distributions were also considered because of their previous success in describing mobile radio signals [64].

In order to compare the goodness-of-fit among the four distribution functions, a minimum chi-squared (χ^2) analysis was made between the hypothesised and experimental PDFs. The χ^2-distribution provides a non-parametric or distribution-free test (i.e. the results do not depend on the shape or parameters of the underlying distribution) for the goodness-of-fit of theoretical models. Comparisons between

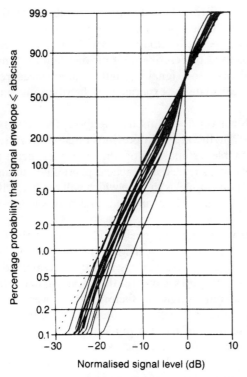

Figure 7.10 Cumulative distributions of the normalised fast fading signal received in several different rural environments: (——) measured results, (- - - -) Rayleigh distribution.

theoretical populations and actual data are made by computing the χ^2-statistic, defined as

$$\chi^2 = \sum \frac{(\text{observed frequency} - \text{theoretical frequency})^2}{\text{theoretical frequency}} \tag{7.11}$$

To estimate the difference between measured and theoretical PDFs, eqn. (7.11) can be rewritten as

$$\chi^2 = \sum_i \frac{N(\hat{p}(\chi_i) - p(\chi_i))^2}{p(\chi_i)} \tag{7.12}$$

where $p(\cdot)$ denotes the theoretical PDF, $\hat{p}(\cdot)$ denotes the estimated PDF, and N is the total number of data samples.

In a minimum χ^2 goodness-of-fit analysis the value of the χ^2-statistic not only indicates the best theoretical model for the measured data, but also provides a quantitative estimate of the goodness-of-fit for any hypothesised distribution through reference to a χ^2-distribution table for the appropriate number of degrees of freedom [65].

In the tests described above, the measured and theoretical PDFs of the logarithmic signal strength values were compared by calculating the number of samples of the fast fading signal within each 1 dB interval from -30 dB to $+10$ dB, i.e.

$i = -30, \ldots, 10$ in eqn. (7.12). This technique, which essentially gives equal importance to all measured values of signal strength, was used because the dynamic range of the fast fading can be greater than 30 dB and an excessively large number of points would have to be used to describe the four distribution functions accurately on a linear scale. Also, since the signal strength is often measured in logarithmic units, it is far simpler to calculate the experimental distributions in these units rather than in linear units.

It is worth pausing for a very brief discussion about the validity of giving equal weighting to all signal strength values. In previous research work [66] (Chapter 4) it was suggested that reliable estimation of the quantiles near the tails of the distribution is of greatest importance in practical planning situations. Thus, it was argued that the most suitable model is the one which will predict, with the least error, the quantiles between 1% and 20% at one end, and between 80% and 99% at the other.

In the analysis, equal weighting was given to all values of the fast fading signal, and therefore no emphasis was given to the tails of the distributions. However, in preliminary χ^2 goodness-of-fit tests, weighting factors were applied to both the tails of the distributions and to values close to the mean signal level. This analysis produced results, for different weighting factors, which were very close to those obtained for equal weighting, and therefore indicated that little was to be gained by emphasising particular sections of the distribution.

As the PDF of the received signal envelope was determined in terms of logarithmic units, it was necessary to translate the theoretical distributions onto a logarithmic scale. This transformation is given in Appendix C for the Rayleigh and Rician distributions, and results can be obtained using the same method for the other two distributions [62].

The PDF for a Rayleigh variable, expressed in decibels, is given by equation (C.6):

$$p(y) = \frac{1}{M\sigma^2} \exp\left[\frac{2y}{M} - \frac{1}{2\sigma^2} \exp\left(\frac{2y}{M}\right)\right] \tag{7.13}$$

If r has a Rician distribution, the PDF is expressed in terms of the parameter K by eqn. (5.61). Alternatively, the PDF of y ($= 20 \log r$) can be expressed by equation (C.11):

$$p(y) = \frac{1}{M\sigma_n^2} \exp\left\{\frac{2y}{M} - \frac{1}{2\sigma_n^2}\left[r_s^2 + \exp\left(\frac{2y}{M}\right)\right]\right\} I_0\left[\frac{r_s}{\sigma_n^2} \exp\left(\frac{y}{M}\right)\right] \tag{7.14}$$

If there is no dominant signal component then $r_s = 0$, the Rician distribution reduces to the Rayleigh distribution and eqn. (7.14) reduces to eqn (7.13).

If r has a Nakagami distribution then the PDF of y can be written as

$$p(y) = \frac{2m^m}{M\Gamma(m)\bar{u}^m} \exp\left[\frac{2my}{M} - \frac{m}{\bar{u}} \exp\left(\frac{2y}{M}\right)\right] \tag{7.15}$$

where $\bar{u}$ is the mean square value of r, with $r = \exp(y/M)$. The parameter m is a measure of the signal variability. If $m = 1$ then eqn. (7.15) reduces to eqn. (7.13), i.e. the PDF for a Rayleigh variable expressed in decibels.

Finally, the PDF for a Weibull variable, expressed in decibels, be written as:

$$p(y) = \frac{w}{M}\left(\frac{b}{V}\right)^w \exp\left[\frac{wy}{M} - \left(\frac{b}{V}\right)^w \exp\left(\frac{wy}{M}\right)\right]$$ (7.16)

where V is the RMS value of the variable y expressed in linear units, w is a measure of the signal variability, and parameter W is defined as $W = w/V$. If $w = 2$ the Weibull distribution reduces to the Rayleigh distribution.

7.6.3 Small-scale signal variations: statistical modelling

In order to investigate which of the four distribution functions best describes the experimental results, the PDFs measured for numerous data files were modelled by each of the four distributions. Since the parameters of the theoretical distribution functions can be varied, giving different χ^2 quantities, they were chosen to give the minimum possible value in each of the four cases. Results obtained using this method are shown in Figure 7.11. The same axes are used to plot the measured and theoretical probability distributions for the four distributions considered. The parameter values that yielded the best-fit or minimum χ^2-statistic were then noted.

Results similar to Figure 7.11 were obtained for different types of terrain, and for various transmitter–receiver separations up to 15 km. To assess the overall performance of each distribution function in these areas, the respective parameter values for the various propagation distances were averaged. In order to estimate the overall accuracy of the four PDFs in each terrain environment, the observed values of χ^2 for the appropriate experiments and the number of degrees of freedom from each individual experiment were added. The total value of χ^2, i.e. $\sum \chi^2$, can then be tested with the total number of degrees of freedom.

Relevant χ^2 quantities were obtained for each environment, and in all the typical rural areas considered, the Rician distribution gave the least overall value of $\sum \chi^2$ and thus the best fit. This indicates that the Rician PDF is consistently the best model in all environments considered. Furthermore, it suggests that in any rural environment there is a high probability of receiving a line-of-sight path, or a dominant signal component. In towns and woodland areas, where there is a lower probability of receiving a line-of-sight path, the Rician distribution still gave the best fit to the measured data.

Examination of the parameter K, which indicates the ratio of the steady signal power to random signal power for the environments concerned, showed that in small villages where a dominant signal component would intuitively be expected, the value of K was a maximum. Furthermore, in both towns and woodland areas, the value of K reduced as the Rician PDF tended towards a Rayleigh distribution. This highlights the flexibility obtained in using the Rician distribution, with the ability of the function to model the signal statistics accurately when a dominant component may or may not be important.

For all terrain environments, the Rayleigh distribution had the highest $\sum \chi^2$ quantity, i.e. the Rayleigh distribution was the least accurate in modelling the statistics of the fast fading envelope. The reason for this is its simplicity and the fact that it has only one parameter. For the Rayleigh distribution, the value of σ^2

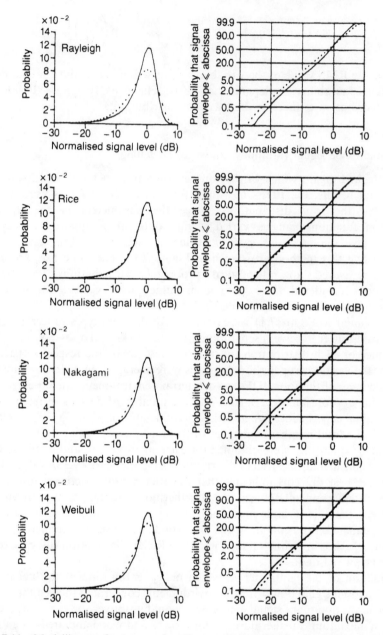

Figure 7.11 Modelling performance of various statistical distributions when tested against normalised fast fading data measured in a rural village: (——) experimental and (- - - -) theoretical.

essentially positions the peak theoretical probability coincident with the peak of the measured data to minimise χ^2. However, the other three distributions have a second parameter that effectively alters the shape of the probability distribution to further reduce the value of χ^2. This feature was also noted by Suzuki [62].

The type of terrain environment had a significant effect on the modelling accuracy of the various distribution functions. For example, the Rayleigh distribution showed a far worse fit to the experimental PDF for rural lanes than for woodland areas. With fewer scatterers intuitively expected in rural lanes, there is a higher probability of a dominant component in the received signal, which causes the signal statistics to be non-Rayleigh.

There did not appear to be any change in the modelling accuracy of the Rician distribution depending on the terrain environment. This is a result of its ability to model the magnitude of any dominant component through a change in the value of K. The analysis revealed that the value of K had an inverse relationship to the scatterer density, intuitively expected in the different terrain environments. In rural villages, for example, there is a significant dominant component, which causes the signal statistics to be non-Rayleigh; the value of K is fairly high and consequently the fit with Rician statistics is quite good; the Rayleigh distribution gives a poor fit to the measured PDF.

As with the Rician distribution, the terrain environment does not affect the modelling accuracy of the Weibull and Nakagami distributions. They too have parameters which are a measure of the signal variability: m for the Nakagami distribution and w for the Weibull distribution. Both these parameters also varied in the same way as the Rician parameter K. But in its overall fit to the measured data, the Weibull distribution was better than the Nakagami distribution.

In relatively open areas, such as villages and rural lanes, there was a noticeable increase in the amplitude of the dominant component in the received signal as the range between the transmitter and receiver decreased. This is clearly due to the fact that as the mobile gets closer to the transmitter, the number of obstructions between the two terminals will decrease, causing an increased probability of receiving a direct line-of-sight path. Therefore, as the range clearly affects the magnitude of the dominant signal component in 'open' areas, its effect on the modelling accuracy of the Rayleigh distribution becomes more obvious. The Rayleigh distribution has only one parameter and cannot accommodate a dominant signal component; its accuracy therefore deteriorates as the range reduces, when the dominant component becomes more significant. It is far better to use any of the other distribution functions to describe the fast fading signal statistics in these areas, especially when the range is small.

The most important conclusion from this analysis is that the Rician distribution function best describes the fast fading signal statistics within rural areas. Furthermore, the Rayleigh model was the least effective of the four, and this has considerable significance since the Rayleigh distribution is commonly used by system planners in the design and evaluation of mobile radio systems.

Extensive analysis of experimental data obtained in non-hilly rural areas showed that the Rician distribution with appropriate parameter selection provides the best model for the fast fading signal. Furthermore, the Rician distribution proved to be more accurate than the Rayleigh, Nakagami or Weibull distributions in modelling the signal statistics in both discrete rural environments and over extensive rural areas.

It may well be appropriate to consider the Rician distribution for modelling signal statistics in rural areas, because within rural areas it is important that the chosen model has the flexibility to accurately describe the situations when a dominant signal component may or may not be present in the received signal.

Table 7.4 Optimum value of the Rice parameter for various classes of terrain environment

Terrain environment	Optimum Rice parameter K	
	Range < 6 km	Range > 6 km
Woodland	-6	-14
Town	-1.2	-6
Village	0.6	-6
Hamlet	2.1	1.3
Rural lane	0.8	0
Minor road	0.9	0
Major road	1.4	-2.6

In this context, Table 7.4 gives optimum values of the parameter K that could be used in specific classes of rural environment as a function of transmission distance. Note that these are global estimates which in many cases can only be used as a starting point for an estimation of channel characterisation and radio system performance.

REFERENCES

1. Hoffman H.H. and Cox D.C. (1982) Attenuation of 900 MHz radio waves propagating into a metal building. *IEEE Trans.*, **AP30**(4), 808–11.
2. Cox D.C., Murray R.R. and Norris A.W. (1983) Measurement of 800 MHz radio transmission into buildings with metallic walls. *Bell Syst. Tech. J.*, **62**(9), 2695–717.
3. Cox D.C., Murray R.R. and Norris A.W. (1984) 800 MHz attenuation measured in and around suburban houses. *AT&T Tech. J.*, **63**(6), 921–54.
4. Durante J.M. (1973) Building penetration loss at 900 MHz. *Proc. IEEE VT'93 Conference*, pp. 1–7.
5. Wells P.I. and Tryor P.V. (1977) The attenuation of UHF radio signals by houses. *IEEE Trans.*, **VT26**(4), 358–62.
6. Akeyama A., Tsuruhara T. and Tanaka Y. (1982) 920 MHz mobile propagation test for portable telephone. *Trans. Inst. Electron. Commun. Eng. Jpn*, **E65**(9), 542–43.
7. Walker E.H. (1983) Penetration of radio signals into buildings in the cellular radio environment. *Bell Syst. Tech. J.*, **62**(9), 2719–34.
8. Cox D.C. (1985) Universal portable radio communications. *IEEE Trans.*, **VT34**(3), 117–21.
9. Rice P.L. (1959) Radio transmission into buildings at 35 and 150 MC. *Bell Syst. Tech. J.*, **38**(1), 197–210.
10. Barry P.J. and Williamson A.G. (1987) Modelling of UHF radiowave signals within externally illuminated multi-storey buildings. *J. IERE*, **57**(6), S231–40.
11. Turkmani A.M.D., Parsons J.D. and Lewis D.G. (1988) Measurement of building penetration loss on radio signals at 441, 900 and 1400 MHz. *J. IERE*, **58**(6), S169–74.
12. Mino N. and Yamada Y. (1985) Pocket bell personal signalling service. *Jpn Telecom. Rev.*, **7**(4), 211–18.
13. Deitz J. *et al.* (1967) *Examination of the feasibility of conventional land mobile operations at 900 MHz*. Federal Communications Commission, Office of the Chief Engineer, Research Division Report R7102.1.
14. Barry P.J. and Williamson A.G. (1991) Statistical model for UHF radio-wave signals within externally illuminated multistorey buildings. *IEE Proc. Part I*, **138**(4), 307–18.
15. Toledo A.F., Turkmani A.M.D. and Parsons J.D. (1998) Estimating coverage of radio transmission into and within buildings at 900, 1800 and 2300 MHz. *IEEE Personal Commun.*, **5**(2), 40–7.

16. Molkdar D. (1991) Review of radio propagation into and within buildings. *Proc. IEE*, **138**(1), 61–73.
17. Hashemi H. (1993) The indoor radio propagation channel. *Proc. IEEE*, **81**(7), 943–68.
18. Acampora A.S. and Winters J.H. (1987) System applications for wireless indoor communications. *IEEE Commun. Mag.*, **25**(8), 11–20.
19. Tsujimura K. and Kuwabara M. (1977) Cordless telephone system and its propagation characteristics. *IEEE Trans.*, **VT26**(4), 367–71.
20. Alexander S.E. (1983) Radio propagation within buildings at 900 MHz. *Electron. Lett.*, **19**, 860.
21. Huish P.W. and Pugliese G. (1983) A 60 GHz radio system for propagation studies in buildings. *Proc. ICAP'83 (IEE Conference Publication 219)*, pp. 181–5.
22. Horikoshi J., Tanaka K. and Morinaga T. (1986) 1.2 GHz band wave propagation measurements in concrete building for indoor radio communications. *IEEE Trans.*, **VT35**(4), 146–52.
23. Akerberg D. (1988) Properties of a TDMA picocellular office communication system. *Proc. IEEE Globecom Conference*, pp. 1343–49.
24. Bultitude R.J.C. (1987) Measurement, characterisation and modelling of indoor 800/900 MHz radio channels. *IEEE Commun. Mag.*, **25**, 5–12.
25. Rappaport T.S. and McGillion C.D. (1987) Characterising the UHF factory multipath channel. *Electron. Lett.*, **23**, 1015–17.
26. Motley A.J. and Keenan J.M.P. (1988) Personal communication radio coverage in buildings at 900 MHz and 1700 MHz. *Electron. Lett.*, **24**(12), 763–4.
27. Owen F.C. and Pudney C.D. (1989) Radio propagation for digital cordless telephones at 1700 MHz and 900 MHz. *Electron. Lett.*, **25**(1), 52–3.
28. Seidel S.Y. and Rappaport T.S. (1992) 914 MHz path loss prediction models for indoor wireless communications in multi-floored buildings. *IEEE Trans.*, **AP40**(2), 207–17.
29. Seidel S.Y. *et al.* (1992) The impact of surrounding buildings on propagation for wireless in-building personal communications system design. *Proc. IEEE VT Conference*, Denver CO, pp. 814–18.
30. Davies J.G. (1997) Propagation of radio signals into and within multi-storey buildings at 900 MHz and 1800 MHz. PhD thesis, University of Liverpool.
31. Violette E.J., Espeland R.H. and Allen K.C. (1988) Millimeter-wave propagation characteristics and channel performance for urban–suburban environments. *NTIA Report 88-239*.
32. Andersen J.B., Rappaport T.S. and Yoshida S. (1995) Propagation measurements and models for wireless communication channels. *IEEE Commun. Mag.*, 42–9.
33. Devasirvatham D.M.J., Banerjee C., Krain M.J. and Rappaport D.A. (1990) Multi-frequency radiowave propagation measurements in the portable radio environment. *Proc. IEEE ICC'90*, pp. 1334–40.
34. Rappaport T.S. (1996) *Wireless Communications: Principles and Practice*. Prentice Hall, Englewood Cliffs NJ.
35. Toledo A.F. and Turkmani A.M.D. (1992) Propagation into and within buildings at 900, 1800 and 2300 MHz. *Proc. IEEE VT Conference*, Denver CO, pp. 633–6.
36. Turkmani A.M.D. and Toledo A.F. (1993) Modelling of radio transmissions into and within multi-storey buildings at 900, 1800, and 2300 MHz. *Proc. IEE Part I*, **140**(6), 462–70.
37. Devasirvatham D.M.J. (1984) Time delay spread measurements of wideband radio signals within a building. *Electron. Lett.*, **20**(23), 950–1.
38. Devasirvatham D.M.J. (1986) Time delay spread and signal level measurements of 850 MHz radio waves in building environments. *IEEE Trans.*, **AP34**(11), 1300–5.
39. Devasirvatham D.M.J. (1987) A comparison of time delay spread and signal level measurements within two dissimilar office buildings. *IEEE Trans.*, **AP35**(3), 319–24.
40. Chuang C.-I. (1986) The effects of multipath delay on timing recovery. *Proc. IEEE ICC'86*, Toronto, Canada, **1**, 55–9.
41. Andersen J.B., Lauritzen S.L. and Thommeson C. (1986) Statistics of phase derivatives in mobile communications. *Proc. IEEE VT Conference*, VETEC '86, 228–31.

42. Bultitude R.J.C., Mahmoud S.A. and Sullivan W.A. (1989) A comparison of indoor radio propagation characteristics at 910 MHz and 1.75 GHz. *IEEE J.*, **SAC7**(1), 20–30.
43. Saleh A.A.M. and Valenzuela R.A. (1987) A statistical model for indoor multipath propagation. *IEEE J.*, **SAC5**(2), 128–37.
44. Ganesh R. and Pahlavan K. (1989) On arrival of paths in fading multipath indoor radio channels. *Electron. Lett.*, **25**(12), 763–5.
45. Rappaport T.S. (1989) Indoor radio communication for factories of the future. *IEEE Commun. Mag.*, **27**(5), 15–24.
46. Rappaport T.S. (1989) Characterisation of UHF multipath radio channels in factory buildings. *IEEE Trans.*, **AP37**(8), 1058–69.
47. Alexander S.E. and Pugliese G. (1983) Cordless communication within buildings: results of measurements at 900 MHz and 60 GHz. *Br. Telecom. Tech. J.*, **1**, 99–105.
48. Seigel S.Y. and Rappaport T.S. (1992) A ray-tracing technique to predict path loss and delay spread inside buildings. *Proc. IEEE Globecom'92*, Orlando FL, pp. 649–53.
49. McKeown J.W. and Hamilton R.L. (1991) Ray-tracing as a design tool for radio networks. *IEEE Networks Mag.*, **5**(6), 27–30.
50. Schaubach K.R. and Davis N.J. (1994) Microcellular radio-channel propagation prediction. *IEEE Antennas and Propagation Mag.*, **36**(4), 25–34.
51. Athanasiadou G.E., Nix A.R. and McGeehan J.P. (1995) A ray tracing algorithm for microcellular wideband propagation modelling. *Proc. IEEE VTC'95 Conference*, Chicago IL, pp. 261–5.
52. Athanasiadou G.E., Nix A.R. and McGeehan J.P. (1997) Comparison of predictions from a ray tracing microcellular model with narrowband measurements. *Proc. IEEE VTC'91 Conference*, Phoenix AZ, pp. 800–4.
53. Rizk K., Wagen J.T. and Gardiol F. (1997) Two-dimensional ray tracing modelling for propagation prediction in microcellular environments. *IEEE Trans.*, **VT46**(2), 508–18.
54. Wang S.S. and Reed J.D. (1997) Analysis of parameter sensitivity in a ray-tracing propagation environment. *Proc. IEEE VTC'97 Conference*, Phoenix AZ, pp. 805–9.
55. Athanasiadou G.E., Nix A.R. and McGeehan J.P. (1998) Investigation into the sensitivity of a microcellular ray-tracing model and comparison of the predictions with narrowband measurements. *Proc. IEE VTC'92 Conference*, Ottawa, Canada, pp. 870–4.
56. Farmer R.A. and Shepherd N.H. (1965) Guided radiation: the key to tunnel talking. *IEEE Trans.*, **VC14**, 93–8.
57. Reudink D.O. (1968) Mobile radio propagation in tunnels. *IEEE VT Group Conference*, San Francisco.
58. Emslie A.G., Lagace R.L. and Strong P.F. (1975) Theory of the propagation of UHF radio waves in coal mine tunnels. *IEEE Trans.*, **AP23**, 192–205.
59. Zhang Y.P., Hwang Y. and Parsons J.D. (1999) UHF radio propagation in straight open-groove structures. *IEEE Trans.*, **VT48**(1), 249–54.
60. Lee W. C.-Y. (1986) *Mobile Communications Design Fundamentals*. Sams, Indianapolis IN.
61. Mockford, S. (1989) Narrowband characterisation of UHF mobile radio channels in rural areas. PhD thesis, University of Liverpool.
62. Suzuki H. (1977) A statistical model for urban radio propagation. *IEEE Trans.*, **COM25**(7), 673–80.
63. Davis B.R. and Bogner R.E. (1985) Propagation at 500 MHz for mobile radio. *Proc. IEE Part F*, **132**(5), 307–20.
64. IEEE Vehicular Technology Society Committee on Radio Propagation (1988) Special issue on radio propagation. *IEEE Trans.*, **VT37**(1).
65. Pearson E.H. and Hartley H.O. (1976) *Biometrika Tables for Statisticians*, Vol. 1. Cambridge University Press, Cambridge.
66. Parsons J.D. and Ibrahim M.F. (1983) Signal strength prediction in built-up areas. Part 2: signal variability. *Proc. IEE Part F*, **130**(5), 385–91.

Chapter 8

Sounding, Sampling and Simulation

8.1 CHANNEL SOUNDING

In the earlier chapters we discussed the characteristics of mobile radio channels in some detail. It emerged that there are certain parameters which provide an adequate description of the channel and it remains now to describe measuring equipment (channel sounders) that can be used to obtain experimental data from which these parameters can be derived. It is often of interest to make measurements which shed some light on the propagation mechanisms that exist in the radio channel but engineers are usually more interested in obtaining parameters that can be used to predict the performance, or the performance limits, of communication systems intended to operate in the channel.

The choice of channel sounding technique will usually depend on the application foreseen for the propagation data. Basically, a choice has to be made between using narrowband or wideband transmissions and whether a time or frequency domain characterisation is required. In what follows we will briefly describe both narrowband and wideband systems and provide an indication of how relevant data can be extracted from measurements. We make only a brief reference to the data processing techniques, particularly in the case of wideband channels; for details the interested reader will need to consult the literature [1–4].

8.2 NARROWBAND CHANNEL SOUNDING

It is clear from the earlier discussion that when the mobile radio channel is excited by an unmodulated CW carrier (i.e. a single tone), large variations are observed in the amplitude and phase of the signal received by a moving antenna. These variations are apparent over quite small distances. A considerable number of mobile radio propagation studies have been undertaken by transmitting an unmodulated carrier from a fixed base station, receiving the signal in a moving vehicle and recording the signal envelope. It is common to use a receiver which provides a DC output voltage proportional to the logarithm of the received signal amplitude, and a suitable receiver calibration therefore produces the signal strength in dBm or, if a calibrated antenna is used, the field strength in dBmV/m.

Figure 8.1 shows a simplified block diagram of a generic receiving and recording system which has the basic features required. The signal envelope at the output of the

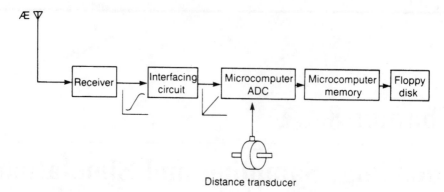

Figure 8.1 Simplified block diagram of a receiver and data logging system for use in the field.

receiver is fed via a suitable interfacing circuit and an ADC into the memory (RAM) of a microcomputer. Distance pulses from a transducer are used to trigger the ADC so that samples are taken at an appropriate rate. Analysis of the stored data can either be carried out in suitable batches as field trials proceed or the stored data can be retained for analysis later. It is not always convenient, or necessary, to initiate sampling using distance pulses and if the system is made portable for use within a room or building, for example, then time sampling is much more convenient.

The phase of the received signal is sometimes of interest and can be measured, relative to a fixed reference, if the signal is demodulated in two quadrature channels. Such receivers have been used by Bultitude [5] for indoor measurements and by Feeney [6] for small-cell measurements outdoors. To measure phase accurately it is essential that the local oscillators in the transmitter and receiver are phase-locked. In the majority of cases this is impracticable but the use of extremely stable sources, such as rubidium oscillators, can provide adequate coherence over quite long periods of time. In this manner, only those phase variations introduced by the propagation channel, and not those due to the transmitter/receiver combination, are measured. Of course, the phase information cannot realistically be studied at the carrier frequency. Translation of the quadrature information to a suitable lower frequency can be carried out by heterodyning to an intermediate frequency; two possibilities exist, either a conveniently low intermediate frequency or a direct conversion to zero-IF.

In the first type of receiver, care is needed in the choice of IF to avoid images, arising from the mixing process, from falling within the passband of the IF filter. This can be achieved using an initial frequency upconversion or by employing image rejection mixers. Two advantages exist for this architecture: the input frequency is not restricted to a narrow RF band and a suitable network analyser can be used to isolate sources of amplitude and phase unbalance in the various signal paths. The zero-IF (direct conversion) receiver requires mixers which have a sufficiently high operating frequency at the RF port, together with a DC-operating IF port. The operating frequency is restricted to a narrow range due to the constraint of maintaining quadrature in the various signal paths. High RF power levels are required to drive the mixers, so that an adequate dynamic range is achieved.

These disadvantages are minimised if the design is limited to one carrier frequency. Other advantages also exist; for example, only one phase-locked stage is required and the single mixing process down to zero-IF provides an inherent detection function. Images are no longer a problem because they are well separated from the wanted information and are easily removed. Any imbalance in the amplitude or phase responses of the two channels can be reduced or eliminated through careful calibration or the use of digital correction techniques. Figure 8.2 shows the dual-channel receiver used by Feeney [6] for propagation and diversity experiments at 900 MHz. A dynamic range of 45 dB was achieved.

8.2.1 A practical narrowband channel sounder

For characterising the channel in respect of its likely effect on narrowband systems it is usually adequate to transmit a CW carrier and to measure the variation in the envelope as the receiver is moved around within a given small area. Almost without exception, equipment designed for this purpose uses a fixed transmitter and a mobile receiver. A data acquisition and analysis unit can easily be incorporated into the receiving system and designs can be tailored to meet any specific requirement, e.g. outdoors or indoors, or in confined spaces. The equipment described below was used by Davies [7] for indoor measurements but it is not restricted in any way and could easily find other applications.

A backpack system was preferred so that the operator could move around freely. This necessitated battery operation with a battery capacity adequate for several hours of operation. The system was specified to have a dynamic range of 80 dB at 1.8 GHz, an automated attenuation control to allow the operator to walk into a room or area and conduct a test without any pretesting routine and a data acquisition system which stored not only the samples of signal strength, but also the setting of the attenuator control. Time sampling was used, the sampling rate being such that 4 or 5 samples per wavelength were taken at normal walking speed. The data acquisition system was designed to enable a large number of samples to be taken, subsequently averaged and the mean value stored.

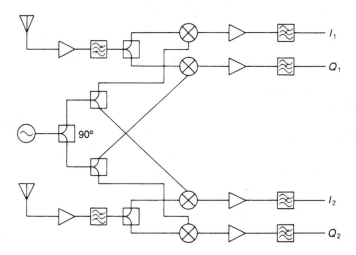

Figure 8.2 Feeney's dual-branch, phase-locked direct conversion receiver.

It was also designed to acquire the average signal level and CDF for a large number of locations. It was intended that the signal strength data should be analysed using a notebook computer, accessed via its printer port. Since it is not possible to insert a standard data acquisition card into such a computer, and since access via the printer port interface is slow, it was necessary for the data acquisition system to have on-board memory so that signal could be sampled and stored for downloading later. The notebook computer effectively controls the acquisition via a specially written program, and allows downloading from the on-board memory to the hard disk for permanent storage.

The transmitter was of conventional design; it used a commercial frequency synthesiser as a signal source, the output being amplified to provide an output power of 3 W, before being fed to a $5\lambda/16$ collinear antenna. A simplified block diagram of the receiver, which is based on a single-conversion superheterodyne architecture, is shown in Figure 8.3. The receiving system is in two parts, a backpack unit and a handset similar in size to a modern cellphone, which incorporates the receiving antenna.

In the receiver, the signal passes through an RF amplifier and bandpass filter before being downconverted to a 10 MHz IF. Further filtering is provided by a crystal filter with very sharp roll-off characteristics and the signal is then fed to a logarithmic IF amplifier/detector which has a dynamic range in excess of 80 dB. The input range of the receiver is controlled by the use of a programmable attenuator having a range of 128 dB in 1 dB steps. The control of this attenuator is automatic via the logging system and its setting is stored.

The efficient logging of data is carried out by the data acquisition unit (DAU). Since it is only required that the mean signal level be recorded, a system was designed to enable the output of the receiver to be sampled and averaged in batches to produce a single value. A diagram of the DAU system is shown in Figure 8.4. The computer interface allows any recorded data to be downloaded to an IBM-compatible computer and stored for later analysis. If it is desired to have approximately 4–5 samples per wavelength at an average walking speed of 1.5 m/s, then the sampling frequency required is approximately 40 Hz at 1800 MHz ($\lambda \approx 17$ cm).

The microprocessor used for this application is the Texas Instruments TMS320-E15. This processor has the advantage that its program memory is contained in the on-chip EPROM of the device, so reprogramming is straightforward. The processor interfaces with several devices, namely an analogue-to-digital converter (ADC), a dynamic RAM, the programmable attenuator and a set of input switches and LCD display located in the handset. Interfacing with a notebook computer is performed by a parallel printer interface on the PC. The DAU can be in one of two modes, *download* or *record*.

In record mode, the system 'hangs' until the user wishes to sample. After initiating a measurement, 128 sample values are taken via the ADC at a sampling frequency of approximately 40 Hz. These 128 values are averaged to produce the mean signal level and if necessary an adjustment is made to the programmable attenuator. The change in attenuation is calculated automatically, using an algorithm which evaluates the mean signal strength, the dynamic range of the system and the current attenuation setting. A further 1024 samples are then taken at the constant sampling rate and the mean signal strength is calculated. Signal levels for the CDF of the collected data are also produced at probabilities of 1%, 50% and 99%. The calculated values are all displayed on the LCD and stored in the dynamic RAM, which also features a small backup battery to enable short-term storage of captured data.

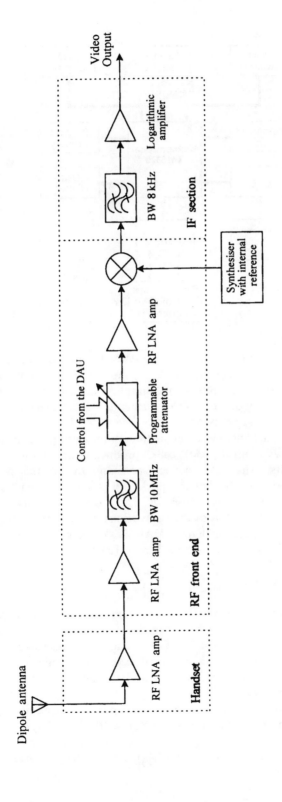

Figure 8.3 Block diagram of the receiver.

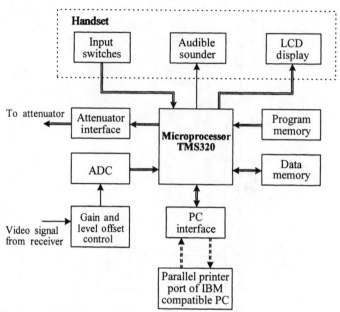

Figure 8.4   The data acquisition unit.

In download (or interface) mode, the user is allowed to interface the system with a computer or manually view the contents of the memory via the LCD on the handset. Using specially written software, full system calibration can also be undertaken. The software also provides testing of all the DAU elements; a useful feature which can be used before any field measurements.

The power source for the backpack signal strength measuring system is provided by a set of nickel–cadmium (NiCd) cells, producing an output voltage of approximately 13 volts. The total current consumption of the backpack is approximately 0.8 amps, so the battery pack will sustain the backpack for a period of up to 7 hours of continuous use. DC–DC converters are used to provide constant output voltages regardless of the fluctuations of the input power source.

The complete receiver system including DAU is shown in Figure 8.5; it has a dynamic range of 80 dB and the noise floor is at −125 dBm.

8.3   SIGNAL SAMPLING

Any record of signal strength has to be analysed in order to obtain the required parameters. The raw information, whether in linear or logarithmic units, has two components which represent the slow and fast fading; the mean value is influenced by the distance from the transmitter. The analysis can be designed to obtain the mean or median value in a certain area and/or to derive information about the first- and second-order statistics of the fading envelope.

If it is desired to obtain information about the depth and duration of fades, it is necessary to sample the signal at a rate appropriate to the task. Expressions for the average level crossing rate and average fade duration of a Rayleigh fading signal have been obtained in eqns (5.43) and (5.47), and Table 5.1 gives values, in

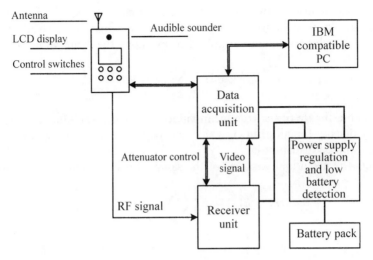

Figure 8.5 The complete receiving system.

wavelengths, with respect to the median value. For example, the average duration of a fade 30 dB below the median value is 0.01λ, and at 900 MHz this value corresponds to a distance of 0.33 cm. Fairly rapid spatial sampling is therefore necessary to ensure such fades are not missed. In practice there is lognormal fading superimposed on the Rayleigh fading, and in order for results to be compared with theory it is necessary to separate the two fading processes by a technique of normalisation.

Clarke's suggestion [8] of normalisation as a method of dealing with a signal in which the underlying process is Rayleigh was discussed in Chapter 5. It has become widely known as the running mean or moving average technique. The result is a new PDF

$$p_n(r_n) = 2r_n \exp\left(-r_n^2\right)$$

which is a Rayleigh process with $\sigma^2 = 0.5$ and an RMS value of unity. The question now arises as to what is a suitable distance for normalisation of experimental data? Parsons and Ibrahim [9] experimented with various windows having widths between 2λ and 64λ, coming to the conclusion that it was reasonable to treat the data as a stationary Rayleigh process for distances up to about 40 m at VHF and about 20 m at UHF. Davis and Bognor [10] investigated the effect of measurement length on the statistics of the estimated fast fading at 500 MHz and showed that as the distance was increased above about 25 m, variations in the local average values appeared. It seems therefore, from experimental evidence, that distances of up to 40 m are suitable at VHF, while there is danger in going above 25 m at UHF. We have seen that rapid sampling is necessary to accurately obtain the second-order statistics of the signal; but following on from the above argument we might ask, in the context of extracting the local mean value, how many samples do we really need within the given measurement length and also, given those samples, with what accuracy and confidence can we estimate the local mean?

8.4 SAMPLED DISTRIBUTIONS

To answer the question about estimation of the local mean, we need to obtain some simple relationships that apply to sampled distributions. We can state quite generally that if the probability density function of a random variable x is $p(x)$ and if x_1, $x_2, \ldots, x_N$ are observed sample values of x, then any quantity derived from these samples will also be a random variable. For example, the mean value of x_i can be expressed as

$$\bar{x} = \frac{1}{N} \sum_{i=1}^{N} x_i$$

and this is an estimate of the true mean value $E\{x\}$; $\bar{x}$ is a random variable and the probability density function $p_1(\bar{x})$, which can be found provided $p(x)$ is known, is called the sampled distribution.

Generally the mean and variance of the sampled distribution can be written

$$\hat{m} = E\{\bar{x}\} = E\left\{ \frac{1}{N} \sum_{i=1}^{N} x_i \right\}$$

$$= \frac{1}{N} E\left\{ \sum_{i=1}^{N} x_i \right\} = E\{x_i\} = m \tag{8.1}$$

and, assuming independent samples,

$$\hat{\sigma}^2 = E\{(\bar{x} - \hat{m})^2\} = E\{(\bar{x} - m)^2\}$$

$$= E\left\{ \left(\frac{1}{N} \sum_{i=1}^{N} (x_i - m) \right)^2 \right\}$$

$$= \frac{1}{N} E\{(x_i - m)^2\} = \frac{\sigma^2}{N} \tag{8.2}$$

8.4.1 Sampling to obtain the local mean value

Theoretical analyses have been published that deal with the question of signal sampling. Early work in this field includes that of Peritsky [11] and Lee [12]. Peritsky investigated the statistical estimation of the local mean power assuming independent Rayleigh-distributed samples, and Lee presented an analysis concerned with estimating the local mean power using an averaging process with a lowpass filter. Their work was based on a statistical estimation of the RMS and mean signal strength in volts, i.e. they assumed a receiver with a linear response. Practical measurements, however, are often taken using a receiver with a logarithmic response; then the signal samples are expressed directly in decibels relative to some reference value and estimates can be made directly from them. If we consider the case of a Rayleigh fading signal, it is possible to determine the number of independent samples N necessary to estimate the mean or median value within a certain confidence interval. The need for independent samples then enables us to relate N to the distance (length of travel) over which these samples should be obtained.

Increasing the sample size can make the estimate more accurate through a knowledge of the effects that sampling rate and measurement length have on the standard deviation of the estimate, but some care is needed. Simply increasing the number of samples is not sufficient since for a small measurement length they will not be independent and may be on an unrepresentative portion of the fading envelope.

Similarly, a long measurement length and a sampling rate that is insufficient to resolve the fading envelope would not adequately represent the local mean or median. It is necessary to have a sufficiently large sample size, and it is also necessary to take the samples over a measurement length that allows an accurate estimation of the required parameters.

Additionally, in the real fading environment, slow fading also exists and this will have an influence if the measurement distance is too large. It is necessary to take this into account in order to arrive at a compromise between measurement length and accuracy in practical measurements.

8.4.2 Sampling a Rayleigh-distributed variable

The relationships between linear and logarithmic samples of a Rayleigh-distributed variable are derived in Appendix B. A widely used parameter is the median value r_M of the logarithm of the signal strength. This can be obtained using $(2k + 1)$ samples and finding the sample above and below which there are exactly k samples. Alternatively, the mean value of the logarithm of the signal strength can be found. This is given by the mean of the dB-record:

$$E\{r_{dB}\} = \frac{1}{N} \sum_{i=1}^{N} 20 \log_{10} r \tag{8.3}$$

Note that $E\{r_{dB}\}$ depends on the values of all the samples of r. The relationship between this value and the value obtained from the mean of a linear receiver, i.e. $E\{r\}$, will be related through the statistics of the signal envelope (Rayleigh in this case) and in general this relationship will not be simple. This is not so for the median value, which is the same sample irrespective of whether the receiver response is logarithmic or linear. The median is widely used in mobile communications, firstly because it does not require a receiver with a characteristic that closely follows a predetermined law (say logarithmic or linear), merely one which can be calibrated with respect to any given reading. Secondly, the 50% cumulative distribution level is meaningful in estimating the quality of service in a given area.

8.5 MEAN SIGNAL STRENGTH

For estimation of mean signal strength in decibels, the distribution of the estimate is not known. The estimate is obtained from the sum of independent samples, and if the number of samples is sufficiently large, the distribution can be approximated by a Gaussian distribution, using the central limit theorem, irrespective of the distribution of the individual samples.

Let us write a standardised variable z, corresponding to a Gaussian variable x as

$$z = \frac{x - m}{\sigma}$$

The probability that z is less than a specified value Z is then

$$\text{prob}[z \leqslant Z] = P(Z) = \int_{-\infty}^{Z} \frac{1}{\sqrt{2\pi}} \exp\left(-\frac{z^2}{2}\right) dz \tag{8.4}$$

$P(Z)$ can be determined by reference to tables.

Now, in terms of the mean signal strength that we are trying to estimate,

$$\bar{z} = \frac{\bar{x} - \hat{m}}{\hat{\sigma}}$$

which, using eqns. (8.1) and (8.2), can be written as

$$\bar{z} = \frac{\bar{x} - m}{\sigma/\sqrt{N}} \tag{8.5}$$

Substituting this in eqn (8.4) we obtain

$$P(Z) = \text{prob}\left[\bar{x} \leqslant \frac{Z\sigma}{\sqrt{N}} + m\right] \tag{8.6}$$

8.5.1 Confidence interval

We are seeking to establish the number of signal strength samples, N, that are necessary in order that we can assert, with a given degree of certainty (often expressed as a percentage), that the mean value of these samples lies within a given range of the true mean. This range is called the *confidence interval* and can be found by confirming that

$$\text{prob}[-Z_1 \leqslant z \leqslant +Z_1] = \int_{-Z_1}^{+Z_1} p(z)\,\mathrm{d}z = 2P(Z_1)$$

We can now extend eqn. (8.6) to obtain

$$\text{prob}\left[\bar{x} - \frac{Z_1\sigma}{\sqrt{N}} \leqslant m \leqslant \bar{x} + \frac{Z_1\sigma}{\sqrt{N}}\right] = 2P(Z_1) \tag{8.7}$$

or alternatively

$$\text{prob}\left[-\frac{Z_1\sigma}{\sqrt{N}} \leqslant m - \bar{x} \leqslant \frac{Z_1\sigma}{\sqrt{N}}\right] = 2P(Z_1) \tag{8.8}$$

Table 8.1 has been compiled using Gaussian statistics and shows the range, in terms of σ, within which a given percentage of values fall. For example, 68% of values fall within $\pm\sigma$.

If we are dealing with samples taken from a receiver with a logarithmic characteristic then we know, from the relationships given in Appendix B, that

Table 8.1 Values of $P(Z_1)$ and confidence intervals

$P(Z_1)$	Range
68%	$\pm\sigma$
80%	$\pm1.28\sigma$
90%	$\pm1.65\sigma$
95.46%	$\pm2\sigma$
99%	$\pm2.58\sigma$

$\sigma = 5.57 \, \text{dB}$. Thus, $5.57/\sqrt{N}$ is the standard deviation of the sample average of N independent logarithmic samples. If we are interested in estimating within $\pm 1 \, \text{dB}$ then $(m - \bar{x}) = 1$ and for 90% confidence Z_1 is given by Table 8.1 as 1.65. Thus the number of independent samples required is obtained from

$$\frac{Z_1 \sigma}{\sqrt{N}} = 1 = \frac{1.65 \times 5.57}{\sqrt{N}} \quad \text{so} \quad N = 85$$

This is different from the number given by Lee [12].

If the samples are taken from a receiver with a linear characteristic then the mean and standard deviation are related to σ by eqns (5.21) and (5.23). Equation (8.8) still applies, but σ_r is now given by (5.23). Again, the mean value is approximately normally distributed and the sample size required to estimate within $2 \, \text{dB}$ ($\pm 1 \, \text{dB}$) with a 90% degree of confidence is given by

$$20 \log_{10}(m_r + 1.65 \hat{\sigma}_r) - 20 \log(m_r - 1.65 \hat{\sigma}_r) < 2$$

which yields $N = 57$. So the required sample size is greater when a logarithmic estimator is used.

It is now necessary to relate these sample numbers to the measurement distances over which they need to be taken. Assuming that, at the mobile, incoming multipath waves arrive from all spatial angles with equal probability, the correlation between the envelopes of signals measured a distance d apart is given by $J_0^2(2\pi d/\lambda)$, and for two adjacent samples to be uncorrelated this gives $d = 0.38 \lambda$. The minimum distances required are therefore approximately 33λ and 22λ for logarithmic and linear sampling, respectively.

Figure 8.6 is an example which shows the 95% confidence interval for the estimation of mean signal strength in decibels. The required sample size clearly

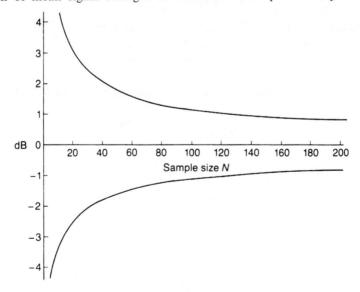

Figure 8.6 Relationship between 95% confidence interval and sample size for estimating mean signal strength (dB) in a Rayleigh fading environment.

depends on how accurately we wish to estimate the local mean. Since the confidence interval decreases very slowly for large N, a smaller confidence interval necessitates a very much larger number of samples and a correspondingly larger measurement distance. If the mobile is close to the base station or is in a radial street where a strong direct path exists, the fading may be Rician rather than Rayleigh and a smaller sample size may then be sufficient. On the other hand, if there are only a few multipaths so that the spatial arrival angle is non-uniformly distributed then longer distances may be necessary.

Figure 8.6 shows that estimation within ± 1 dB with 95% confidence requires about 125 samples. The corresponding measurement length at 900 MHz is 48λ, i.e. about 16 m. Experimental evidence has shown that 20–25 m is the maximum distance before the effects of slow fading become apparent so, for 95% confidence, ± 1 dB represents a practical limit on the accuracy with which the mean value can be measured.

8.6   NORMALISATION REVISITED

Before leaving the subject of signal sampling, we briefly clarify two different approaches to normalisation that appear in the literature. Some authors describe the local mean power of the fast fading as being lognormally distributed, whereas others use the lognormal distribution to describe the local mean signal voltage. These, quite clearly, are different assumptions and the implications can be explained as follows [13].

If the local mean power of the fast fading is lognormally distributed, the probability density function is

$$p(\overline{s^2}) = \frac{10}{\overline{s^2}\sigma_p \ln 10\sqrt{2\pi}} \exp\left(-\frac{(10\log\overline{s^2}-m_p)^2}{2\sigma_p^2}\right) \tag{8.9}$$

where m_p is the mean of the slow fading component (dB)
  σ_p is the standard deviation (dB)
  $\overline{s^2}$ is the mean power of the fast fading component

But if the local mean voltage is lognormally distributed then the PDF is

$$p(\overline{s}) = \frac{20}{\overline{s}\sigma_v \ln 10\sqrt{2\pi}} \exp\left(-\frac{(20\log\overline{s}-m_v)^2}{2\sigma_v^2}\right) \tag{8.10}$$

where m_v is the mean of the slow fading component (dB)
  σv is the standard deviation (dB)
  $\overline{s}$ is the mean of the fast fading voltage

Now, if the fast fading is Rayleigh distributed,

$$\overline{s^2} = \frac{4}{\pi}\overline{s}^2$$

Closed-form relationships between m_p, σ_p and $m_v\sigma_v$ can be obtained through the PDF transformation property:

$$p(\overline{s^2}) = p(\overline{s}) \left| \frac{\mathrm{d}\overline{s}}{\mathrm{d}\overline{s^2}} \right|$$

thus

$$p(\overline{s^2}) = \frac{10}{\overline{s^2}\sigma_v \ln 10\sqrt{2\pi}} \exp\left(-\frac{10\log\overline{s^2} - (m_v - 10\log\pi/4)^2}{2\sigma_v^2}\right) \qquad (8.11)$$

Equations (8.9) and (8.11) must be equivalent, hence

$$m_v - 10\log\frac{\pi}{4} = m_v + 1.049 = m_p$$

$$\sigma_v = \sigma_p \qquad (8.12)$$

The standard deviation is therefore the same whether the voltage or power is assumed to have a lognormal distribution. The difference between the two means is 1.049 dB, i.e. normalisation using mean power will yield a slow fading mean (dB) that is 1.049 dB greater than the mean (dB) obtained if normalisation is undertaken using the mean voltage.

8.7 WIDEBAND CHANNEL SOUNDING

In Chapter 6 it was shown that parameters such as the average delay, the delay spread and the coherence bandwidth are useful ways to characterise wideband radio channels and they provide relevant information for system designers. The scattering function can give an insight into the propagation mechanism. We now describe how these parameters can be measured.

The channel models in Chapter 5 [8,14] have been extended to consider the correlation between two spaced frequencies in the presence of time-delayed multipath, but in order to verify the models, either single-tone measurements have to be repeated at various frequencies over the band of interest, or an alternative sounding technique has to be used. A primary limitation of the single-tone sounding technique is its inability to illustrate explicitly the frequency-selective behaviour of the channel. In order to surmount this difficulty, a spaced-tone sounding method can be used, in which several frequencies (often two in practice) are transmitted at the same time.

The earliest measurements employing this technique were reported in 1961. There appears to be slight confusion in the literature as to who carried out these measurements: Clarke [8] credits Ossanna [14], but Gans [15] credits Hoffman, with Ossanna carrying out computational work. Although unpublished, this work formed the basis of support for Clarke's and Gans' theoretical scattering models for predicting the frequency coherence of multipath channels. Comparisons with frequency correlation functions obtained from wideband measurements [16] in urban New York City were also used to substantiate the theoretical models. However, the echo power-delay profiles were assumed to have a smooth exponential distribution as a function of time delay. Although this assumption is valid in some instances, there are times when the echo power profile contains echoes with significant energy arriving at large excess time delays. When this occurs the frequency correlation function is highly oscillatory and becomes a multivalued function [2,17]. Ambiguities in determining the frequency coherence of the channel can arise, depending on the separation of the transmitted tones. This limitation can be overcome by repeating the experiment and varying the frequency separation.

A study was carried out in the UK [18] using frequency separations between 50 kHz and 200 kHz. By sequentially stepping the tones across a band of frequencies, measurements of the channel frequency transfer function were obtained. This method provided a wideband measurement using relatively simple and inexpensive narrowband equipment, but it had two major drawbacks. Firstly, stepping a synthesiser over a large bandwidth in small steps is time-consuming, even using modern fast switching designs. Secondly, it is impossible to make mobile measurements using such a system due to the frequency stepping technique. Therefore, no Doppler shift and hence no angle-of-arrival information can be obtained, which precludes identification of significant single scatterers.

As an alternative to changing the frequency in discrete steps, a swept frequency (chirp) method can be used to excite the mobile channel. Although chirps are quite popular in high-resolution radars and HF ionospheric links [19], and they can be adapted for mobile use [20], they have not yet been used extensively in studies of mobile radio channels.

8.8   WIDEBAND SOUNDING TECHNIQUES

Channel sounding using a number of narrowband measurements (simultaneously or sequentially) is attractive from the viewpoint of equipment complexity, but has clear limitations. It is usually preferable to employ a genuine wideband sounding technique in which the transmitted signal occupies a wide bandwidth. Several methods are possible.

8.8.1   Periodic pulse sounding

When a pseudo-impulse (i.e. a short duration pulse) is used to excite the mobile propagation channel, the received signal represents the convolution of the sounding pulse with the channel impulse response. In order to observe the time-varying behaviour of the channel, periodic pulse sounding must be employed. The pulse repetition period has to be sufficiently rapid to allow observation of the time-varying response of individual propagation paths, while also being long enough to ensure that all multipath echoes have decayed between successive impulses. Figure 8.7 illustrates the technique, in which the duration of the pulse determines the minimum discernible path difference between successive echo contributions, while the repetition rate determines the maximum unambiguous time delay i.e. the

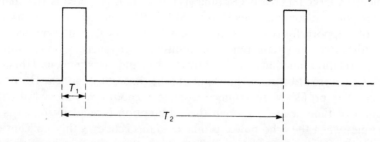

Figure 8.7   Periodic pulse sounding: T_1 = minimum echo-path resolution, T_2 = maximum unambiguous echo-path delay.

maximum distance for which an echo contribution can be unambiguously resolved. Periodic pulse sounding of the channel provides a series of 'snapshots' of the multipath structure, with successive snapshots forming a 'motion picture' representation of the multipath propagation between transmitter and receiver (either or both of which can be mobile).

The first reported study of the impulse response of the mobile radio propagation channel was by Young and Lacy [16] in urban New York City, at 450 MHz using a sounder with a pulse duration of 0.5 μs (equivalent spatial resolution = 150 m). A further study was carried out by Turin [3] in San Francisco using essentially the same method. Impulse response measurements were obtained using a 0.1 μs duration pulse (i.e. ~ 30 m spatial resolution) at carrier frequencies of 488, 1280 and 2920 MHz. In later studies by Van Rees [21,22] in Leidschendam, The Hague, impulse response measurements were obtained by transmitting a 10 W peak power pulse, at 910 MHz, every 100 μs from a moving vehicle. Pulse durations of 50, 100 [21] and 200 ns [22] were used, corresponding to spatial resolutions of 15, 30 and 60 m respectively.

All three systems used an envelope detection technique, so the phase information was discarded. But the phase information contains the angles of arrival of the echo paths in the form of Doppler shifts, and because this information was discarded, it was impossible to identify the sources of significant single scattering. The Doppler shifts can, of course, be determined by coherently demodulating the quadrature components of the received signal. Possibly the major limitation of the periodic pulse sounding technique is its requirement for a high peak-to-mean power ratio to provide adequate detection of weak echoes. Since, in general, pulsed transmitters are peak power limited, a possible way of overcoming this constraint is to use a sounding method which provides pulse compression.

8.8.2 Pulse compression

The basis for all pulse compression systems is contained in the theory of linear systems [23]. It is well known that if white noise $n(t)$ is applied to the input of a linear system, and if the output $w(t)$ is cross-correlated with a delayed replica of the input, $n(t - \tau)$, then the resulting cross-correlation coefficient is proportional to the impulse response of the system, $h(\tau)$, evaluated at the delay time. This can be shown as follows:

$$E[n(t)n^*(t - \tau)] = R_n(\tau) = N_0\delta(\tau) \tag{8.13}$$

where $R_n(\tau)$ is the autocorrelation function of the noise, and N_0 is the single-sided noise power spectral density. The system output is given by the convolution relationship

$$w(t) = \int h(\xi)n(t - \xi)\,d\xi \tag{8.14}$$

so the cross-correlation of the output and the delayed input is given by

$$E[w(t)n^*(t - \tau)] = E\left[\int h(\xi)n(t - \xi)n^*(t - \tau)d\xi\right]$$

$$= \int h(\xi)R_n(\tau - \xi)d\xi$$

$$= N_0h(\tau) \tag{8.15}$$

Therefore, the impulse response of a linear system can be evaluated using white noise, and some method of correlation processing.

In practice it is unrealistic to generate white noise, and as a result, experimental systems must employ deterministic waveforms which have a noise-like character. The most widely known examples of such waveforms are probably maximal length pseudo-random binary sequences (*m*-sequences), alternatively known as pseudo-noise (PN) sequences. These have proved extremely popular in communications, navigation and ranging system [24], since they are easily generated using linear feedback shift registers, and they possess excellent periodic autocorrelation properties [25], as illustrated in Figure 8.8.

8.8.3 Convolution matched-filter

One method of effecting pulse compression is to use a filter which is matched to the sounding waveform. This is known as the convolution matched-filter technique, and has been used in a study at 436 MHz [2] using an experimental surface acoustic wave (SAW) device to realise the matched filter. The principle is illustrated in Figure 8.9.

Because the SAW filter is matched to the specific *m*-sequence used in the transmitter, there is no requirement in this technique for local regeneration of the *m*-sequence at the receiver in order to produce the pulse compression. It can therefore be termed an asynchronous sounding technique and has many advantages in terms of cost and complexity. In addition the system operates in real time because the output of the matched filter is a series of snapshots of the channel response and amounts to a one-to-one mapping of time delays in the time domain. There are, however, several disadvantages which limit its appeal for channel sounding.

Firstly, the real-time information cannot be recorded without expensive equipment, and the consequent requirement for bandwidth reduction prior to recording necessitates the addition of special-purpose circuitry. Secondly, the performance of practical SAW devices is limited by deficiencies in the devices themselves. Specifically,

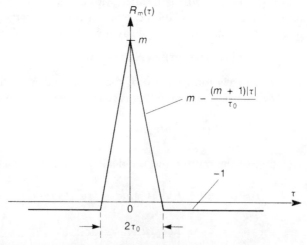

Figure 8.8 Periodic autocorrelation function of a maximal length pseudo-random binary sequence: τ = time delay, τ_0 = chip rate (clock period).

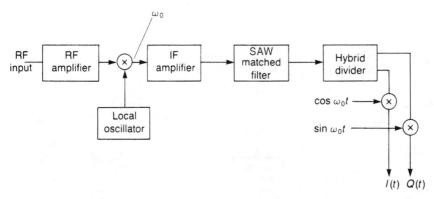

Figure 8.9 Principle of pulse compression using a matched filter.

long sequences are difficult to obtain and the generation of spurious acoustic signals gives rise to phenomena such as multiple reflection, bidirectional re-radiation and scattering of the surface acoustic waves. Also, since the devices are fabricated using standard photolithographic techniques, the placement accuracy in the mask-making process produces errors in the positioning of the interdigitated transducers. As excitation of the transducers is dependent on the accurate spatial position of the interdigitated structures, a degradation in performance arises. The combination of these effects causes time sidelobes to appear in the output of the matched filter and results in a reduced sensitivity to weak echoes.

8.8.4 Swept time-delay cross-correlation

As an alternative to convolution, it is possible to design a receiver in which the signal processing is based on correlation. Real-time correlation processing (equivalent to the convolution process previously described) would require a bank of correlators with infinitesimally different time delay lags, but clearly this is unrealistic.

In practice, correlation processing is often achieved with a single correlator, using a swept time-delay technique in which the incoming signal is correlated with an *m*-sequence identical to the sequence used at the transmitter, but clocked at a slightly slower rate. Time scaling (bandwidth compression) is inherent in this process; the scaling factor is determined by the difference in clock rates at transmitter and receiver. The essential blocks in such a receiver are illustrated in Figure 8.10. In an equivalent implementation, instead of clocking the receiver *m*-sequence at a slightly slower rate it is possible to use the same clock frequency but to reset the sequence after (*m* + 1) bits, so the two sequences at the transmitter and receiver pass each other on a step-by-step basis rather than drifting slowly and continuously.

The earliest impulse response measurements of the mobile radio channel using a swept time-delay cross-correlation (STDCC) sounder were obtained by Cox [1] in New York City at 910 MHz. In these experiments a 511-bit *m*-sequence, clocked at 10 MHz, was used to phase-reversal modulate a 70 MHz carrier. This modulated signal was then translated to the sounding frequency by mixing with an 840 MHz local oscillator, and was amplified to produce an average radiated power of 10 W. The signal was radiated from an omnidirectional antenna mounted at a fixed base

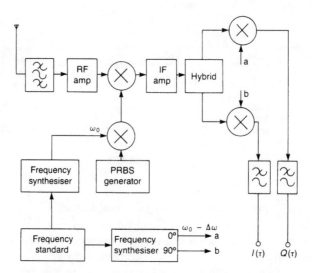

Figure 8.10 Principle of pulse compression using a cross-correlation process.

station site. All frequencies used in the transmitter were derived from a stable 5 MHz frequency standard. In the mobile receiver an identical 5 MHz standard was also used to derive all frequencies.

Figure 8.11 shows a receiver schematic in which, following front-end amplification and filtering, the received signal is translated down to 70 MHz by mixing with an 840 MHz local oscillator. The 70 MHz IF signal is then split in a wideband quadrature hybrid and applied to two correlators. In each correlator, an identical m-sequence to that formed in the transmitter, but clocked at a slightly slower rate (9.998 MHz), phase-reversal modulates a 70 MHz carrier. This signal is then multiplied with the IF signal from the quadrature hybrid. A lowpass integrating filter completes the cross-correlator.

The difference in the clock rates for the two m-sequences, Δf, determines the bandwidth of the cross-correlation function, in this case 2 kHz (i.e. 10 MHz − 9.998 MHz). This corresponds to a time scaling factor of 5000, which means that the features of 5000 individual responses are contained within each delay profile obtained at the output of the cross-correlator. Since it was not anticipated that path delays would exceed 15 μs, the slower receiver m-sequence was reset every 75 ms ($5000 \times 15\,\mu$s). At a constant speed of 1.4 m/s the vehicle would have travelled a spatial distance of approximately one-third of a wavelength of the transmitted carrier in this time. As a result, the 5000 individual responses are unlikely to have appreciably altered in their multipath structure. The validity of this argument was confirmed by Bajwa [2] as a result of observing the output of a matched-filter receiver. The bandwidth reduction inherent in this system easily allows data recording with conventional analogue tape recorders for later off-line analysis.

Demodulating the received signal in quadrature demodulators permits extraction of the Doppler shifts associated with each time-delayed echo. Accurate timing information was obtained by synchronising identical 10 MHz m-sequences in the transmitter and receiver, and by using stable frequency standards. This enabled, for the first time, the simultaneous measurement of time delays and Doppler shifts in multipath mobile radio channels.

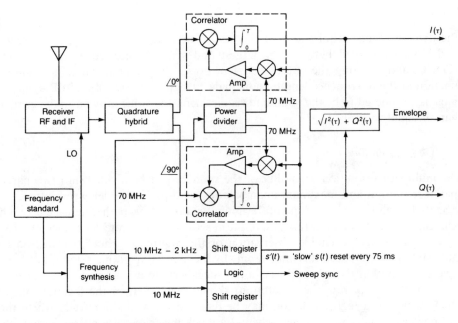

Figure 8.11 Channel sounder receiver as used by Cox.

There have been several further studies made, in the mobile radio [7,26–30] and microwave [31] fields, using the swept time-delay cross-correlator method. The measuring equipment in all these studies was, in essence, a replica of the system used by Cox, although some sounders employed envelope detectors instead of quadrature demodulators, because only investigations of the received envelope were required.

Although the information directly obtainable from these channel sounders is not exactly the same, they are all eventually equivalent. For example, the outputs of the periodic pulse sounder, the matched-filter convolution sounder and the swept (or stepped) time-delay cross-correlator are all the same and equal to $P(\tau)$. If a chirp technique is used then the output, after Fourier transform processing (an intrinsic requirement [19]) is also equal to $P(\tau)$.

8.9 SYSTEM REQUIREMENTS

The recent rapid growth in private mobile radio schemes, particularly cellular radio-telephony, has increased the need for accurate methods of assessing, and/or predicting, the performance of these radio systems. From a systems engineering standpoint, modulation schemes, data rates, diversity techniques, coding formats and equalisation techniques are of principal concern; from the standpoint of radio propagation modelling, the principal concern is to relate multipath phenomena to the local environment. An ideal channel sounder would be able to satisfy both standpoints simultaneously; however, due to the method of operation of practical channel sounders such as the STDCC, there is an interrelationship between the measured parameters (e.g. delays and Doppler shifts) such that to effect an improvement in one parameter may cause a degradation in another. This will become clearer in the following sections.

Dynamic range

The dynamic range requirement of the system depends on how large a difference needs to be observed between the largest and smallest received echoes. For an STDCC, and ignoring the effect of system noise, the dynamic range is a simple function of the m-sequence length and equals $20 \log_{10} m$. Hence, if a 30 dB dynamic range is considered to be the minimum requirement, the value of m has to be greater than or equal to 31.

Multipath resolution

The multipath resolution capability of the sounder can be divided into two parts: spatial resolution and maximum unambiguous echo-path time-delay resolution.

Spatial resolution is a measure of the minimum discernible path difference between echo contributions, and is a function of the m-sequence clock rate. The clock rate has to be high enough to enable observation of the multipath echoes (which lead to intersymbol interference); it is limited only by the highest operating rates of available logic gates. Within these bounds, i.e. a few megahertz to a few hundred megahertz, the choice of clock rate (i.e. resolution) should depend upon the location of the experiment, e.g. a high resolution may be required for an indoor study where the scatterers are very close together, whereas a much lower resolution would probably suffice for a study in rural, mountainous areas.

The maximum unambiguous echo-path time-delay which can be measured with an STDCC system is given by $m\tau_0$, the product of the length (in bits) and the clock period of the m-sequence. This must be sufficiently long to ensure that no echoes are detectable after this time.

Scaling factor

As stated in Section 8.8.4, the STDCC works by correlating two identical m-sequences that are produced at slightly different clock rates. This difference produces time scaling (bandwidth compression) of the cross-correlation function, where the scaling factor is the ratio of the highest clock rate to the frequency difference.

The choice of time scaling factor k may be thought arbitrary, depending only upon the final bandwidth required for data recording. However, Cox [1] found that severe distortion was produced in the cross-correlation function if k was set too low.

Doppler-shift resolution

To identify the location of scatterers, or scattering centres, it is necessary to determine the angles of arrival of echo paths in the form of Doppler shifts. The limit to which Doppler-shift information can be resolved depends upon the following factors:

- Vehicle velocity (v) and stability
- Carrier frequency (f_c) and stability
- Length (m) and clock period (τ_0) of the m-sequence
- The scaling factor (k) of the swept correlator

The maximum Doppler shift experienced by a mobile receiver moving with velocity v is given by

$$f_D = \frac{v f_c}{c} \qquad (8.16)$$

where c is the velocity of electromagnetic waves in free space. However, the maximum Doppler shift that can be measured using an STDCC is given by

$$f_D = \frac{1}{2km\tau_0} \qquad (8.17)$$

Comparing equations (8.16) and (8.17) gives

$$v = \frac{c}{2km\tau_0 f_c} \qquad (8.18)$$

Equation 8.18 shows that, for k, τ_0 and f_c fixed, v is inversely proportional to m. Therefore, although doubling m may be beneficial in resolving long time delays, it would mean halving the vehicle speed to permit equivalent Doppler-shift resolution. This may prove impractical due to the lower vehicle speed required. For $k = 5000$, $m = 127$, $\tau_0 = 0.1\,\mu s$ and $f_c = 900\,MHz$, the vehicle velocity would have to be less than or equal to 2.61 m/s ($\sim$ 6 mph). Increasing the m-sequence length to 255 would require a vehicle speed of less than 1.3 m/s (3 mph). The overall frequency resolution, however, will depend upon the stability of the frequency sources and the ability to maintain a constant vehicle speed throughout the measurement period.

8.9.1 Accuracy of frequency standards

The accuracy of time measurement and frequency generation depends on the performance of the frequency standards employed in the transmitter and receiver systems [32]. Furthermore, in a coherent system, their performance in relation to each other is paramount over their performance relative to a primary master source such as a caesium atomic standard. For perfect coherent signal demodulation, the injected carrier must be identical in both phase and frequency to that of the received signal. Phase synchronism, however, is impossible due to the random location of the mobile receiver, hence the need for quadrature detection.

Assuming that identical frequency multipliers are employed at the transmitter and receiver, the degree to which frequency synchronism can be achieved depends upon the magnitude of any frequency offset between the transmitter and receiver standards, and their stability.

Any small frequency difference between the standards causes a slow drift between the transmitter and receiver systems, and this sets two performance bounds for the channel sounder. Firstly, the drift causes a relative shift in timing, so that echoes with the same path delay no longer occupy the same time resolution cell. Therefore, there will be a maximum time of field trial operation before resynchronisation of the sounder is required. Secondly, the slow drift determines the lowest Doppler-shift frequency that can be unambiguously resolved. This has a bearing on how accurately the sounder can measure echo contributions arriving with angles close to 90° relative to the direction of motion.

A measurement period can be defined as the time it takes for a drift of a single time resolution bin (e.g. $\tau_0 = 0.1 \, \mu s$). If this period is to be of the order of 30 min, the frequency difference between the standards has to be of the order of 5.6×10^{-11} ($0.1 \, \mu s/30$ min) and the stability has to be good enough to maintain this difference over the 30 min period.

8.9.2 Phase noise in signal sources

Assuming that perfect frequency synchronism exists between transmitter and receiver, the outputs of two quadrature demodulators define a received vector with a constant amplitude and a fixed, arbitrary phase angle. However, this statement assumes that all frequency sources are ideal and produce outputs that are constant in both amplitude and frequency, whereas in practice all signal sources exhibit random perturbations in both amplitude and phase. The spurious amplitude modulation is usually very small and is generally ignored [33], but the phase modulation (phase noise) is important since it leads to a degradation in system performance, particularly in low-data-rate communications and Doppler radars.

The effect of phase noise in an STDCC system is to induce random amplitude fluctuations in the quadrature signal components. This can be best understood by considering the system to be both phase and frequency synchronous, such that all the received energy appears in the in-phase channel. The effect of any phase jitter is to cause random perturbations in the phase of the received vector. For narrowband phase modulation, the amplitude of the in-phase component will 'appear' fixed; however, a small component now appears in the quadrature channel. The result of this jitter will be to cause slight broadening of the measured Doppler spectral components.

8.10 A PRACTICAL SOUNDER DESIGN

Several studies have been undertaken using the STDCC method with sounders identical in form to that of Cox [1]. However, it is possible to improve the sounder design to obtain a reduction in circuit complexity.

The first change is in the transmitter, and involves removing the IF stage and directly modulating the RF carrier with the pseudo-random code. This has the advantages of obviating the need to synthesise the IF and eliminating the need to filter the RF signal in order to remove the unwanted sideband following the upconversion. The second change is in the receiver. Cox's sounder was essentially a direct implementation of the cross-correlator idea; that is, the received signal was translated to IF, split in a wideband quadrature hybrid, and finally demodulated in two correlators. An alternative approach is to multiply the received signal by the slower *m*-sequence at the same time as translation to IF. Demodulating two cophasal components of the IF signal with quadrature sinusoids, and applying the products to two lowpass filters results in outputs identical to Cox's. This approach, however, requires the multiplicative part of the cross-correlation process to be performed in a single place. Additionally, and more importantly, carrying out the multiplication coincident with RF-to-IF translation results in a reduced IF bandwidth from twice the clock rate to twice the difference in clock rates. The need for wideband components in the IF stage is thereby eliminated, and accuracy is improved since only the IF oscillator needs to be split into quadrature components.

Beyond this, however, for many applications such as characterising the channel inside buildings it is desirable to improve the time-delay resolution capability and furthermore, although the measurement of the Doppler spectrum is a desirable feature, it places very stringent requirements on the stability and phase noise specifications of the signal sources. Doppler measurements permit the identification of significant scatterers and scattering centres, but only receivers in fast-moving vehicles and railway trains are really subjected to performance degradation by Doppler effects. Hand-portable equipment carried by users on foot is almost completely unaffected. Some recent designs of channel sounder have therefore abandoned the implementation of Doppler-shift measurements, allowing them to incorporate further simplifications in design by using, for example, a logarithmic IF amplifier/envelope detector and high-stability crystal oscillators rather than rubidium frequency standards. Furthermore, in order to make rapid measurements in the field, data acquisition systems, which include on-board memory and AGC circuits, have been incorporated into receiver designs.

A simplified block diagram of the transmitter used in a recent design [30] is shown in Figure 8.12. A 30 MHz clock is used to drive a PRBS generator, and the resulting 511-bit sequence is used to phase-reversal modulate an 1800 MHz carrier. The output signal is passed through a filter having a 60 MHz bandwidth to reduce interference outside that band and after amplification is radiated using a discone antenna.

The receiver is shown in Figure 8.13. An attenuator having discrete steps is included for AGC purposes and a 29.992 MHz clock is used to drive the PRBS generator, which is identical to the PRBS generator in the transmitter. This produces an 8 kHz 'slip rate' and gives rise to a time scaling and hence bandwidth compression ratio of $30/(30-29.992) = 3750$. For a PRBS of length 511 bits, the power-delay profile duration of 17.1 μs is therefore recorded, after time scaling, in $(17.1 \times 10^{-6} \times 3750) = 64.125$ ms. The receiver IF is 10 MHz. Because the multiplicative element of the cross-correlation process is carried out during the frequency downconversion, the IF bandwidth is restricted to 19 kHz (just greater than 2 × the 'slip rate' of 8 kHz) by a crystal filter having a very sharp roll-off. The output is then passed to a logarithmic IF amplifier/envelope detector which has an 80 dB dynamic range and a 2 MHz bandwidth centred on 10 MHz. This also facilitates an easy implementation of the AGC design.

The data acquisition system (DAS) is designed to interface with the printer port of a notebook computer. The system is required to sample and store the output from the envelope detector (analogue) and the AGC (digital). An ADC is required to digitise the output of the envelope detector with a minimum sampling rate of 16 kHz to satisfy the Nyquist criterion. Due to the rate at which data needs to be collected, it is necessary to design an acquisition system with on-board memory.

8.10.1 Data processing

A block diagram of the data acquisition system (DAS) is shown in Figure 8.14. Information is stored in two halves of a 1 Mb × 16 DRAM memory, one half holding the sampled power-delay profiles and the other half holding, in corresponding memory locations, the setting of the AGC attenuator during the collection of any particular profile. The operation of the DAS and subsequently the downloading and analysis of data is controlled by a data acquisition and analysis software package mounted on a notebook computer interfaced with the receiver via its printer port. There is no need for an external ADC card within the computer, or any expansion slots. The receiver and the notebook computer are connected via a 25-way cable.

The associated software package has built-in analysis routines which compute all the CCIR-recommended small-scale time domain descriptors of the power-delay profiles, namely the average delay, delay spread, delay interval at 9, 12 and 15 dB below the maximum and delay windows at 50%, 75% and 90% of the total power. The package is

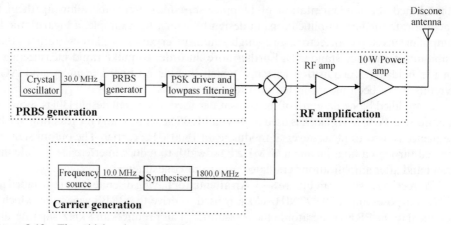

Figure 8.12   The wideband transmitter.

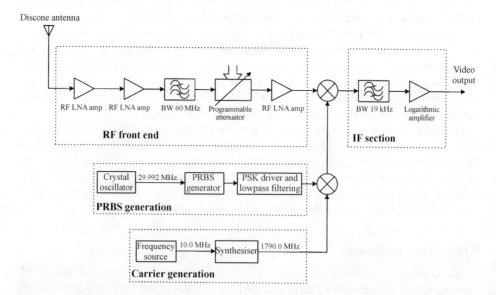

Figure 8.13   The wideband receiver.

also capable of averaging a user-specified number of profiles and will then calculate the average delay and the delay spread of that averaged profile. It is entirely menu driven with pop-up and pull-down menus and has extensive online help facilities.

With the program running, the operator uses the pull-down menu to select the measurement parameters to be set in the computer for the experiment about to be undertaken. These include the sampling rate (selectable by the user in the range 20–100 kHz) to be used during recording of the power-delay profiles. At a sampling rate of 40 kHz, a 26 s record can be stored and this equates to over 400 profiles when a profile is recorded in 64 ms.

The logic circuits needed to control the data acquisition process are implemented in the form of two programmable logic devices (PLDs) and communication between the computer and the DAS establishes the settings required. The main PLD clock at 16 MHz is suitably divided down in accordance with the required sample rate and a counter is set for the number of samples to be recorded. The measurement process is initiated from the computer, but because the sounder uses two PRBS with different clock rates, no synchronisation is possible. Moreover, because no synchronisation signal is provided to the data acquisition system by the sounder, sampling may not begin at the start of a profile. To overcome this problem, a feature has been included in the software to allow the user to locate, graphically, the position corresponding to the start of the first profile. The user can display a part of the recorded waveform equivalent to one profile duration and is prompted to move a blinking cursor to an appropriate position. Once selected, the sample number of this position is written to the data file to be used when analysing the data. No data analysis can take place until this has been completed.

Sampling of the measured video signal is via an 8-bit ADC; conversion takes place simultaneously with the clocking of data into the on-board memory. This is achieved

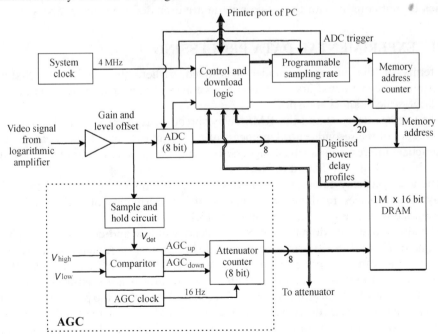

Figure 8.14 The data acquisition system.

by using the memory address clock produced from the PLD that performs the 'RAM controller' function. Suitable time delays and data latches are arranged to provide synchronisation.

For AGC purposes, an auto-ranging RF attenuator is inserted in the signal path. This has an insertion loss of 4 dB and the attenuation can be switched over a range of 64 dB in 4 dB steps. The way in which the attenuator is controlled is as follows: the video signal from the logarithmic amplifier is fed to a quasi-peak-follower circuit, the output of which is applied to a voltage level circuit connected as a window comparator. This provides an indication of whether an increase or decrease of attenuation is required to maintain the receiver within its 30 dB linear dynamic range. The clocking of the AGC up/down signal is controlled by an AGC clock which produces a pulse train with a period of approximately 70 ms. This is deliberately chosen to be slightly greater than the time used to record one power-delay profile (64 ms). The AGC can be disabled using a switch on the receiver for calibration procedures that need to be undertaken before a measurement campaign. The value of inserted attenuation is recorded concurrently with the measured signal data and stored in the memory as indicated above.

At the end of each measurement run, all information stored in the on-board memory is downloaded to the hard disk of the computer and all 'good' profiles are analysed. In this context a 'good' profile is one which meets two criteria:

- No change in attenuator setting has taken place during the recording of the profile
- The peak-to-spurious is at least 18 dB

When parameters are being computed for an 'averaged' profile, all individual profiles which do not comply with these two criteria are excluded from the analysis.

8.11   EXPERIMENTAL DATA PROCESSING

In terms of the two-stage model mentioned earlier, the small-scale channel descriptors are evaluated first, followed by averages of these parameters to estimate the large-scale channel statistics.

The average delay D and the RMS delay spread S, respectively the first and the square root of the second central moments of $P(\tau_i)$, are two time domain parameters of practical interest to systems designers. In terms of measured quantities, they are given by equations (6.75) and (6.76).

The average delay causes ranging errors in phase ranging systems, whereas the delay spread places fundamental limits on the performance of wide bandwidth transmissions over non-equalised channels [15,17].

The alternative time domain parameters, delay window and delay interval, given by eqns (6.77) and (6.79), can also be established from experimental data. Digital transmissions can be expected to produce satisfactory performance (i.e. low BER) with a carrier-to-interference ratio of about 10 dB, and it is recommended by the CCIR that delay intervals for thresholds 9, 12 and 15 dB below the peak value should be measured. Likewise, delay windows for 50, 75 and 90% of the total energy are suggested. It is important to recognise the existence of noise and spurious signals in the measuring system and to set an appropriate threshold for measurements. A safety margin of 3 dB is recommended to ensure the integrity of results and it is further suggested that only delay profiles in which the peak-to-spurious ratio exceeds 15 dB (excluding the 3 dB safety margin) are used to compute statistical parameters.

Several studies have been undertaken in an attempt to establish a 'figure of merit' for a given channel in terms of one or more of these measurable parameters [34].

8.11.1 Frequency domain characterisation

The frequency correlation function is a measure of the correlation between two spaced carrier frequencies. This function is easily evaluated from the values of $P(\tau_i)$ through fast Fourier transform (FFT) techniques [35]. The coherence bandwidth, defined as the maximum frequency difference for which two signals have a specified value of correlation, is a frequency domain parameter that is useful for assessing the performance of various modulation or diversity techniques. No definitive value of correlation has been established for the specification of coherence bandwidth but values of 0.9 ($B_{0.9}$), and 0.5 ($B_{0.5}$) are the two most popular. The resolution in the frequency domain, however, is related to the pulse repetition frequency (PRF) of the spread-spectrum sounding signal, which is defined as

$$\text{PRF} = \frac{1}{m\tau_0} \tag{8.19}$$

For $m = 127$ and $\tau_0 = 0.1\,\mu s$ the PRF is 78.74 kHz.

In his study in New York City, Cox [1] used a 511-bit m-sequence and a chip period of 0.1 μs, which provided a frequency resolution of ~19.6 kHz. The smallest values of $B_{0.9}$ and $B_{0.5}$ he reported [17] were 20 kHz and 55 kHz respectively. Obviously, the degree of confidence in these small coherence bandwidths, which are the most critical in terms of error performance, must be low. Furthermore, if the frequency resolution is insufficiently fine, detail may be lost in the estimation of the frequency correlation function, resulting in erroneous values for $B_{0.9}$ and $B_{0.5}$. In essence, this is the same problem that afflicts the spaced-tone sounding technique.

One obvious method of counteracting this problem is to increase the length of the m-sequence, thereby maintaining the same time resolution. However, there are penalties to be paid for adopting this approach, thus limiting the maximum practical value of m.

An alternative and more elegant solution has been proposed. Since it has been assumed that all distinguishable echoes in the power-delay profile occur within a certain time-delay window, e.g. 12.7 μs [2], then if the length of the m-sequence is doubled, the new power-delay profile will contain exactly the same information up to a delay of 12.7 μs with only the system noise floor extending to 25.4 μs. However, the new frequency resolution capability will improve from 78.74 kHz to 39.37 kHz. In practice, therefore, the frequency resolution capability can be improved by increasing the length of the time-delay window off-line, i.e. after completion of the field trials, by taking the system noise floor and extending it in time, prior to using the FFT.

As opposed to the case where m is physically increased, the only penalty of increasing the length of the time-delay window off-line is the increased time of computation. In reference 29 the power-delay profile was extended up to 204.8 μs, thus providing a frequency resolution of ~ 4.9 kHz. The value of 204.8 μs, for the sequence length, was obtained by continually increasing the length from 12.7 μs, until it was felt that the increase in computation time outweighed any further improvement in the estimates of coherence bandwidth. Figure 8.15 shows an average power-delay profile obtained under extreme multipath conditions and Figure 8.16 shows normalised frequency correlation functions derived from this profile with and without the use of off-line profile lengthening. The coherence bandwidth obtained without zero padding clearly underestimates the frequency selectivity of the channel.

8.11.2 Large-scale characterisation

The small-scale descriptors presented above are essentially measures of the channel response at 'single' locations. Obviously, systems engineers must design communication systems that will operate satisfactorily in a large variety of geographical locations; they therefore require measures of the variability in the small-scale channel descriptors over the large-scale area. Specifically, they need to know for what percentage of locations a specific level of performance can be maintained; that is, they require the cumulative distribution functions (or just distribution function) of each parameter. These can be obtained easily from sets of small-scale characteristics.

8.11.3 Summary

It appears that the swept time-delay cross-correlator method is the nearest to an optimum sounding technique. However, subtle changes to the receiver layout originally proposed by Cox result in a sounder with a simpler architecture. The interrelationship between all the system factors is such that great care must be taken when specifying an STDCC system, if it is to produce meaningful results. Zero-padding can be used in the calculation of the frequency correlation function, in order to achieve sufficiently fine frequency resolution. This is particularly important when the average power-delay profile contains significant echoes at large excess time delays.

8.12 RADIO CHANNEL SIMULATION

Testing radio communication systems in the field is time-consuming and expensive since there are uncertainties in the statistical variations actually encountered. Extensive trials

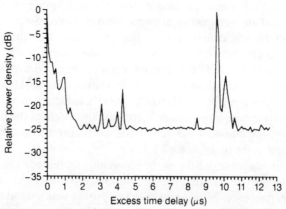

Figure 8.15 Average power-delay profile under severe multipath conditions: average delay $D = 4.595\ \mu s$, delay spread $S = 5.123\ \mu s$.

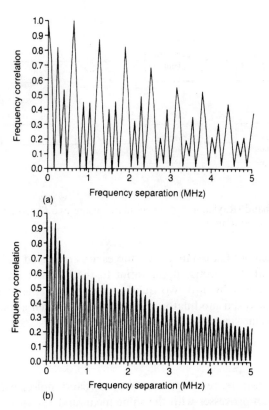

(a)

(b)

Figure 8.16 Normalised frequency correlation function of the average power-delay profile in Figure 8.15: (a) no profile extension, $B_{0.9} = 31.2\,\text{kHz}$, $B_{0.5} = 105\,\text{kHz}$; (b) with profile extension, $B_{0.9} = 11.8\,\text{kHz}$, $B_{0.5} = 27.1\,\text{kHz}$.

therefore have to be undertaken to ensure the results are truly representative of all the conditions likely to be encountered in practice. It is clearly attractive to test systems in the laboratory since conditions can then be tightly controlled, but it is very important to ensure that all the relevant properties of the signal can be adequately simulated. The major decision that has to be taken is whether to use a hardware or software simulation or a combination of both. It is also necessary to decide, in the light of the intended application, whether simulation of the channel as a frequency-selective medium is necessary (wideband simulation) or whether a simpler non-frequency-selective, multipath simulation (narrowband simulation) is sufficient.

We have seen earlier that the characteristics of mobile radio channels, although complex in nature, can often be adequately represented by known statistical distributions. Simulation therefore amounts to producing, in the laboratory, signals that have appropriate statistical properties. Once a simulator is available it can be used not only for testing existing systems, but also as a design tool in the development of new systems, coding and modulation schemes, and in the evaluation of equalisation and diversity techniques.

8.12.1 Hardware simulation of narrowband channels

Several simulators that reproduce the Rayleigh-distributed fast fading encountered in mobile radio channels have been based on the block diagram shown in Figure 8.17

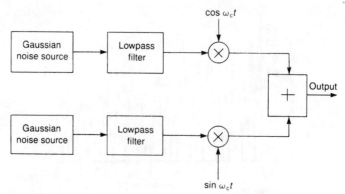

Figure 8.17   Narrowband (Rayleigh) fading simulator using two Gaussian noise sources and quadrature amplitude modulation.

[36–38]. Two independent Gaussian noise sources are connected to identical lowpass shaping filters so that the spectra at the input to the balanced modulators are the same. An RF source is split into two quadrature components and applied to the other ports of the balanced modulators, the outputs of which are added together.

In essence this circuit simulates the equation

$$n(t) = x(t) \cos \omega t - y(t) \sin \omega t \tag{8.20}$$

which is a well-known representation of narrowband noise; $x(t)$ and $y(t)$ are independent Gaussian processes with the same mean and variance. The output $n(t)$ has an envelope which is Rayleigh distributed and a phase which is uniformly distributed in the interval $(0, 2\pi)$, as required.

Simulation of the fading spectrum appropriate to mobile radio is obtained by properly shaping the spectrum of the two noise sources, i.e. by choosing appropriate characteristics for the two shaping filters. It is important to remember, in this context, that although the spectrum of a Gaussian process is affected by filtering, the PDF is not, so the process at the output of the shaping filter remains Gaussian. The required spectrum depends on the assumption made about the angle-of-arrival statistics and the radiation pattern of the receiving antenna, but for isotropic scattering and an omnidirectional antenna the spectrum is represented by eqn. (5.15) and is illustrated in Figure 5.9.

It is impossible to design a filter that truly follows the shape represented in Figure 5.9, so approximations have to be sought. In early implementations [36] active analogue filters were used to produce a suitable characteristic with a cut-off frequency equal to the maximum Doppler frequency. A practical simulator would need several such filters to simulate different vehicle speeds but digital filters can provide built-in flexibility. Tests on simulators of this kind show the output signal envelope to be a close approximation to a Rayleigh distribution. The phase is uniformly distributed and the level crossing rates and average fade durations agree well with theoretical predictions.

The slow lognormal fading that characterises mobile radio channels can be incorporated into simulators of this type by an additional unit. Such units are usually based on generating a signal $x(t)$ with Gaussian statistics and subsequently obtaining a lognormal signal $L(t)$ using the transformation

$$L(t) = 10^{x(t)/20}$$

Practical fading simulators often have two fading channels with a facility to set the correlation coefficient between them. This aids investigations of diversity reception techniques and the effects of co-channel and adjacent channel interference.

An alternative hardware implementation of a Rayleigh fading simulator is possible using a model presented by Jakes [39, Ch 1]. Again, the basic idea is to generate two quadrature signals as represented by eqn. (8.20) and to add them together to produce a signal with a Rayleigh envelope and uniform phase. The mathematical model leads to the implementation shown in Figure 8.18.

Here N_0 low-frequency oscillators with angular frequencies equal to the Doppler shifts $\omega_m \cos(2\pi n/N)$, $n = 1, 2, \ldots, N_0$, together with one oscillator at frequency ω_m, are used to generate signals that are added together and modulated on to quadrature carriers. The amplitudes of all oscillators are the same (say unity) with the exception of the oscillator at ω_m, which has relative amplitude 0.707. The phases β_n are appropriately chosen so that the PDF of the resultant phase approximates to a uniform distribution. Figure 8.18 shows the relationships that exist with $N_0 = 8$. The

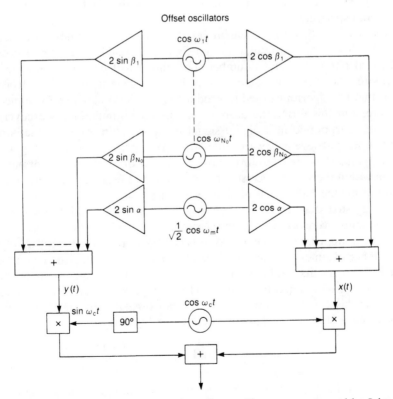

Figure 8.18 Rayleigh fading simulator using offset oscillators, as proposed by Jakes.

proper amplitude and phase relationships are provided by amplifiers with gains of $2\cos\beta_n$ or $2\sin\beta_n$.

It is apparent from the diagram that

$$x(t) = 2\sum_{n=1}^{N_0} \cos\beta_n \cos\omega_n t + \sqrt{2}\cos\alpha\cos\omega_m t \qquad (8.21)$$

$$y(t) = 2\sum_{n=1}^{N_0} \sin\beta_n \cos\omega_n t + \sqrt{2}\sin\alpha\cos\omega_m t \qquad (8.22)$$

where $\beta_n = \pi n/N_0$, $\omega_n = \omega_m \cos(2\pi n/N)$, $\omega = 2\pi v/\lambda$ and $N = 2(2N_0 + 1)$.

The phase of the output $n(t)$ has to be random and uniformly distributed in the range $(0, 2\pi)$. To achieve this it is necessary to ensure that $\langle x^2 \rangle \approx \langle y^2 \rangle$ and $\langle xy \rangle \approx 0$. It is interesting that Jakes defines β_n in two slightly different ways ($\pi n/N_0$ in his Fig. 1.7-2 and $\pi n/(N_0 + 1)$ at the top of p. 72). It is not a question of one being correct and the other incorrect, but they do lead to slightly different results. If $\pi n/N_0$ is used with $\alpha = \pi/4$, then $\langle x^2 \rangle = \langle y^2 \rangle$ and $\langle xy \rangle \approx 0$; alternatively if $\pi n/(N_0 + 1)$ is used with $\alpha = 0$, then $\langle x^2 \rangle \approx \langle y^2 \rangle$ and $\langle xy \rangle = 0$. In practice the effect on the distribution of $n(t)$ is insignificant, although the second definition which requires $\alpha = 0$ is probably easier to implement. In a practical embodiment, excellent agreement has been obtained between the theoretical and experimental envelope distributions, autocorrelation functions and spectra.

It is very cumbersome to implement Jakes' model in the physical form shown in Figure 8.18. It is time-consuming and costly and to build a number of oscillators with identical amplitudes, and a number of amplifiers with carefully controlled gains. Furthermore, the resulting equipment is likely to be rather bulky. It is clear, however, that the algorithms used to produce $x(t)$ and $y(t)$ are readily implemented in software, and this forms the basis for a computer simulation. Moreover, if the software is incorporated in a digital signal processor chip, we have the basis of a flexible hybrid (software/hardware) simulator that combines the advantages of both forms of simulation. Indeed it is possible to incorporate a lognormal fading algorithm within the same processor to obtain a versatile and flexible instrument for the study of mobile radio transmission techniques.

Jakes suggested that in applications where a number of independent Rayleigh fading simulators were needed, e.g. in the simulation of a frequency-selective channel modelled by a tapped delay line, it was unnecessary to replicate the set-up of Figure 8.18 a number of times. He suggested that a *single set* of oscillators could be used, provided arrangements could be made to ensure that a number of independent outputs $n(t)$ were obtained from them. The arrangement suggested by Jakes, and illustrated by his Fig 1.7-7, was to give the nth oscillator in the jth simulator an additional phase shift $\beta_{nj} + \gamma_{nj}$, where one possibility was

$$\beta_{nj} = \frac{\pi n}{N_0 + 1} \quad \text{and} \quad \gamma_{nj} = \frac{2\pi(j - 1)}{N_0 + 1}$$

An interest in using the equations that govern the operation of the simulator as part of a software simulation has, however, cast some doubt on the validity of this approach [40]. Although it appeared that the programming involved in simulating a wideband channel would be considerably simplified if the above equations were used to generate several independent Rayleigh fading signals from one set of oscillators, in practice it was found that quite high correlations (>0.6) existed.

The same investigation revealed what appears to be an error in Jakes Fig. 1.7-7; the values of phase shift should use β_{nj} in all cases, instead of $\beta_{nj}/2$. Faced with this result, it was decided to implement the software simulator using large, arbitrarily chosen time delays instead of the suggested phase shifts; in other words $x(t)$ and $y(t)$ were calculated using $\cos \omega_n(t + \Delta)$ where Δ is fairly large. The justification for this is that the autocorrelation function of $n(t)$ is approximated by a zero-order Bessel function of the first kind and the value of this function is very small for large values of delay.

8.13   WIDEBAND CHANNELS

8.13.1   Software simulation

A software simulation of a wideband channel can be based on the model proposed by Turin *et al.* [3]. The model assigns statistical distributions to the perceived features of the propagation medium; it was investigated by Suzuki [41] and refined by Hashemi [42]. The multipath medium is modelled as a linear filter with a complex-valued impulse response given by

$$h(t) = \sum_{k=0}^{\infty} A_k \delta(t - t_k) \exp(\mathrm{j}\phi_k) \tag{8.23}$$

This model is quite general and can be used to obtain the response of the channel to any signal $s(t)$ by convolving it with $h(t)$. It represents the channel in terms of a set of amplitudes $\{A_k\}$, echo arrival times $\{t_k\}$ and phases $\{\phi_k\}$; $h(t)$ represents attenuated, delayed and phase-shifted echoes of a transmitted pulse.

Hashemi [42] assumed a priori that the signal phases $\{\phi_k\}$ are uniformly distributed in the interval $(0, 2\pi)$. In order to determine the statistical properties of the amplitude sequence $\{A_k\}$ and the arrival time sequence $\{t_k\}$, he envisaged a hypothetical experiment in which a vehicle travelling along a city street takes samples of the channel impulse response at various points along the route. Such data was available from Turin's experiments [3] at frequencies of 488, 1280 and 2920 MHz. The envelopes of the signals received in a moving vehicle were recorded in the form of photographs from an oscilloscope display. These photographs were optically scanned and a series of $\{A_k, t_k\}$ pairs were obtained for each echo profile. Experiments were conducted in a heavily built-up area, the centres of a medium-sized city and a medium-sized town, and in residential suburbs. This data formed the basis for the work of Suzuki and Hashemi.

The basis of simulation is now clear. A series of impulse responses are generated according to equation (8.23) using characterisation parameters that change from one profile to the next. The time-varying impulse response of the channel is constructed from a number of successive profiles. Clearly the objective is to produce a simulation program that generates profiles having statistics that conform very closely to those empirically determined in respect of the correlation between the variables of spatially adjacent profiles, the temporal correlation of variables within the same profile, and small and large area fluctuations of the channel statistics.

Simulation of arrival times

Because the buildings and other obstacles that give rise to echoes of the transmitted signal are randomly located, it is tempting to describe the arrival times in terms of a Poisson distribution. This hypothesis, however, did not conform with the observed results, and Turin suggested an alternative second-order model, in the form of a modified Poisson process that was further developed and refined by Suzuki and Hashemi. It is termed the Δ–k model and embraces the fact that echoes may arrive in groups from closely spaced buildings.

The original Δ–k model has two states, S1 in which the mean arrival rate of echoes is $\lambda_0(t)$ and S2 in which the mean arrival rate is $k\lambda_0(t)$. The process starts in state S1; if an echo arrives at time t, a transition is made to S2 for a time $(t, t + \Delta)$, i.e. the mean arrival rate is changed for the next Δ seconds, where k and Δ are parameters to be chosen. If $k > 1$ the probability that an echo will occur in the next Δ seconds is increased; the converse is the case if $k < 1$. The value of k determines whether the echoes cluster together or spread out. Hashemi refined the model by using discrete time intervals Δ of 100 ns, which were called *delay bins*. He also attempted to describe the spatial correlation of arrival time sequences between adjacent profiles. It is assumed that no more than one echo exists in any delay bin. The mean or underlying echo arrival rate λ_0 assumes the value λ_j for the jth bin.

The model is illustrated in Figure 8.19. The probability that an echo exists in bin 1 is λ_1, so the probability of no echo is $(1 - \lambda_1)$. If an echo exists in the $(j-1)$th bin then the probability of an echo in the jth bin is $k\lambda_j$. As an example the probability of having echoes in bins 1, 2 and 4 but not in bin 3 is $\lambda_1 k\lambda_2(1 - k\lambda_3)\lambda_4$. In order to fit the Δ-k model to the experimental data, the underlying probabilities λ_j need to be determined from the empirical probabilities determined from the experimental data.

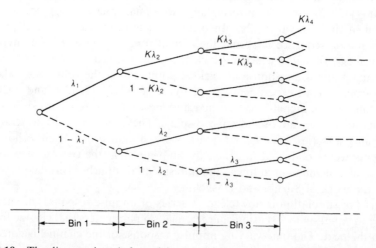

Figure 8.19   The discrete-time Δ–k model.

Having done this, Hashemi produced 'probability of occupancy' curves (i.e. probability of observing a path in any given bin) and 'path number' distributions (i.e. probability of observing n echoes in N bins). Typical examples are shown in Figure 8.20, which indicates a very close correspondence between simulation and experiment.

Simulation of amplitude and phase distributions

Turin originally concluded that over large global areas the signal amplitudes followed a lognormal distribution. Further analysis by Suzuki, however, led to a modification of this for the earlier echoes which appeared to follow a Nakagami distribution. Nevertheless, because of computational and other difficulties, Hashemi decided to use the lognormal distribution to generate all amplitudes. Factors which influenced this choice included the need to simulate correlation between successive echo amplitudes in the same profile (temporal correlation) and correlation between amplitudes in successive profiles (spatial correlation). It was assumed a priori that phases are uniformly distributed in the interval $(0, 2\pi)$.

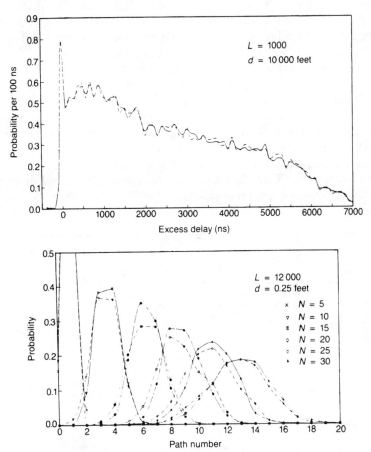

Figure 8.20 Experimental (——) and simulation (———) curves for (a) probability of occupancy and (b) path number, the probability of observing n echoes in N bins (after Hashemi).

For the first profile, the means and variances were generated to fit lognormal distributions having parameters estimated from the experimental data. Using these means and variances the amplitude (dB) of the echo in the first bin was generated from a normal distribution, and the amplitude for the jth bin was generated from a conditional normal distribution, the condition being the amplitude of the $(j-1)$th path. The temporal correlation between the jth and $(j-1)$th bins was made a decreasing function of the difference in arrival times of these two bins. For subsequent profiles a modified procedure was invoked in order to introduce spatial correlation (not relevant in the first profile). For the mth profile the mean and variance for any specific bin were generated according to empirically determined lognormal distributions if there was no echo in the same bin of the $(m-1)$th profile; or if such a path existed, using a conditional lognormal distribution, the condition being the amplitude of the corresponding echo in the $(m-1)$th profile.

After the means and variances had been calculated in this manner, the echo amplitudes for the mth profile were calculated in the following way. For the first bin a bivariate normal distribution was used, taking into account the spatial correlation with the $(m-1)$th profile. Generating the amplitude for an echo in the jth bin is more complicated because both spatial and temporal correlation have to be taken into account. This was achieved using a three-dimensional normal distribution. This aspect of the simulation was evaluated by producing large numbers of echo strength (path strength) distributions; some examples are shown in Figure 8.21. Again, agreement between simulation and experiment is very close.

Finally, Hashemi combined his arrival time, amplitude and phase simulations into a program that ran on a large mainframe computer. He used it successfully to simulate the fading of a CW signal. Investigation has verified that the simulation produced results representative of a narrowband fading signal and could be used to demonstrate the frequency-selective nature of a wideband channel [43].

The model relies heavily on the realistic simulation of echo arrival times, and the modified Poisson sequence appears to yield logical results. The probability of occupancy

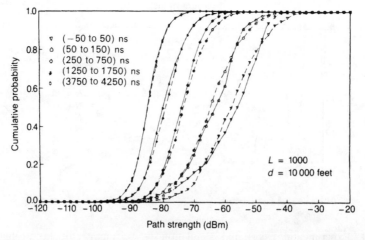

Figure 8.21 Some path strength distributions for selected excess delay intervals (after Hashemi).

curves show that a line-of-sight path is much more likely in a residential area than in a heavily built-up area, and this is intuitively the case. The technique relies on the generation of random numbers, which confines the model to use in software simulations.

8.13.2 Hardware simulation

The tapped delay line representation in Figure 6.3 provides the basis of a hardware simulator [44,45]. An original implementation incorporates both lognormal fading and weighting to give the generic model shown in Figure 8.22. The questions that arise relate to the appropriate correlation between the Rayleigh and lognormal modulators attached to the various taps, and the relationship between the various weighting factors.

Experimental evidence shows that, after normalisation to remove slow fading effects, the small-scale amplitude variations, particularly for paths with delays less than 1 μs, can be very accurately modelled by a Rayleigh distribution. When plotted in decibels, the large-scale variations approximate to a normal distribution, although departures are apparent for paths with longer time delays [29]. This finding tends to confirm the assumption made by Turin [3].

Figure 8.23 shows average correlation coefficients ($\pm \sigma$) between the small-scale amplitude fluctuations in neighbouring time-delay bins as given by Demery [29]. It shows very clearly that these fluctuations are almost completely uncorrelated, pointing to the need for independent Rayleigh modulators attached to each tap of the delay line, and it confirms that the small-scale signal variations are consistent with the GWSSUS model. As far as the large-scale amplitude fluctuations are concerned, the average correlation coefficients measured for neighbouring cells are shown in Figure 8.24.

In contrast to the small-scale fluctuations, there is significant correlation between adjacent time-delay bins. Although the correlation coefficients for separations greater than 1 time-delay bin are larger than their small-scale counterparts, the values are less than 0.5; this indicates that these fluctuations are only weakly correlated. The amplitude fluctuations in each time-delay cell can therefore be well

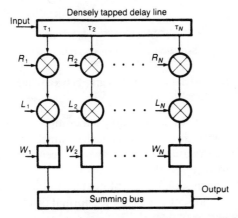

Figure 8.22 Schematic of the tapped delay line simulator for wideband multipath channels: the R_i are independent zero-mean complex Gaussian modulators, the L_i are zero-mean lognormal modulators, and the W_i are weighting factors.

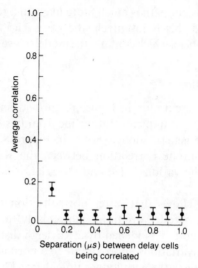

Figure 8.23 The average correlation coefficients between the small-scale amplitude fluctuations in neighbouring time-delay cells.

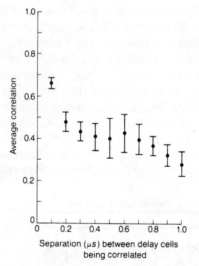

Figure 8.24 The average correlation coefficients between the large-scale amplitude fluctuations in neighbouring time-delay cells.

approximated as uncorrelated Rayleigh fading superimposed on partially correlated lognormal fading. For simplicity, in a practical channel simulator it is probably adequate to provide for correlation between two adjacent taps only. It remains to establish the weighting to be applied to each tap.

Figure 8.25 shows the mean signal levels plotted as a function of excess time delay together with curves that represent $\pm \sigma$ of the lognormal distribution within that cell. It can be seen that the mean signal level decreases continuously with increasing time delay, which is as expected. Figure 8.25 also provides information that is important when implementing the model with a limited number of taps, because it explicitly

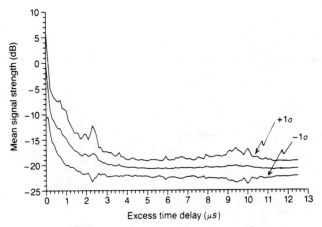

Figure 8.25 The large-scale mean signal strength (i.e. weighting factor) versus excess time delay.

indicates the values of delay that are most essential for an accurate channel simulation. To highlight this point, assume that 12 taps are available in the simulator. Figure 8.25 shows that the majority of significant echoes arrive with delays less than $2\,\mu s$. It might be reasonable therefore to assign 6 taps to cover this time-delay period. The remaining taps can then be assigned to provide delays of 2, 2.3, 3.6, 5.9, 9.4 and $10.0\,\mu s$; the selection is based on examining Figure 8.25 to identify echoes of significant amplitude or time-delay bins where the standard deviation is large, i.e. where the signal variability is greatest. Using fixed time delays in a channel model conflicts somewhat with the concept of a real channel, but identifying significant delay cells using information derived from graphs such as Figure 8.25 is considerably easier, and more realistic, than computing them from Poisson-distributed random numbers. The weighting to be applied to each tap can then be derived, and the weighting can be implemented using attenuators or amplifiers depending upon the reference point used and the required value of signal level.

Practical channel sounders based on a tapped delay-line representation of the channel have been built and tested. Caples *et al.* [44] described a design operating at an intermediate frequency (230–370 MHz) in which the delay line was implemented using a SAW device. The line produced six delayed versions of the incident signal with delays up to $9.3\,\mu s$, any three of which could be selected. Four multipath components were thereby available. The outputs from the selected taps were modulated by independent Rayleigh modulators only; there were no lognormal modulators. Weighting of each output was necessary to set the required level and to compensate for losses in the SAW delay line. Tests showed that the simulator performance was in close agreement with theoretical predictions and experimental observations.

Modular systems exemplified by Figure 8.26 have also been developed. The RF unit downconverts the incoming signal to an intermediate frequency of 35 MHz and upconverts the output of the delay section which is the major subsystem. The delay section itself consists of a chain of fixed $5\,\mu s$ delay lines (SAW devices) with fading modules as shown. Each fading module contains another chain of four $1\,\mu s$ delay lines (also SAW devices); Rayleigh fading is achieved using complex modulators.

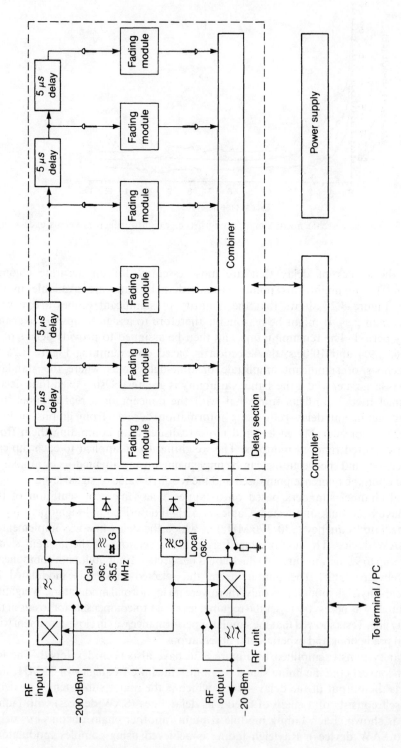

Figure 8.26 Block diagram of a modular wideband fading simulator.

Again the Rayleigh fading modulators are independent, but there is no lognormal component. The spectrum of the Rayleigh components simulates a uniform spatial arrival angle – the 'classical' distribution with both generation and filtering accomplished using a digital signal processor. The architecture allows the simulation of delay profiles with a minimum delay tap spacing of $1\,\mu s$ and a maximum determined by the number of $5\,\mu s$ sections installed. It has a usable bandwidth of 3.5 MHz.

The major limitations of these fading simulators are the absence of lognormal fading on each tap and the coarse time-delay resolution. The first limitation is easily rectified, because it is possible to incorporate both Rayleigh and lognormal fading within the same digital signal processor. The coarse time-delay resolution presents no problems in principle, because delay lines with closely spaced taps are readily available. The real problem is the actual number of delay paths that are to be used. Each path requires its own modulator (Rayleigh plus lognormal) and weighting network. To produce a realistic simulation, it is desirable to have a large number of paths, but this is very expensive. It seems from the work of Demery [29] that 12 taps would provide an extremely accurate delay profile, and in practice 8 would probably suffice.

REFERENCES

1. Cox D.C. (1972) Delay-doppler characteristics of multipath propagation at 910 MHz in a suburban mobile radio environment. *IEEE Trans.*, **AP20**(9), 625–35.
2. Bajwa A.S. and Parsons J.D. (1982) Small-area characterisation of UHF urban and suburban mobile radio propagation. *Proc. IEE Part F*, **129**(2), 102–9.
3. Turin G.L., Clapp F.D., Johnston T.L., Fine S.B. and Lavry D.A. (1972) Statistical model of urban multipath propagation. *IEEE Trans.*, **VT21**(1), 1–9.
4. Ibrahim M.F. and Parsons J.D. (1983) Signal strength prediction in built-up areas. Part 1: median signal strength. *Proc. IEE Part F*, **130**(5) 377–84.
5. Bultitude R.J.C. (1987) Measurement, characterisation and modelling of indoor 800/900 MHz radio channels for digital communications. *IEEE Commun. Mag.*, **25**(6), 5–12.
6. Feeney M.T. (1989) The complex narrowband mobile radio channel. PhD thesis, University of Liverpool.
7. Davies J.G. (1997) Propagation of radio signals into and within multi-storey buildings at 900 MHz and 1800 MHz. PhD thesis, University of Liverpool.
8. Clarke R.H. (1968) A statistical theory of mobile radio reception. *Bell Syst. Tech. J.*, **47**(6), 957–1000.
9. Parsons J.D. and Ibrahim M.F. (1983) Signal strength prediction in urban areas. Part 2: signal variability. *Proc. IEE Part F*, **130**(5), 385–91.
10. Davis, B.R. and Bogner R.E. (1985) Propagation at 500 MHz for mobile radio. *Proc. IEE Part F*, **132**(8), 307–20.
11. Peritsky M.M. (1973) Statistical estimation of mean signal strength in a Rayleigh-fading environment. *IEEE Trans.*, **COM21**(11), 1207–13.
12. Lee W.C.-Y. (1985) Estimate of local average power of a mobile radio signal. *IEEE Trans.*, **VT34**(1), 22–7.
13. Turkmani A.M.D. Unpublished work.
14. Ossanna J.F. (1964). A model for mobile radio fading due to building reflections: theoretical and experimental fading waveform power spectra. *Bell Syst. Tech. J.*, **43**, 2935–71.
15. Gans M.J. (1972) A power-spectral theory of propagation in the mobile radio environment. *IEEE Trans.*, **VT21**(1), 27–38.
16. Young W.R. and Lacy L.Y. (1950) Echoes in transmission at 450 megacycles from land-to-car radio units. *Proc. IRE*, **38**, 255–8.

17. Cox D.C. and Leck R.P. (1975) Correlation bandwidth and delay spread multipath propagation statistics for 910 MHz urban mobile radio channels. *IEEE Trans.*, **COM23**(11), 1271–80.
18. Matthews P.A. and Molkdar D. (1987) Wideband measurements of the UHF mobile radio channel. *Proc. ICAP'87 (IEE Conference Publication 274)*, Pt 2, pp. 73–6.
19. Salous S. (1986) FMCW channel sounder with digital processing for measuring the coherence of wideband HF radio links. *Proc. IEE Part F*, **133**(5), 456–62.
20. Salous S., Nikandrou N. and Bajj N. (1995) An ASIC solution for mobile radio channel sounders. *Proc. IEEE Int. Conf. on Electronics, Circuits and Systems*, Amman, Jordan, pp. 451–5.
21. Van Rees J. (1986) Measurements of impulse response of a wideband radio channel at 910 MHz from a moving vehicle. *Electron. Lett.*, **22**(5), 246–7.
22. Van Rees J. (1987) Measurements of the wideband radio channel characteristics for rural, residential, and suburban areas. *IEEE Trans.*, **VT36**(1), 2–6.
23. Papoulis A. (1965) *Probability, Random Variables and Stochastic Processes*. McGraw-Hill, New York.
24. Simon M.K., Omura J.K., Scholtz R.A. and Levitt B.K. (1985) *Spread Spectrum Communications* (3 vols). Computer Science Press, Rockville MD.
25. Sarwate D.V. and Pursley M.B. (1980) Crosscorrelation properties of pseudorandom and related sequences. *Proc. IEEE*, **68**(5), 593–619.
26. Nielson D.L. (1978) Microwave propagation measurements for mobile digital radio application. *IEEE Trans.*, **VT27**(3), 117–31.
27. Devasirvatham D.M.J (1986) Time delay spread and signal level measurements of 850 MHz radio waves in building environments. *IEEE Trans.*, **AP34**(11), 1300–5.
28. Sass P.F. (1983) Propagation measurements for UHF spread spectrum mobile communications. *IEEE Trans.*, **VT32**(2), 168–76.
29. Demery D.A. (1989) Wideband characterisation of UHF mobile radio channels in urban areas. PhD thesis, University of Liverpool.
30. Nche C. (1995) UHF propagation measurements for future CDMA systems. PhD thesis, University of Liverpool.
31. Linfield R.F., Hubbard R.W. and Pratt L.E. (1976) *Transmission channel characterisation by impulse response measurements*. US Department of Commerce, Office of Telecommunications Report OT76-96.
32. Kartaschoff P. (1978) *Frequency and Time*. Academic Press, New York.
33. Robins W.P. (1982) *Phase Noise in Signal Sources*. Peter Peregrinus, London.
34. Ladki M. (1991) The determination of a figure of merit for the wideband mobile radio channel. PhD thesis, University of Liverpool.
35. Elliot D.F. and Rao K.R. (1982) *Fast Transforms: Algorithms, Analyses, Applications*. Academic Press, New York.
36. Arredondo G.A., Chriss W.H. and Walker E.H. (1973) A multipath simulator for mobile radio. *IEEE Trans.*, **VT22**(4), 241–4.
37. Comroe R.A. (1978) All-digital Rayleigh fading simulator. *Proc. Nat. Electron. Conf.*, **32**, 136–9.
38. Ball J.R. (1982) A real-time fading simulator for mobile radio. *Radio and Electronic Engineer*, **52**(10), 475–8.
39. Jakes W.C. (ed.) (1974) *Microwave Mobile Communications*. John Wiley, New York.
40. Ladki M. Unpublished work.
41. Suzuki H. (1977) A statistical model for urban radio propagation. *IEEE Trans.*, **COM25**(7), 673–80.
42. Hashemi H. (1979) Simulation of the urban radio propagation channel. *IEEE Trans.*, **VT28**(3), 213–25.
43. Natarajan N. (1989) Software-based wideband channel simulator. MSc dissertation, University of Liverpool.
44. Caples E.L., Massad K.E. and Minor T.R. (1980) A UHF channel simulator for mobile radio. *IEEE Trans.*, **VT29**(2), 281–9.

Chapter 9

Man-made Noise and Interference

9.1 INTRODUCTION

The performance of any communication system depends on the characteristics of the transmission medium and can often be improved by using techniques which successfully exploit these characteristics, for example by using an optimum modulation method. The important characteristics for the communications engineer are the frequency and time responses of the channel, and the magnitude and nature of the noise. The channel responses have been discussed in earlier chapters; we now deal with the problem of noise. There are two basic reasons for a study of noise. Firstly there is a need to understand the nature of the noise in order to devise methods by which it can be characterised. Knowledge of the sources of noise may also lead to methods by which it can be suppressed. Secondly there is a vital need to be able to predict the performance of communication systems that have to operate in noisy environments.

A mobile radio system is beset with noise from various sources, each having different characteristics. Firstly there is receiver noise which is Gaussian in nature and arises from the receiving system itself. Receiver noise is usually expressed in terms of nkT_0B, where n is the factor by which the total receiver noise exceeds ambient noise. Atmospheric noise may also be present, but it decreases rapidly with frequency and is generally negligible in the VHF range. Galactic noise is also insignificant in the VHF band as it is well below the background noise. By far the most important source of noise in mobile communication is the noise radiated by electrical equipment of various kinds. This noise, commonly termed *man made noise*, is impulsive in nature and therefore has characteristics quite different from Gaussian noise. It can be detected at frequencies up to 7 GHz [1] and the magnitude of various noise sources as a function of frequency is shown in Figure 9.1. The characterisation of Gaussian noise is fairly straightforward, but impulsive noise is a quite different matter.

There are several potential sources of impulsive noise which could play a role in mobile communication systems. The radio is often installed in a vehicle, itself a source of noise due to its own ignition and other electrical systems, and the vehicle commonly operates in urban, suburban and industrial areas where it is close to other noisy vehicles. There are various extraneous sources of noise such as power lines and

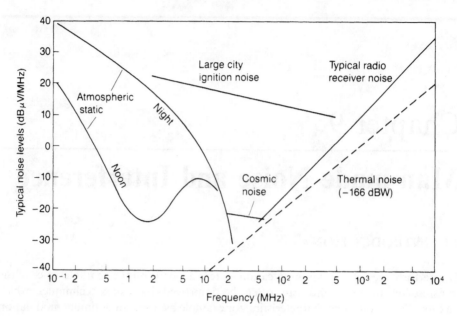

Figure 9.1 Typical average noise levels in a 6 kHz bandwidth.

neon signs, industrial noise from heavy-current switches, arc welders and the like, and noise from various items of domestic electrical equipment. These may or may not be significant contributors in any specific situation. In practice the level of man-made noise varies with location and time [2,3], so from a limited series of observations it is only possible to derive typical values and obtain some estimate of the variability. Some years ago it was established that in urban areas the impulsive noise generated by motor vehicles was a major source of interference to mobile radio systems, particularly in the lower part of the VHF band. The ignition system was the main source [4], although there were also contributions from ancillary electrical equipment [5]. Nowadays, although motor vehicles make much greater use of electronic equipment, suppression methods have been greatly improved and the problem seems much less severe.

Throughout the literature, the terms *Gaussian* and *impulsive* are used to denote two distinct types of noise. Only the power spectral density of Gaussian noise is affected by linear filtering; the probability density function remains Gaussian. The in-phase and quadrature components of narrowband Gaussian noise are independent, as are the envelope and phase distributions. For any other type of noise, both the power spectral density and the probability density function are changed by filtering; the in-phase and quadrature components, although uncorrelated, are not independent. In the general case, the envelope and phase of random noise are independent, the phase being uniformly distributed in the interval $(0, 2\pi)$.

In general terms we may consider an impulse as a transient that contains an instantaneous uniform spectrum over the frequency band for which it defined; a uniform spectrum requires that all frequencies are present and they must be of equal strength over the frequency band concerned. Impulsive noise is the combination of successive impulses which have random amplitudes and random time spacings; these factors may sometimes be such that adequate separation of successive impulse responses by a narrowband receiver is not possible.

Thermal noise can produce an annoying 'hiss' on a voice channel, but does not significantly degrade intelligibility unless its RMS value is relatively high. Impulsive noise causes clicks which, although disturbing, may be tolerable. The degradation of the channel is not easily defined and is usually based on some kind of subjective assessment; indeed the quasi-peak measurement (see later) has been shown to have some correspondence with the subjective assessment of degradation on AM radio and television [6]. Conceptually, digital transmissions are easier to deal with since the bit error rate (BER) provides a good quantitative indication of how well the communication system reproduces the transmitted information. The BER produced by thermal noise is readily established for various kinds of modulation system and the analysis is available in several textbooks. We will discuss the methods for expressing the properties of impulsive noise, and the extent to which they provide information that is directly useful in predicting performance degradation in communication systems.

9.2   CHARACTERISATION OF PULSES

Impulse generators find widespread use as calibration sources for measuring instruments such as spectrum analysers and receivers. These generators are calibrated in terms of a quantity known as *spectrum amplitude*, which is commonly used to characterise broadband signals. The units of spectrum amplitude are volts per hertz or more commonly microvolts per megahertz. It is defined [7] in terms of the magnitude of the Fourier transform $V(f)$ of a time domain signal function $v(t)$ as

$$S(f) = 2|V(f)|$$

where

$$V(f) = \int_{-\infty}^{+\infty} v(t) \exp(-j2\pi ft)\, dt \tag{9.1}$$

Alternatively we can write

$$v(t) = \int_{-\infty}^{+\infty} V(f) \exp(j2\pi ft)\, df \tag{9.2}$$

which is the inverse Fourier transform.

The decibel expression, dB relative to 1 microvolt per megahertz (dBμV/MHz) is also in common use and is defined as

$$S(\text{dB}) = 20 \log_{10} \left[\frac{S\,(\mu\text{V}/\text{MHz})}{1\mu\text{V}/\text{MHz}} \right] \tag{9.3}$$

Note that $V(f)$ is complex, as shown by eqn. (9.1), and therefore spectrum amplitude may not be sufficient to describe the signal completely. Phase information may sometimes be needed, but for many purposes $|V(f)|$ is a very useful quantity.

9.2.1   Spectrum amplitude of a rectangular pulse

We consider the rectangular pulse shown in Figure 9.2 as an example. This has an amplitude A from $t = 0$ to $t = \tau$ and is zero elsewhere. It can be written

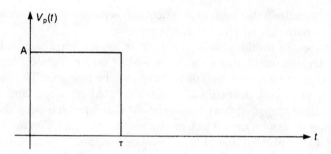

Figure 9.2 Single rectangular baseband pulse.

$$V_p(t) = \begin{cases} A & 0 < t < \tau \\ 0 & \text{elsewhere} \end{cases}$$

The Fourier transform can be obtained as

$$V_p(f) = A \int_0^\tau \exp(-j2\pi ft)\, dt$$

$$= \frac{A}{2\pi f}[\sin 2\pi f\tau - j2 \sin^2 \pi f\tau]$$

which is a complex quantity.

It is easy to show that

$$S(f) = 2|V_p(f)| = 2A\tau \left| \frac{\sin \pi f\tau}{\pi f\tau} \right| \tag{9.4}$$

and a graph of this function is given in Figure 9.3.

The spectrum amplitude $S(f)$ has a $(\sin x)/x$ variation. A feature of this and many other impulsive-shaped pulses, e.g. the triangular pulse, is the fact that the spectrum amplitude approaches $2A\tau$ at low frequencies. $A\tau$ is simply the area under the $v(t)$ curve and is known as the *impulse strength* [8].

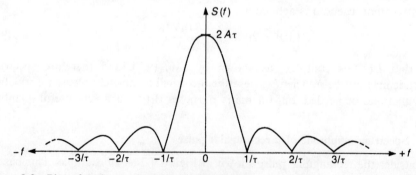

Figure 9.3 Plot of $S(f)$ as a function of frequency.

9.2.2 Impulse generators

Commercial impulse generators usually produce a uniform train of rectangular pulses with extremely short duration, typically 0.5 ns. The spectrum amplitude may extend up to 80 dBμV/MHz and repetition rates are either controllable or locked to the line frequency. The analysis for a single pulse will not be extended here to the case of a finite train of pulses, but note that for any physical (i.e. realisable and non-infinite) signal, $S(f)$ is not constant with frequency. It varies in a manner determined by the signal waveform. Impulse generators are often specified in terms of the spectrum amplitude at low frequency and it is then implicitly assumed that this value may be used up to a frequency where $S(f)$ has fallen to a value 3 dB lower. However, no physical signal has a truly 'flat' spectrum amplitude for all frequencies; this would imply infinite energy.

For impulsive noise, the exact pulse shapes are of little importance since the information cannot be put to practical use. Any two pulses are indistinguishable if they provide the same spectrum amplitude over the frequency range of interest.

9.3 CHARACTERISATION OF IMPULSIVE NOISE

One unfortunate aspect of impulsive noise characterisation is the apparent lack of any agreement about the techniques of measurement and the relative usefulness of the parameters for characterisation. The situation becomes more complex due to non-stationarity in the noise data, as any statistical model evolved may not be valid for all situations. One possible approach is to associate the measured parameters with the environmental conditions under which the measurements were carried out, and for noise in urban areas these conditions include traffic density, distance of the monitoring antenna from the traffic flow, and its position with respect to traffic lights, etc. Once this association has been established, the problem of non-stationarity may be minimised and the performance of a communication system may be predictable (using certain approximations) given the general environmental conditions under which the system is operating.

9.3.1 Measurement parameters

To provide a starting point for discussing how to characterise impulsive noise, we adopt a very simple physical model in which the pulses are very narrow and are described by

$$A_{\mathrm{T}}(t) = \sum_{m=1}^{k} A_m \delta(t - t_m) \tag{9.5}$$

We do not lose generality by assuming positive-going pulses only, since after an impulse has passed through a bandpass filter operating at RF it is not possible to determine its original polarity. Therefore at the output of a bandpass filter, the waveform, the detailed shape of which depends on the filter bandwidth, will have the form shown in Figure 9.4(a). The carrier phase will be random and the detected

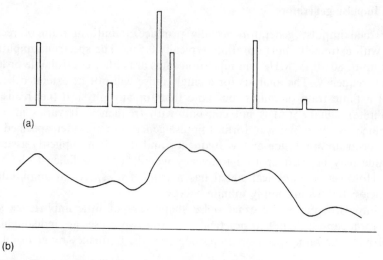

Figure 9.4   (a) Elementary model of impulse noise and (b) detected waveform.

waveform will be as shown in Figure 9.4(b). There may be secondary responses, which depend on on the impulse response of the filter; there may also be overlapping, as between the third and fourth pulses.

We are normally mainly interested in the magnitude (envelope) of the noise, and the following list gives some parameters which can be measured and which provide relevant information:

- Mean or average voltage
- Peak voltage
- Quasi-peak voltage
- RMS voltage
- Impulsiveness ratio
- Amplitude probability distribution (APD)

  - level crossing rate (LCR)
  - pulse duration distribution (PDD)
  - pulse interval distribution (PID)

- Noise amplitude distribution (NAD)

  - pulse height distribution (PHD)

This list is not exhaustive, but it illustrates some of the quantities that can be measured; it remains to establish the relative usefulness of these quantities for various purposes. The first five parameters in the list characterise noise in terms of a single parameter and are useful principally in detecting the presence of radio noise emissions or in specifying regulatory levels of noise. For example, peak voltage is useful to determine whether or not a particular area or object is a source of radio noise but it is of no use in characterising ignition noise, which is postulated to consist of numerous peaks of random amplitude.

Impulsiveness ratio is useful in comparing the noise from two different kinds of source, e.g. car ignitions and power lines, and it may indicate that one is more or less impulsive than the other. It is used as a measure of impulsiveness at HF and below, and its value is used to define a certain shape APD [9]. However, it is unlikely to be useful on its own as a measure of ignition interference because in that case the APD varies with both traffic density and traffic pattern and its shape cannot be adequately described by a single parameter. Average or mean is useful for giving a general indication of level of background noise in any given area and allows a comparison of industrial, urban, suburban and rural regions.

The most common single parameter used for characterisation is the *quasi-peak* value, and the use of the quasi-peak voltmeter has grown to such an extent that most national radio administrations specify it as the method of measurement for regulatory levels of conducted or radiated noise. The original reason for the development of the quasi-peak meter was to try to compensate for an undesirable feature of the peak detector: its response is insensitive to pulse rate for pulse rates greater than a few pulses per second. Note that the terms 'peak' and 'quasi-peak' actually refer to the characteristics of a detector, usually placed at the output of an IF amplifier. These terms therefore describe a detector function rather than an inherent characteristic of the impulsive noise.

The specifications for quasi-peak meters to be used in different frequency ranges are contained in various CISPR publications [10–12] but an essential feature is that they have a charge time constant much longer than that appropriate for a peak detector and a discharge time constant much shorter, so their response is a function not only of impulse strength but also of impulse rate. The quasi-peak method recognises that the noise output from a receiver has a joint probability distribution involving amplitude and time, and this is a fundamental point.

There are two other methods of characterisation which recognise time as an important factor in the characterisation of impulsive noise. The *amplitude probability distribution* (APD) is usually plotted on Rayleigh-scaled graph paper and shows the percentage of time for which the noise at the output of a receiver exceeds any particular level. The ordinate is usually expressed in decibels above kT_0B. The reason for choosing kT_0B and this type of graph paper is explained in Appendix A, where it is shown that detected receiver noise has a cumulative probability distribution which plots as a straight line on this type of paper. In order to interpret APD it is necessary to know the characteristics of the receiver used to make the measurement, since bandwidth and filter response can affect the APD shape.

A typical APD is shown in Figure 9.5, together with the receiver noise line from a receiver with a noise figure of 10 dB. There are two distinct regions to this curve, one of low slope at low amplitude levels (the background noise) and the other of high slope at high amplitude levels (the impulsive noise). The APD gives the 'first-order' statistics of impulsive noise; that is, it allows a determination of the overall fraction of time for which the noise exceeds any particular value. It gives no information about how this time was made up, i.e. whether the value was exceeded by one pulse, or ten or a hundred. This kind of information is given by the *noise amplitude distribution* (NAD).

The NAD is a method of presenting impulse noise data in a form which gives much more information than provided by, say, the quasi-peak detector. It provides a

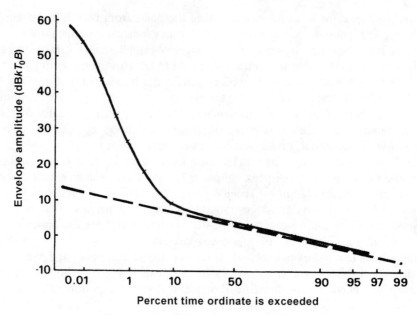

Figure 9.5   A typical amplitude probability distribution (APD) curve: (- - - -) receiver noise.

method of estimating the noise at the input of the receiver rather than at the output; this estimate is independent of the bandwidth and largely independent of the characteristics of the measuring equipment. The NAD concept also provides an empirical method for determining the susceptibility of analogue [13] and digital [14] communications receivers to impulse noise.

The information given by the NAD is the number of pulses per second which exceed a given strength (or more exactly, contain more than a given energy). It is presented in graphical form; the ordinate is spectrum amplitude (μV/MHz or dB above 1μV/MHz) and the abscissa is average pulse rate. There are advantages in expressing the amplitude in μV/MHz; firstly it is the unit normally found on impulse generators, and secondly since it is normalised with respect to bandwidth it allows a direct comparison with results for other bandwidths. The NAD is not a probability distribution, it is a system-independent measure which was originated as a means to extract information from radio noise in a form which allows evaluation of the effect of that noise on land mobile communication systems [13,15]. The different methods of characterising impulsive noise are compared in Tables 9.1 to 9.3; they were originally presented by Shepherd [16].

9.4   MEASURING EQUIPMENT

Having identified the parameters it is desirable to measure, we can now specify the measuring equipment. Ideal measuring instruments and measuring systems have characteristics which do not influence the quantity to be measured. This ideal situation may not always be realisable, especially when the quantity is as complex as impulsive noise, and it is then important to be able to estimate the effects of the

Table 9.1 Evaluation of different types of impulsive noise measurement receivers (detector circuits)

Evaluation	Detector function					
	Peak	Quasi-peak	Average	RMS	Impulse count (NAD)	Envelope time distribution (APD)
Usual units measured	dBμV/MHz	μV	μV	μV	dBμV/MHz vs i/s	dBkT vs % time
Standard	IEEE (1)	CISPR	None (1)	None (1)	CCIR (2) IEC (3)	CCIR (2)
Equipment complexity	Simple	Moderately simple (4)	Moderately simple (4)	Moderately complex (4)	Moderately complex	Highly complex
Accuracy of measurement	Operator dependent	Equipment dependent (5)	Equipment dependent (6)	Equipment dependent (7)	Operator dependent	Operator dependent (8)

Notes
1. CISPR specifies relationship with quasi-peak but makes no recommendation for use.
2. Noise data obtained by an approximation of the envelope detector method of measurement appears in CCIR Report 322-1 based on a bandwidth of 200 Hz.
3. The IEC has approved methods of measurement of degradation of receiver performance due to impulse noise. The method approved requires that the NAD be used as a method of presentation of noise data. Documents approved at the interim meeting of CCIR SG8 include the NAD as a method of presentation of noise data.
4. Field intensity meters with log IF amplifiers cannot measure parameter accurately without increased complexity.
5. Quasi-peak meters are calibrated at 25 or 100i/s while accuracy at other pulse rates depends on design and tolerance of components used in manufacture.
6. Measurements are usually near or below noise level of measurement receiver.
7. Can be calibrated at different pulse rates and amplitude, but it requires considerable time.
8. Accurate for only one bandpass filter envelope amplitude.

Table 9.2 Evaluation of different methods of presentation of results (impulsive noise)

Evaluation	Method of presentation						
	Peak	Quasi-peak	Average	RMS	NAD	EAD	APD
Usual type of presentation	Single parameter	Single parameter	Single parameter	Single parameter	dB/log	dB/log	dB/%
Efficiency of presentation	Good	Good	Good	Good	Good (2)	Good	Poor (1)
Difficulty of presentation	Simple	Simple	Simple	Simple	Moderately complex (2)	Moderately complex (3)	Highly complex (3)
Range of accuracy	Low impulse rates only	Middle impulse rates only	High impulse rates only	Middle impulse rates only	All impulse rates	Accurate for specific filter (4)	Accurate for specific filter (4)

Notes
1. When the APD is given on Rayleigh graph paper, 90% of the important data is squeezed into 10% of the paper. Good for only one filter. Can be converted by approximation to other bandwidths.
2. The NAD need not be converted from one bandwidth to another. It is usually easy to distinguish between impulse noise data and receiver noise data.
3. Applies for a specific bandpass filter envelope amplitude, which is almost never furnished with the presentation.
4. If the impulse response of the measurement equipment is known, it is possible to make an approximate conversion to the NAD.

Table 9.3 Evaluation of accuracy in determining degradation of receiver performance due to impulsive noise

Evaluation	Peak	Quasi-peak	Average	RMS	NAD	EAD	APD
Method of use	None (1)	None (1)	None (1)	None (1)	NAD overlay	None (1)	By prediction
Types of receiver	All (2)	All (2)	All (2)	All (2)	All	All (4)	Data (3) communication
Accuracy, 1 error	9 dB	10 dB	(5)	9 dB	2.5 dB	(5)	(5)

Notes
1. No information is available as to how data is used to determine degradation of performance to a receiving system.
2. By calculating the signal-to-interference ratio.
3. It is reported that the APD can be used for predicting errors to data communication systems. It is not known whether V_d must also be known.
4. When a method becomes available.
5. No data available.

measuring equipment on the parameters to be measured. We briefly consider the various characteristics of measuring equipment, and their effects, so that the requirements of equipment designed to make realistic measurements can be specified. It is particularly important to consider the effects of bandwidth, sensitivity, dynamic range, impulse response and type of detector.

9.4.1 Bandwidth

Man-made noise occurs mainly in the form of narrow pulses a few nanoseconds wide, and it would require a measuring system with a very wide bandwidth to produce a characterisation applicable to all possible system bandwidths. Such an instrument would be expensive and difficult to use, and in a congested radio spectrum the concept is totally unrealistic because coherent interference would invariably be present within the bandwidth of the instrument.

The decision about a suitable bandwidth is influenced by the fact that provided the measurement bandwidth is considerably larger than the proposed communication system bandwidth, the noise characterisation is likely to be valid. The decision therefore rests on the way in which the radio spectrum is used in the frequency range of interest. For PMR and other systems in the VHF and lower UHF bands, frequency allocations are in fairly narrow bands, and a measurement bandwidth of a few hundred kilohertz seems adequate. Indeed, the CISPR specification for a quasi-peak voltmeter in the frequency range 25–1000 MHz calls for a filter bandwidth of 120 kHz, and this seems entirely appropriate except perhaps for GSM systems at 900 MHz. Consideration of other factors also leads to a similar conclusion. For example, a larger bandwidth leads to a requirement for an increased dynamic range.

9.4.2 Dynamic range

The dynamic range required of the measurement system is probably the most difficult parameter to assess, since it depends on both the bandwidth and the centre frequency of measurement.

The mean magnitude of VHF man-made noise has been measured at various frequencies, and has been found to decrease as the measurement frequency is increased [17,18]. In the UK this mean level is given by

$$N = 67 - 28 \log_{10} f \qquad (9.6)$$

where N is in $\mathrm{dB}kT_0B$ and f is in MHz. This is about 10 dB lower than the value measured in the USA [19].

Examination of the literature reveals that the probability of the noise exceeding 60 dBkT_0B is about 10^{-5} in the HF range [20], and measurements of APD made at 50 MHz using a receiver bandwidth of 10 kHz indicate a dynamic range of over 80 dB between the 0.0001 and 99 percentiles. For bandwidths greater than 10 kHz, the required dynamic range will be larger because the peak amplitude of the impulse response at the receiver output is proportional to the energy in the input impulse. In fact, for any two bandwidths B_2 and B_1 the required dynamic range ratio is $10 \log(B_2/B_1)$. This, however, is only an approximation since it does not take into account any change in dynamic range requirements caused by the overlapping of impulse responses, an effect which increases as the bandwidth is reduced. So, if the bandwidth is reduced, the full reduction in dynamic range given by the above expression cannot be realised since some allowance has to be made for the increased overlapping which is likely to occur. Considering these points, it seems that if it is necessary and desirable to cover the probability range 0.0001% to 99%, a dynamic range near 80 dB will be required. However, in practice it is usual to compromise and use relatively inexpensive equipment with a 50–60 dB dynamic range capability.

9.4.3   Receiver sensitivity and noise figure

The receiver sensitivity is a measure of the lowest signal that can be detected. When measuring impulsive noise it is also necessary to cope with very high peaks, and the lowest detectable signal may be restricted by limitations in the dynamic range. Put another way, in a measuring system designed to cope with the highest peaks, the lower signals may be obscured by receiver noise unless the dynamic range is very large. The problem is made worse by the fact that the amplitude of the highest peak is uncertain, but there are indications in the literature that in the bandwidths appropriate for mobile radio systems, a reasonable estimate of the peak level is $80 \text{ dB} k T_0 B$. If we are not interested in the very low levels of noise, which occur with high probability, then very high sensitivity is not necessary and a commercially available receiver with a noise figure of about 6 dB and a dynamic range of 50–60 dB will be adequate [21].

The receiver front-end gain should be such that its sensitivity is not degraded by local oscillator and mixer noise, and receivers with front-end gains in excess of 20 dB are readily available. The overall gain of the receiver depends on the minimum detectable signal and the detector characteristics. Non-linearities in the detector should be avoided, thus the knee voltage of a detector should provide the output for the minimum detectable signal. The necessary gain can therefore be calculated, and in practice an overall predetector gain of about 80 dB is required.

9.4.4   Impulse response

Impulse response is probably the most important parameter of impulsive noise measurement systems. Limitations in the other parameters such as dynamic range and bandwidth still allow results to be obtained, but limitations in system impulse response result in fundamental modifications to the noise parameters themselves. The ideal impulse response for a measurement system is one in which there is no ringing, such as may be obtained from a Gaussian filter. The presence of ringing, or time sidelobes, may seriously affect the capability of the system to resolve low-strength impulses which follow very high-strength impulses. The various noise parameters are affected differently and the manner in which they are modified depends on the statistical properties of the noise. The extent to which inaccuracies exist in the measurements depends on the probability of occurrence of high strength impulses, and the way in which the individual parameters are affected is discussed below.

Noise measurements

If there is a time sidelobe in the impulse response of a measuring instrument, the measured probability of exceeding lower levels will be enhanced. This arises because there will be a contribution from pulses at that level and an additional contribution from the sidelobe response caused by previous high-level impulses. The extent of the additional contribution may be estimated from a knowledge of the actual impulse response and the relationship between the numbers of high- and low-level impulses (the noise slope). As an example, consider an impulse response with a secondary response 30 dB below the main peak.

If the noise slope is low, i.e. the probability of occurrence of impulses with a given amplitude is of the same order as those with an amplitude 30 dB lower (this is very unlikely to occur in practice), then the measured probability of exceeding the lower level is enhanced by a factor approaching 2, and there is significant error at the lower levels. If the noise slope is high – say the probability of exceeding the given higher level is 1% of the probability of exceeding the lower level – then the measured probabilities at the lower levels are enhanced by a much smaller amount and the errors are much less. It can therefore be seen that noise slope plays an important part in determining the required impulse response, and if the noise slope is high an inferior impulse response can be tolerated.

In practice, Shepherd [20] measured the probability of VHF automotive impulsive noise exceeding $25\,\mathrm{dB}kT_0B$ as 0.05 and of exceeding $65\,\mathrm{dB}kT_0B$ as 10^{-5}. We can therefore see that in practice, although the APD will be affected by impulse response, the errors caused by a time sidelobe about 30 dB below the main response are unlikely to be serious.

Level crossing rate

The effect of a poor impulse response is to cause secondary responses generated by ringing, and these are indistinguishable from responses to genuine low-level impulses. There is therefore an increase in the crossing rates at low threshold levels; this increase depends on the crossing rate at very high levels and the number of significant time sidelobes. Nevertheless, provided the increase at low levels is masked by the number of genuine impulses at these levels, the error in measurement is not significant. However, the level crossing rate (LCR) for high threshold levels at the output is not affected by the poor impulse response and in most cases it is the high-threshold data which is of greatest interest.

Pulse duration and pulse interval measurements

Another effect of the time sidelobes present in the impulse response is to increase the measured pulse duration and decrease the time interval between pulses. Again the low threshold levels are affected, but assuming a well-behaved noise slope the errors are not significant.

9.5 PRACTICAL MEASURING SYSTEMS

A practical measuring system for man-made noise can be built around commercially available equipment. We will briefly discuss the measurement of APD and NAD.

9.5.1 Measurement of amplitude probability distribution

If APD is to be the principal measure of noise amplitude then it is possible to use a commercial receiver having suitable characteristics. The detected noise can be fed into a computer via a fast ADC; the results are then stored and analysed off-line to obtain various parameters of interest. Note, however, that the receiver selected must be used in a fixed-gain mode, i.e. with no AGC, and that any limitations in dynamic range and impulse response must be taken into account.

There are several ways to achieve an improvement in dynamic range, the most attractive being to insert an attenuator either between the antenna and the receiver or in the IF stages when measuring high-level noise. Alternatively the noise can be processed in two parallel branches with different gains. These options are illustrated in Figure 9.6. Effectively they all amount to looking at the noise through two 'windows'; the parallel-branch method has the advantage of looking through the two windows simultaneously but the disadvantage of a much larger capital cost. From an economic and technical viewpoint, the best strategy is to place the attenuator in the IF section of the receiver. This is attractive because the receiver front-end often has a sufficiently high dynamic range, and the use of IF attenuation only causes a marginal degradation in the receiver noise figure.

The effect of inserting an IF attenuator is to increase the measurement capability for high-level signals, and this can be seen as follows. If an attenuator having an attenuation X is inserted in the IF section of the filter (Figure 9.6(a)) then the output

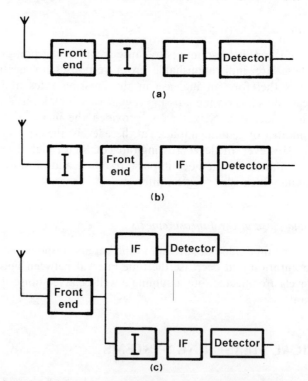

Figure 9.6 Noise measuring systems: (a) attenuator following RF section, (b) attenuator preceding RF section, (c) dual-branch receiver.

signal is reduced by a factor X and the input signal necessary to saturate the receiver output is therefore increased by X. However, provided the front-end gain is high, the receiver noise figure is almost unaffected and the insertion of the attenuator has effectively taken the absolute values of the signals which can be received and moved them up by an amount X without reducing the dynamic range significantly. In other words, by using a variable attenuator we can effectively slide up and down a 'window' with a width equal to the unmodified dynamic range of the system and hence select any convenient portion of the input signal for reception. The limitation is that the inserted attenuation must not approach the value of the front-end gain otherwise the effective front-end gain is reduced with a consequent degradation in the receiver noise figure. Provided this limit is not approached, the absolute range over which signals can be received is equal to the inherent dynamic range of the receiver plus the inserted attenuation.

When a receiver is subjected to impulsive noise with a dynamic range far in excess of the inherent receiver dynamic range, we first consider the situation when there is no inserted attenuation. In this case the high-amplitude impulses tend to saturate the receiver, and although this is not serious in itself, time sidelobes in the receiver impulse response will appear in the range being measured and can cause problems. However, if the number of pulses due to time sidelobes is swamped by the number of genuine pulses at these lower levels, measurement errors will be insignificant. For example, if a time sidelobe exists 30 dB below the mean peak of the receiver response, and at this level the impulse rate is 100 times the rate at a level 30 dB higher, the measurement error is 1%. When there is inserted attenuation the limiting factor is noise generated by those parts of the system which follow the attenuator. This is overcome by making measurements using two 'windows' which overlap slightly and then ignoring information from the lower end of the upper window.

Basically the impulse response problem can only be solved by using suitable filters, and Gaussian filters are ideal. The point in the system where these filters should be inserted is a matter for further consideration. Single-conversion receivers often have a high IF (21.4 or 10.7 MHz is typical) and it is difficult to design narrowband Gaussian filters at these frequencies. The two possible approaches to the problem are either to convert the signal to an IF not greater than about 10 times the required bandwidth and then use Gaussian filters, or to use wider bandwidth filters at the existing IF and to improve the impulse response at baseband by a suitable lowpass filter. In normal circumstances the second approach is more attractive. A measuring system as described would normally be calibrated in $dBkT_0B$ using a suitable Gaussian noise source.

If a measuring system as in Figure 9.6 is used to measure and record noise, the detected noise envelope is a unidirectional time-varying voltage with a shape that depends on several factors, including the noise input, the receiver impulse response and the receiver bandwidth. The amplitude of the noise can be represented by the amplitude probability distribution (APD) which is obtained by fixing a particular level (say L) and then summing all the individual durations Δt_k above that level to find the total time for which the level is exceeded.

The appropriate point on the APD curve can then be plotted from

$$\text{prob}[A > L] = P(L) = \frac{1}{T}\sum_{k=1}^{n}\Delta t_k \qquad (9.7)$$

where T is the total length of the record. Other points can be plotted in a similar way. The APD only gives information about the overall percentage of time for which any particular level is exceeded. There is no information about how the time was made up, i.e. how many pulses exceeded the value in question.

In order to provide more detail, the APD can be supplemented by a graph showing the number of times any particular level is crossed. If, in a record of length T seconds, the level L is crossed n times (in a positive-going direction) then the average crossing rate at that level is n/T crossings per second. Average crossing rate (ACR) is usually expressed in the form of a graph of amplitude against crossing rate, on a logarithmic scale.

ACR curves give only first-order information about the number of times each level is crossed, and when taken in conjunction with APD curves they can only produce information about the average pulse duration and the average pulse interval. Further information about the pulse duration can be provided by a graph of the pulse duration distribution (PDD), i.e. the probability that any given duration is exceeded. To do this it is necessary to count the individual pulse durations at a given amplitude level. The probability that any particular duration Δt is exceeded is then obtained by finding the total number of pulses that exceed Δt and dividing by n, the total number of pulses (or level crossings) at that level. So, at any level L, the probability that the pulse duration is greater than Δt is

$$P_d(\Delta t) = \frac{\text{number of pulses with duration} > \Delta t}{\text{total number of pulses at level } L} \tag{9.8}$$

In a similar way, further information about the intervals between pulses is given by a graph of the pulse interval distribution (PID). To obtain this, all the individual time intervals between successive positive-going crossings at a given level are counted and the probability that any interval Δt is exceeded is calculated by finding the total number of intervals that have a length greater than Δt and dividing by the total number of intervals. Thus, at any particular level L, the probability that a pulse interval is greater than Δt is

$$P_i(\Delta t) = \frac{\text{number of intervals of length} > \Delta t}{\text{total number of intervals at level } L} \tag{9.9}$$

The most straightforward method of obtaining these parameters is to access the digitised stored data and to find the parameters of interest using software.

9.5.2 Measurement of noise amplitude distribution

Equipment designed to measure NAD should be capable of detecting all the noise impulses likely to affect a communications receiver, and be able to distinguish between successive impulses which occur at the highest likely rate. At the very minimum, for conventional land mobile radio systems, this implies a dynamic range of 60 dB and an RF bandwidth of 30 kHz. It is highly desirable that the equipment has a well-behaved impulse response so its effect on the measured results is minimal.

Basically the equipment consists of two parts: a high-gain receiver having as its output a band-limited detected signal, and a pulse height analyser to ensure each noise pulse is simultaneously compared with several preset threshold levels. Each threshold detector is coupled to a digital counter and store, which accumulates the total number of pulses that exceed the particular threshold level. This process is illustrated in Figure 9.7 for a system with six thresholds. Practical realisations have been described in the literature [22,23]; later instruments use sophisticated digital techniques so they can also measure other parameters such as APD, pulse intervals and pulse durations.

Calibration of the equipment in terms of impulse strength at the input is straightforward using an impulse generator. The reference level to each threshold detector is set to maximum; the impulse generator is set to the lower limit of the required measurement range and connected to the receiver input. The reference level

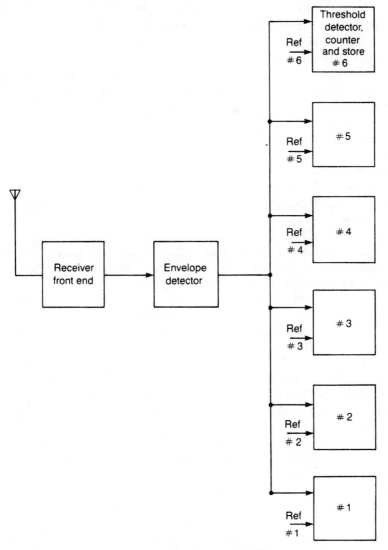

Figure 9.7　Measuring noise amplitude distribution (NAD).

of branch 1 is then adjusted until counter 1 just starts to count. The output of the impulse generator is then increased to the next threshold level (say 10 dB higher) and the reference level for detector 2 is adjusted until counter 2 starts to count. This procedure is repeated for all levels until the instrument is completely calibrated.

In use the instrument is connected to an appropriate antenna (in urban areas this is often a vehicle antenna), all counters are set to zero, and a measurement is made for typically 15 min. The readings of the various counters at the end of this time allow the NAD to be plotted. It is essential to measure at a frequency which is free from coherent interference, and if a suitable receiver is available it is wise to monitor the channel during the measurement period.

9.6   IMPULSIVE NOISE MEASUREMENTS

Measurements of impulsive noise measured at roadside locations must be interpreted with some care since the level of noise depends on the volume of traffic and the influence of road features such as traffic lights and junctions. However, the results are useful in two different ways. Firstly, they can be used to test the validity of existing mathematical models representing impulsive noise, or as a basis for proposing new models. Secondly, the data can be used to predict the performance degradation of communication systems subjected to impulsive noise, using methods based on APD or NAD.

Figures 9.8 to 9.10 show noise data measured at various frequencies between 40 MHz and 900 MHz [24] using a receiver with an IF bandwidth of 20 kHz. In suburban locations, traffic generally flows quite freely without significant influence from road features such as roundabouts or traffic lights. The traffic density varies between 12 and 20 vehicles per minute; a typical value is 15 vehicles per minute. In urban areas, traffic flows in a constant stream at a much steadier rate than in suburban locations; a typical value is 30 vehicles per minute. The APD curves for the two types of location reflect this increase in traffic density, the difference being between 2 and 4 dB at all frequencies. It is also clear that the overall level of noise decreases with an increase in frequency; for example, Figure 9.8(a) shows that the value exceeded with probability 10^{-4} decreases from $56 \, \mathrm{dB} k T_0 B$ at 40 MHz to $30 \, \mathrm{dB} k T_0 B$ at 900 MHz.

At 40 MHz the median value of the curve representing background noise lies approximately 12 dB above the corresponding point on the line representing receiver noise. To avoid congestion this line is not shown in Figure 9.8(a); however, the level $11 \, \mathrm{dB} k T_0 B$ is exceeded by the receiver noise for only 10^{-2} (1%) of the time, whereas it is exceeded by external noise for about 80% of the time. It is therefore safe to conclude that at this frequency the external noise is very dominant. On the other hand, at 400 and 900 MHz the external noise is much lower, and receiver noise is dominant for most of the time. A further comparison between receiver and man-made noise may be made using Figure 9.9(a) by noting that at 40 MHz, man-made noise in urban areas exceeds $11 \, \mathrm{dB} k T_0 B$ for 84% of the time. For 1% of the time at 80 MHz, receiver noise exceeds $11.5 \, \mathrm{dB} k T_0 B$ (receiver noise figure 4 dB) whereas the external noise exceeds this level for about 65% of the time. At the higher frequencies, receiver noise is still very important.

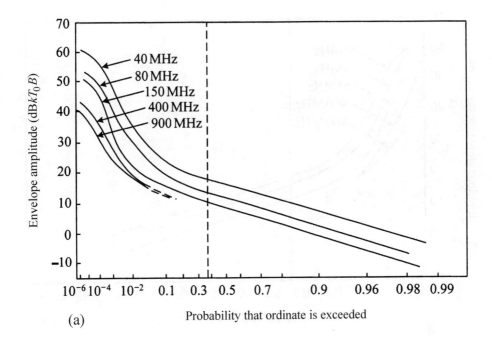

(a)

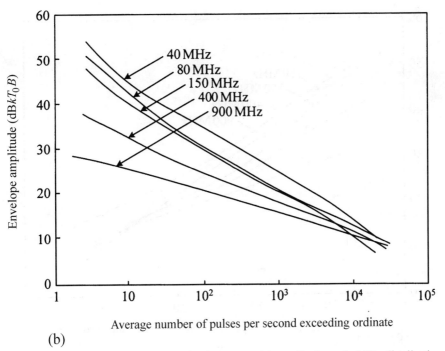

(b)

Figure 9.8 Measured noise data in suburban areas: (a) amplitude probability distribution, (b) average crossing rate.

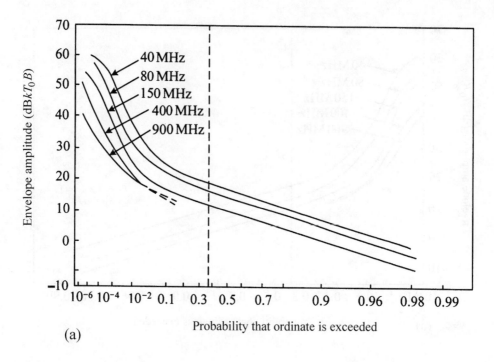

(a)

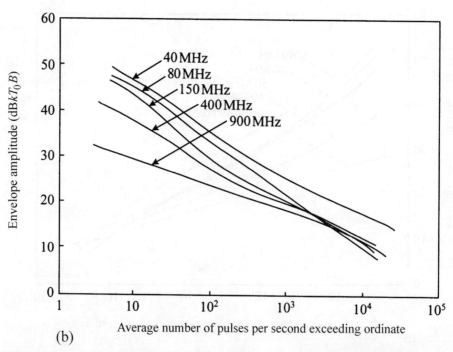

(b)

Figure 9.9 Measured noise data in urban areas: (a) amplitude probability distribution, (b) average crossing rate.

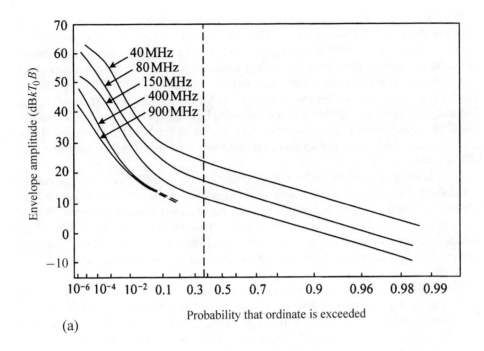

(a)

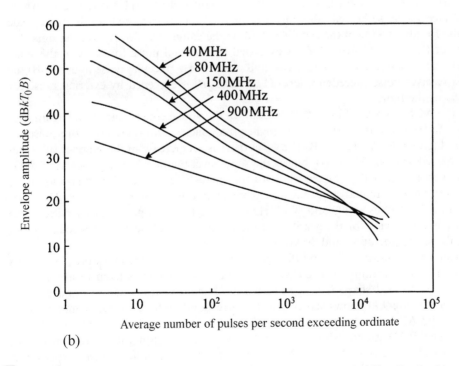

(b)

Figure 9.10 Measured noise data in a city centre: (a) amplitude probability distribution, (b) average crossing rate.

The ACR curves in Figure 9.9(b) also indicates the higher level of noise measured in urban areas. There does not appear to be any significant change in the shape of the graphs with respect to the corresponding curves in Figure 9.8(b) for suburban areas. However, in both cases there is an evident tendency for the curves to become lower and flatter as frequency increases. The number of pulses with high amplitude is greater in urban areas but not excessively so. At the 25 dBkT_0B level, there are about 70 pulses per second at 900 MHz that exceed this level compared with 12 in suburban areas. However, there is no corresponding increase at the lower frequencies. There are still about 300 pulses per second exceeding this level at 150 MHz, with just over 1000 at 40 MHz.

A misleading idea of the influence of traffic in the central business area can be obtained if the only factor taken into account is the number of vehicles passing the observation point. In the city centres, vehicles move slowly under the control of road features such as traffic lights, and although the apparent traffic density is low, 14–24 vehicles per minute in Figure 9.10, there are always a large number of vehicles, moving and stationary, sufficiently close to the receiver to influence the measured noise parameters.

The APD curves in Figure 9.10(a) illustrate this in a convincing way. The noise level here is higher than at the other two types of location. The level at 10^{-4} exceedence probability is now 61 dB at 40 MHz, 39 dB at 400 MHz and 34 dB at 900 MHz, all these values being 2–3 dB above the value measured at urban locations even though the apparent traffic density is significantly lower. In city-centre areas at 40 MHz, man-made noise is completely dominant since the 11 dBkT_0B level, which the receiver noise exceeds for only 1% of the time, is exceeded by the man-made noise for about 93% of the time. At 80 MHz the value exceeded by receiver noise for 1% of the time (11.5 dBkT_0B) is exceeded by external noise for 65% of the time, similar to the figure for urban areas, and at 150 MHz (receiver noise figure 3 dB) the 1% receiver noise exceedence level (12.5 dBkT_0B) is exceeded by external noise for 38% of the time.

The ACR curves in Figure 9.10(b) confirm the above conclusions. Once again they tend to be lower and flatter at the higher frequencies, and there are more pulses at very high levels. At the 25 dBkT_0B level there is an increase in the number of pulses at 900 MHz, from 70 to 100, but at the same level the number of pulses at 40 MHz has increased to 3500. The tendency for the curves to cross at a few thousand pulses per second, apparent at suburban and urban locations, is also present here. The slope of the ACR curves at 400 and 900 MHz is similar to the slope at other locations but there is a tendency for the graphs at lower frequencies to be slightly steeper, about 12 dB per decade at 80 and 40 MHz.

Curves illustrating PDD and PID have been published in the literature [24]. In this case it is not meaningful to derive an 'average' set of curves from measurements, since the detailed shape is very sensitive to changes in traffic density and pattern. Instead Figure 9.11 shows sets of graphs corresponding to test measurements which produced APD and ACR curves very close to the typical suburban curves in Figure 9.8. The PDD graphs show that pulses with a duration exceeding 100 μs (the reciprocal of the post-detection bandwidth) are not uncommon, even at the higher threshold levels, and this is indicative of overlapping receiver impulse responses due to the arrival of groups of pulses (noise bursts) too closely spaced in time to be

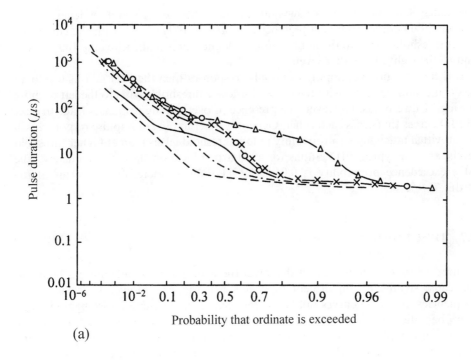

(a)

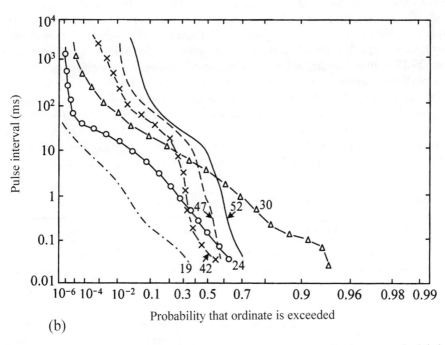

(b)

Figure 9.11 Pulse duration and pulse interval distributions in a suburban area; the labels on the curves show the level in decibels relative to $\mathrm{dB}kT_0B$.

adequately separated by a narrowband receiver. There is a general tendency for the various PDD curves to bunch together close to the $100\,\mu s$ level, with the curves for lower thresholds lying to the left of those for higher thresholds below the $100\,\mu s$ level, and to the right above that level.

If there were no overlapping of impulse responses then the PID graphs would all be distinct and separate, with the curves for lower thresholds lying to the left of those for higher thresholds. However, the presence of overlapping responses leads to a set of PID curves that cross each other. Generally, overlapping of impulse responses will occur with a much higher probability at low threshold levels than at higher threshold levels, and the typical PID graphs for a suburban location show this tendency. The 50% exceedence probability is 0.04 ms at $11\,\mathrm{dB}kT_0B$, increasing to 30 ms at the $50\,\mathrm{dB}kT_0B$ level.

9.7   DISCUSSION

Figures 9.8 to 9.10 show that, at the lower frequencies, man-made noise is high and can cause significant impairment to system performance. However, at the higher frequencies, the APD curves fall below the receiver noise curve (noise figure $9.5\,\mathrm{dB}$) at probabilities less than 10%, even in city-centre areas, and at these frequencies the receiver noise plays a much more important role. There is therefore a strong incentive to design a UHF receiver with a low noise figure because there is much to be gained by doing so. The same is not true for receivers at 40 MHz intended for use in urban areas. The amplitude levels measured at the 10^{-6} probability level (i.e. the values that would be indicated by a peak-reading meter) decrease steadily with frequency. However, they are always significantly more than the maximum contribution expected from receiver noise, even at 900 MHz; the difference is about 20 dB in urban areas. A very misleading impression can therefore be obtained from peak measurements, and their usefulness is severely limited.

Notwithstanding the different locations and the different traffic densities involved, the general level of man-made noise decreases as the measurement frequency is increased. The amplitude characteristics in suburban, urban and city-centre areas are summarised in Table 9.4 and this effect is apparent at all probability levels. The ACR curves tend to be well separated at low pulse rates but cross over each other in the

Table 9.4   Envelope amplitude ($\mathrm{dB}kT_0B$) in different environments at a variety of frequencies

Frequency (MHz)	City centre		Urban area		Suburban area	
	$P = 10^{-4}$	$P = 0.5$	$P = 10^{-4}$	$P = 0.5$	$P = 10^{-4}$	$P = 0.5$
40	61	22	58	17	56	16
80	56	14	55	14	50	12
150	48	10	48	10	45	8
400	39	< receiver noise	38	< receiver noise	36	< receiver noise
900	34		32		30	

region 10^3 to 10^4 pulses per second; the curves at higher frequencies have a lower slope. A major contribution to the noise at lower frequencies is therefore in the form of fairly infrequent high-amplitude pulses. At higher frequencies the variation in pulse heights is much smaller and high-amplitude pulses do not make a major contribution.

To avoid confusion, the individual receiver noise lines have not been drawn on the APD graphs. However, it is shown in Appendix A that the receiver noise at the various frequencies can be represented by a family of straight lines, all having the same slope and passing through a point corresponding to 36.8% cumulative probability on the abscissa (dotted in the APD figures) and the appropriate receiver noise figure, expressed in $\mathrm{dB}kT_0B$ on the ordinate.

The establishment of cellular radio-telephone systems in the 900 MHz band made it important to measure noise levels and propagation characteristics. As far as propagation is concerned, for base-to-mobile operation the excess loss over predictions based on the plane earth equation varies from 35 to 45 dB in city areas and is typically 35 dB in suburban areas. Propagation losses at 900 MHz are 6–9 dB greater than at 455 MHz and about 15 dB greater than at 168 MHz. However, the external noise levels are lower, so a well-designed low-noise-figure receiver should ensure better overall system performance at 900 MHz, compared with 450 MHz, than propagation figures alone would suggest.

To summarise, in the range 40–900 MHz there is a monotonic decrease in the level of detected noise envelope as the frequency of measurement is increased. However, there is no evidence of any significant change in the statistical character of the noise with frequency. The APD curves all have the same general shape, although the detail varies with factors such as frequency, traffic density and traffic pattern. The ACR curves show that the noise is composed of a relatively large number of low-level pulses together with a smaller number of high-level pulses; and the slope of these curves is steeper at low frequencies than at high frequencies. External noise is very dominant at low frequencies but not in the UHF range. The PDD and PID curves show that overlapping receiver responses are quite common, but they also occur with decreasing probability as the measurement frequency is increased.

9.8 PERFORMANCE PREDICTION TECHNIQUES

Assessing the performance of radio receivers in the presence of Gaussian noise is fairly straightforward because good quality wideband Gaussian noise sources are readily available and testing methods are well established. The situation is quite different for impulsive noise. It would be possible to test mobile radio receivers by installing them at a suitable roadside location where interference is generated by passing vehicles, but in general such ad hoc procedures are clearly unsatisfactory. There is an obvious need to establish experimental performance assessment techniques which yield meaningful measures of receiver system performance and which can be used either in the laboratory or at the production stage.

Not all the parameters listed in Section 9.3.1 find use for directly evaluating communication system performance. No attempt has been made, for example, to use the peak detector as a measure of communication system degradation, because of its obvious limitations. As examples of the difficulties, it has been reported that different types of noise from power lines, giving the same peak reading, often result in widely different assessments of the degradation to TV reception. Also, the response of a peak detector to the noise, even from a single source, corresponds very poorly with the resulting degradation to the sensitivity of a radio receiver.

But quasi-peak measurements have been used; indeed one of the reasons for developing the quasi-peak meter was to provide a measurement parameter, with appropriate weighting, for measuring the degradation in performance of radio broadcasting receivers. However, with the trend to use higher radio frequencies it has become obvious that the information provided by a quasi-peak meter is very limited and there is a need to use other parameters for the prediction of system performance.

For performance prediction, the most important methods of characterising noise are those which recognise time as an important factor, i.e. APD and NAD. Each method has its advantages and disadvantages. The APD represents the properties of the detected noise envelope and is therefore a function of the receiver characteristics such as bandwidth and impulse response. There is a lack of generality in such procedures, but measurements made at the IF output do at least have the advantage that they represent the total noise (ambient + impulsive + receiver) presented to the demodulator and may be readily used. The lack of generality is to some extent minimised in PMR systems at VHF and UHF, where the necessity to meet stringent specifications has led to the development of a range of IF filters having impulse responses that are essentially very similar.

The NAD, on the other hand, represents the noise process at the receiver input and models it in terms of a series of pseudo-impulses. The disadvantage therefore is that ambient and receiver Gaussian noise is not included, but this may not be too restrictive if impulsive noise is the dominant feature. The major advantage of using NAD is that methods based on it utilise separate formulations of the receiver and impulse noise characteristics; this means that in principle they can be applied to any receiver whose characteristics are known.

9.8.1 Assessment of receiver performance using APD

A common intermediate step between noise measurement and BER prediction is noise modelling; the models can be empirical or related to the physical noise mechanisms involved. Noise models based on the APD usually depend on fitting some form of mathematical expression to the noise curve [25,26]; the common approach thereafter is to predict BER by integrating the expression over the noise range, using numerical techniques. The expressions derived for BER cannot be reduced to compact forms and are left as integral expressions. The integrations are then evaluated numerically for certain combinations of parameters.

However, in all the BER prediction models that use APD, the measured characteristic can in principle be used to obtain the BER directly, without the need to go through the intermediate stage of mathematical modelling. This brings the advantage that there is no loss of accuracy through fitting a mathematical expression, and there is a clear saving of time. The measured APD can be stored with a high degree of accuracy in a computer, and the BER can be predicted using digital processing techniques [27].

BER prediction techniques based solely on APD are not always sufficient to make an accurate determination of error performance. However, for communication systems such as conventional VHF land mobile radio, where channel bandwidths are narrow, filter characteristics are well defined and data rates occupy the bandwidth efficiently, accurate predictions can be made using APD alone. In most cases the prediction procedure amounts to evaluating a one-dimensional integral of the form

$$P_e = \int_a^b f_r(r)P[e|(r/s)]\,dr \tag{9.10}$$

Here $f_r(r)$ is the probability density function of the noise envelope and $P[e|(r/s)]$ is the conditional probability of error in the receiver given the noise envelope r and the signal envelope s.

In practice, for different modulation techniques, $P[e|(r/s)]$ can be obtained theoretically or approximated. The integral can be evaluated by exploiting the inherent precision in measuring the APD using a large number of histogram bins in the computer and expressing equation (9.10) as a finite summation over the histogram bins.

9.8.2 Assessment of receiver performance using NAD

As a first step in assessing the susceptibility of a radio receiver to impulsive noise, it is useful to measure the degradation in receiver performance caused by impulses having a given strength and a given repetition rate. It has been suggested that this may be expressed in the form of impulse noise tolerance curves (sometimes termed isodegradation or isosensitivity curves). For FM land mobile radio receivers, Shepherd [6] developed a set of isodegradation curves using the system shown in Figure 9.12. The first step is to determine the carrier level required to obtain 20 dB quieting of the audio, with no external noise. An X dB isodegradation curve is then obtained by increasing the carrier level by X dB and using a variable-rate impulse generator to find what combinations of impulse strength and repetition rate will yield the same 20 dB quieting of the audio. These combinations will form a smooth curve as in Figure 9.13, which shows a family of isodegradation curves for a VHF/FM receiver. If the receiver is subjected to an impulse strength of 45 dB above 1 μV/MHz at its input and at a rate of 16 impulses per second, then an additional 5 dB of carrier power is required to provide 20 dB quieting of the audio, i.e. the receiver sensitivity is

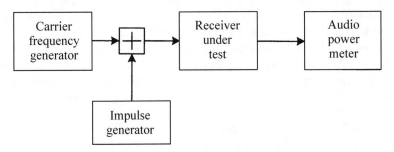

Figure 9.12 Obtaining isodegradation curves for an FM receiver.

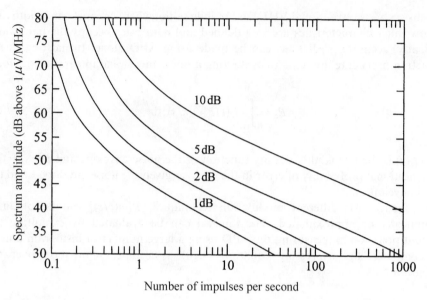

Figure 9.13   Family of isodegradation curves for a VHF/FM receiver.

reduced by 5 dB. If the impulse noise source operates at 4 impulses per second then the degradation is only 2 dB.

For data communication receivers it is dubious whether the 20 dB audio quieting criterion is the best basis for comparison. It has the advantage of being easily measured, but the signal level at which it occurs varies from one receiver to the next, so performance measured relative to the 20 dB quieting level is not an absolute measure. Even in speech communication systems, no systematic correspondence has been established between it and a subjective assessment of receiver performance. It therefore seems sensible to establish, for each type of communication system, a performance criterion that has direct relevance to the type of system being considered [28]. In data communication systems, BER is the obvious measure. Figure 9.14 shows a measuring system for a data communication system. A pseudo-random data stream is used to modulate a carrier, and after corruption by noise from the impulse generator, the signal passes into the receiver. Following demodulation the output is passed into an error detector where it is compared on a bit-by-bit basis with the original data stream and errors are counted.

The experimental procedure is to obtain a set of curves at a certain input carrier level by setting the spectrum amplitude of the impulse generator to a given value and plotting BER as a function of impulse rate. Figure 9.15 shows a family of such curves for a commercial VHF/FM receiver subjected to a periodic pulse train. They verify the work of Bello and Esposito [29] and Engel [30], who suggested that when the incoming pulses are uniformly spaced in time, then for a constant spectrum amplitude, the BER is linearly proportional to the average number of impulses per second. By repeating the experiment at several different input carrier levels it is possible to produce several sets of curves similar to those in Figure 9.15.

There are two ways in which the results can be presented. Firstly, on a single set of curves as in Figure 9.15, we can choose our performance criterion (say BER = 10^{-4})

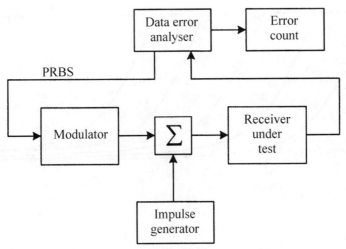

Figure 9.14 Obtaining isodegradation curves for data receivers.

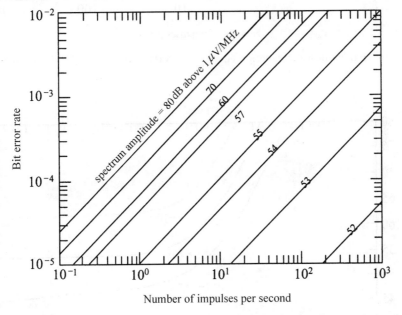

Number of impulses per second

Figure 9.15 Experimental results for a VHF receiver obtained using the experimental system in Figure 9.14.

and by drawing a horizontal line we can determine what combinations of spectrum amplitude and impulse rate give rise to that value of BER. Figure 9.16 shows this form of presentation, in which carrier level is the parameter. Alternatively if we have several sets of curves, similar to Figure 9.15 plotted at various carrier levels, then from each set we can obtain a curve relating spectrum amplitude to average impulse rate, at a predetermined value of BER. Figure 9.17 shows a family of curves produced in this way; note that BER is now a parameter. This form of presentation is almost directly analogous to the method used by Shepherd [6]; the difference is that

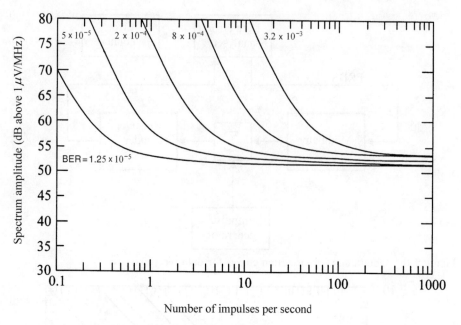

Figure 9.16 Family of isodegradation curves for a VHF receiver using FSK modulation at

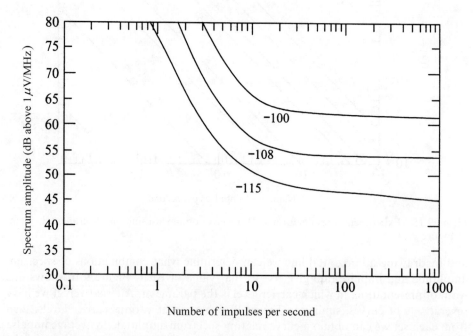

Figure 9.17 Family of isodegradation curves for the same receiver as in Figure 9.16; the labels on the curves represent the signal level in dBm (BER $= 4 \times 10^{-3}$).

the primary performance measure has been set at BER $= 4 \times 10^{-3}$ rather than 20 dB quieting.

The presentation in Figure 9.16 has several practical advantages and is therefore greatly preferred. Firstly, it clearly shows a threshold level below which the noise impulses have negligible effect. Secondly, if we were to conduct an experiment to measure the BER due to a certain noise source, either in the field or in the laboratory, then the simplest way to proceed would be to fix the input carrier level, add the noise at the receiver input, and measure the BER. In practice, therefore, carrier level is the obvious parameter.

An early document relating to the NAD concept [31] also proposed an empirical method for determining the susceptibility of the radio receiver to impulsive noise. For FM radio receivers the NAD 'overlay' technique involves overlaying the measured NAD curve for a given noise source on the isodegradation curves for the receiver under consideration, in the manner shown in Figure 9.18. It was originally suggested that the point of tangency between the NAD and one of the isodegradation curves indicates the degradation to be expected from that source. In Figure 9.18, which shows two NAD curves overlaid on a set of isodegradation curves, the expected reduction in sensitivity is approximately 2 dB.

Note that it is not the relatively small number of high-amplitude noise pulses that cause the maximum degradation, nor indeed is it the large number of low-amplitude pulses. The greatest contribution comes from pulses about 10 dB below the maximum, and they occur at a rate of a few per second. Nevertheless, it is prudent to be a little cautious in using this criterion, because although the tangency point is a good indicator of performance and may be useful in, say, comparing the degradation caused to different receivers by the same noise source, no one has yet produced a theory that establishes the precise relationship between the degradation indicated by the tangency point and the degradation actually measured under

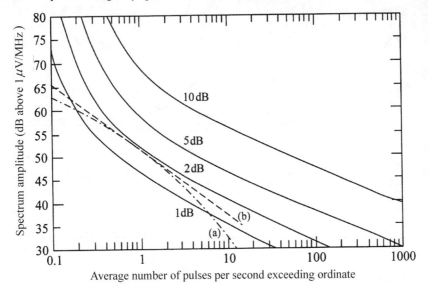

Figure 9.18 Measured NAD curves and isodegradation curves can be overlaid to obtain an estimate of system degradation.

experimental conditions. It would appear intuitively that the relative slopes of the two sets of curves must have an influence. For example, the NAD curve marked (b) in Figure 9.18 is likely to cause greater degradation than curve (a), even though the tangency point is the same; this is because a larger number of high-amplitude impulses are present.

One possible limitation associated with the overlay technique is that the isodegradation curves are plotted for a train of impulses uniformly spaced in time, whereas the NAD for the measured noise source is likely to consist of pulses with random arrival times. An accurate prediction is only possible if there is negligible overlapping of impulse responses at the receiver IF output. The expected effect is small but the question has been investigated [28] by arranging for a pulse train having a Poisson distribution of arrival times to be applied to the 'external trigger' socket of an impulse generator used in the experimental configuration of Figure 9.14. The only observed effects were at high average impulse rates, where there is a significant probability, due to overlapping responses, that one data symbol is affected by more than one noise pulse. The method of measurement using pulses uniformly spaced in time is therefore adequate for almost all practical purposes.

For data communication systems the overlay technique can be extended and used in association with isodegradation curves plotted as in Figure 9.16. The method is illustrated in Figure 9.19 and it can be seen that for the given receiver at a signal level of −108 dBm the point of tangency between the NAD and the isodegradation curves corresponds to an error of approximately 2×10^{-4}.

Experiments have established an empirical relationship between the degradation indicated by the tangency point (termed the tangential degradation) and

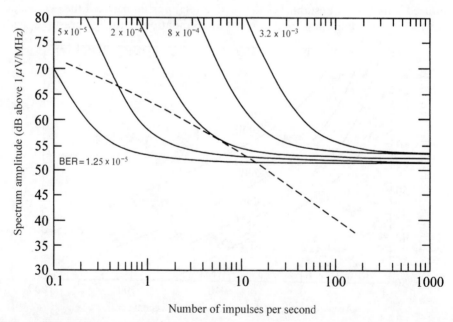

Figure 9.19 The overlay technique can be applied to a data receiver having the isodegradation curves in Figure 9.16.

experimentally measured error rates at different radio locations. The results indicate that a multiplication factor in the range 2 to 2.5 is appropriate depending on whether the mobile is stationary or moving in traffic.

9.9 INTERFERENCE

In planning land mobile radio systems, a fundamental requirement is to obtain a specified minimum grade of service over the intended coverage area. We have seen in earlier chapters that because the signal is subject to multipath propagation its amplitude at any given location can only be expressed in statistical terms, so it is necessary to express the grade of service either in terms of a *service reliability* which is the probability of achieving satisfactory reception, or as an *outage probability* which is the probability of unsatisfactory reception.

Initially it is necessary to define what is meant by 'satisfactory reception'. Clearly we need the value of the wanted signal to be above some minimum value determined by the receiver threshold and noise considerations. In the present context this is a coverage criterion [32]; in other words, given the signal statistics we can calculate, for any specified location, the probability that the minimum signal level is exceeded. As an example of this, Jakes [33] calculates, for a circle of given radius (which could represent a cell in a radio-telephone system), the percentage of locations within the circle that have a mean signal level above a specified value s_0, given the percentage of locations on the circumference that have a mean level above s_0. Assuming the received signal strength to be proportional to d^{-4} and a lognormal variation in mean signal strength over a small area, with $\sigma = 8\,\mathrm{dB}$, about 96% of locations within the circle have a mean signal above s_0 when 90% of locations on the circumference are above the same level.

Alternatively we can consider the signal level at a specific point, i.e. take the Rayleigh fading into account, so that the total signal is Suzuki distributed. In these circumstances with $\sigma = 5\,\mathrm{dB}$, if the base station power is sufficient to produce an adequate signal at 90% of locations on the cell boundary then an adequate signal also exists at 50% of locations at a distance of twice the cell radius, 10% of locations at 3 times, 5% of locations at 3.6 times, and 1% of locations at 4.5 times [32]. Figure 9.20 illustrates the implications of this for a cellular radio system where, for 7-cell clusters the reuse distance is 4.6 times the cell radius; there is an obvious potential for co-channel interference.

The question of coverage will be treated much more extensively in Chapter 11 but we now turn to the second consideration relevant to what has been termed 'satisfactory reception'. In all mobile radio systems, particularly cellular schemes, there are always other base stations operating on the same frequency as the wanted station. Co-channel interference exists as a result of frequency reuse, and satisfactory reception therefore requires that at the receiving point, the signal received from the wanted transmitter exceeds that from unwanted transmitters by a specified factor known as the *protection ratio*; the required value of the protection ratio depends on the nature of the transmissions. This is a carrier-to-interference criterion and outage on this basis has been dealt with, in some detail, by French [34].

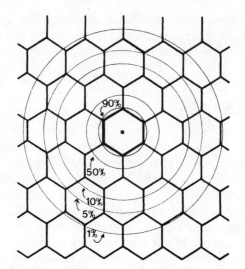

Figure 9.20 The probability of coverage at various distances from a cell in which 90% of locations at the cell edge are covered, i.e. have an adequate 'wanted' signal strength.

In summary, there are two criteria for satisfactory reception. If we consider a receiver located in the small area shown in Figure 9.21 then the coverage criterion demands that the signal s_w from the wanted transmitter T_w exceeds some threshold level s_0, i.e. $s_w \geqslant s_0$. If this criterion is not satisfied then an outage exists; the probability of this is the probability that $s_w < s_0$:

$$p_{out}^c = \int_{-\infty}^{s_0} p_w(s_w) \, ds_w \tag{9.11}$$

Figure 9.22 shows that under fading conditions, even if the mean value of s_w exceeds s_0 by several decibels, there is still a significant probability of outage.

The carrier-to-interference criterion implies that in terms of Figure 9.21, s_w must exceed s_i by the protection ratio R (dB), i.e. $s_w \geqslant s_i + R$. On this basis the probability

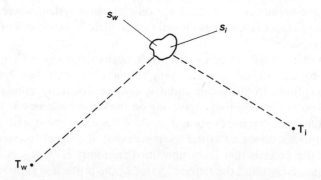

Figure 9.21 Wanted and interfering signals in a small area, arising from transmitters T_w and T_i.

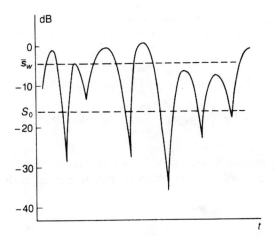

Figure 9.22 Although $\bar{s}_w$ exceeds the threshold s_0 by several decibels, there are still times when an outage occurs.

of outage is the probability that $s_i > (s_w + R)$, weighted by the PDF of s_w and integrated over all possible values of s_w, i.e.

$$p_{out}^i = \int_{-\infty}^{+\infty} p_w(s_w) \left[\int_{s_w-R}^{\infty} p_i(s_i) \, ds_i \right] ds_w \qquad (9.12)$$

Figure 9.23 shows that if both wanted and interfering signals fade, then even if the mean value of s_w exceeds that of s_i by R, there are many times at which the instantaneous values differ by far less than R, and an outage exists.

These two criteria are conceptually straightforward. They have been stated independently, but in practice they need to be satisfied simultaneously; that is, we

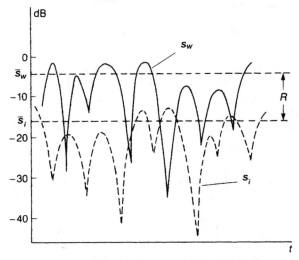

Figure 9.23 Although $\bar{s}_w$ and $\bar{s}_i$ differ by the protection ratio R, there are many times when the instantaneous values differ by far less than R in the presence of fading.

need both an adequate signal and an adequate signal-to-interference ratio (SIR). In the context of Figure 9.23, the probability of outage is the probability of failing to meet both criteria

$$= 1 - \text{probability of meeting both criteria}$$

$$= 1 - \int_{s_0}^{\infty} p_w(s_w) \left[\int_{-\infty}^{s_w - R} p_i(s_i) \, ds_i \right] ds_w \tag{9.13}$$

Of course, in any specific situation, one criterion may dominate to such an extent that it effectively becomes the sole criterion, but in general both must be satisfied together.

9.9.1 Single interferer

As an example we consider the case of a single interfering transmission when both wanted and interfering signals are subject to fading and shadowing, i.e. have Suzuki statistics. The rather simpler case of shadowing only, i.e. lognormal statistics, was treated many years ago by Sachs [35].

The PDF of the signal can be obtained by superimposing the Rayleigh and lognormal distributions, i.e.

$$p(s) = \int_{-\infty}^{+\infty} p_r(s'/\bar{s}') p_l(\bar{s}) \, d\bar{s} \tag{9.14}$$

where the subscripts r and l refer to the Rayleigh and lognormal distributions. s' and $\bar{s}'$ are in linear units, s and $\bar{s}$ are in decibels (i.e. $\bar{s} = 20 \log_{10} \bar{s}'$ and $s = 20 \log_{10} s'$) and

$$p_r(s'/\bar{s}') = \frac{\pi s'}{2\,\bar{s}'^2} \exp\left(-\frac{\pi s'^2}{4\,\bar{s}'^2} \right) \tag{9.15}$$

$$p_l(\bar{s}) = \frac{1}{\sigma\sqrt{2\pi}} \exp\left(-\frac{(\bar{s} - m)^2}{2\sigma^2} \right) \tag{9.16}$$

Equation (9.17) represents the PDF of both s_w and s_i. From eqns. (9.11), (9.12) and (9.13) we can obtain

$$p_{\text{out}}^c = 1 - \frac{1}{\sqrt{\pi}} \int_{-\infty}^{+\infty} \exp(-y^2) \exp\left(-\frac{\pi}{4} 10^{-(\alpha + \sigma y\sqrt{2})/10} \right) dy \tag{9.17}$$

$$p_{\text{out}}^i = \frac{1}{\sqrt{\pi}} \int_{-\infty}^{+\infty} \exp(-y^2)(1 - 10^{(\tau - 2\sigma y)/10})^{-1} dy \tag{9.18}$$

Thus the outage probability is

$$p_{out} = p_{out}^c - \frac{1}{\pi} \int_{-\infty}^{+\infty} \int_{-\infty}^{+\infty} \exp(-y_i^2 - y_w^2)$$

$$\times \exp\left\{ -\frac{\pi}{4} [1 + 10^{\{\tau + \sigma\sqrt{2}(y_w - y_i)\}/10}] 10^{-(\alpha + \sigma y_w \sqrt{2})/10} \right\}$$

$$\times [1 + 10^{\{\tau + \sigma\sqrt{2}(y_w - y_i)\}/10}]^{-1} dy_i \, dy_w \qquad (9.19)$$

where $\alpha = m_w - s_0$, the margin in decibels, by which the mean level of the wanted signal exceeds the threshold; $\tau = (m_w - m_i) - R$, i.e. the margin by which the difference between the mean levels of the wanted and interfering signal exceeds the specified protection ratio. For simplicity, σ has been taken as the same for s_w and s_i. We note that (9.18) is the same as the equation given by French [34].

Results obtained by numerically evaluating eqn. (9.19) are given in Figure 9.24. In Figure 9.24(a) the case $\tau = \infty$ corresponds to p_{out}^c; in Figure 9.24(b) the case $\alpha = \infty$ corresponds to p_{out}^c. When $\alpha \gg \tau$ interference considerations predominate; when $\alpha \ll \tau$ coverage conditions predominate. When $\alpha \cong \tau$, p_{out} is significantly different to either p_{out}^c or p_{out}^i.

As an example we consider a base station transmitter with ERP = 53 dBm, antenna height 100 m, communicating with a mobile having a receiver threshold of -110 dBm, antenna height 2 m, gain 0 dB. We assume that interference arises from a base station 30 km away, having the same parameters as the wanted station. The outage probability along a line joining the two stations is shown in Figure 9.25 as a function of distance from the wanted station. The calculations have been based on $\sigma = 6$ dB and the 'plane earth plus clutter factor' model has been used with $\beta = 45$ dB. These values are typical of 900 MHz transmissions in urban areas. Close to the wanted station the outage probability is very low but it increases further away. Note that p_{out}^c and p_{out}^i are comparable everywhere, and that p_{out} is significantly greater than either of them.

The 'shadowing only' case (lognormal statistics) is of practical interest if diversity reception, or an equivalent technique, is used to mitigate the effects of Rayleigh fading. Results calculated for this situation are also shown in Figure 9.25. The outage probability is much lower in this case as a direct result of the absence of the deep and rapid signal variations associated with Rayleigh fading.

9.9.2 Multiple interferers

In mobile radio systems, particularly cellular systems, it is likely that interference will arise simultaneously from a number of co-channel transmitters. In this case, as far as the carrier-to-interference criterion is concerned, it is necessary for the wanted signal power to exceed the sum of the powers of the interfering signals by an amount equivalent to the protection ratio R (dB). The question then arises as to the treatment of multiple interferers. There are two obvious approaches. One is to find the statistics of a single 'equivalent' interferer; this involves finding the distribution of the sum of the short-term interfering signal powers. If this can be done then eqn. (9.13) can be used to determine the outage probability. Alternatively, if the

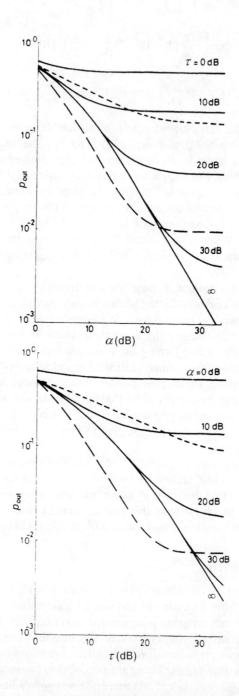

Figure 9.24 (a) Values of p_{out} plotted against α at $\tau = 20$ dB. (b) Values of p_{out} plotted against τ at $\alpha = 20$ dB. In each case (——) $\sigma = 6$ dB, (– – –) $\sigma = 0$ dB, (- - - -) $\sigma = 12$ dB.

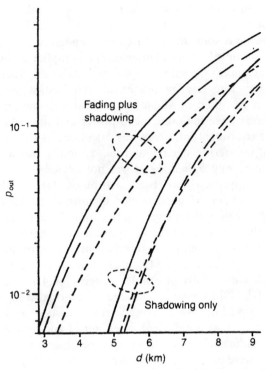

Figure 9.25 Outage probabilities for the case described in the text: (———) p_{out}, (– – –) p_{out}^c, (- - - -) p_{out}^i.

distribution of the wanted signal $p(s_w)$ and those of the individual interfering signals $p(s_{i1})$, $p(s_{i2})$, . . ., $p(s_{in})$ are all known then the outage probability can be expressed as

$$p_{out} = 1 - \int_{s_0}^{\infty} p(s_w) \int_0^{s_w/r} p(s_{i1}) \int_0^{(s_w/r)-s_{i1}} p(s_{i2}) \ldots \int_0^{(s_w/r)-\sum^{n-1} s} p(s_{in}) \, ds_{in} \ldots ds_w$$

(9.20)

In this equation the outermost integral recognises that the wanted signal s_w must exceed some minimum value s_0, i.e. it expresses the coverage criterion, whereas the inner n integrals recognise that s_w must simultaneously be greater than the instantaneous sum of the interfering signal powers by a margin of the protection ratio r (here r is in linear units). This 'multiple integral' method was the approach used by Sowerby [36].

In general, the statistics of the single equivalent interferer can only be approximated. For example, Engel [37] treated multiple interferers each having Rayleigh statistics by approximating them to a single Rayleigh interferer with a mean equal to the sum of the means of the actual interferers. Lundquist and Peritsky [38] used a similar method to treat the case of six co-channel interferers surrounding a wanted cellular radio base station. Muammar and Gupta [39] represented a number of interferers in terms of a normal distribution. However, they found the mean voltage and variance of the equivalent interferer by summing signal voltages rather than powers, a procedure which is only valid if the interferers are coherent. Consequently, their results are pessimistic.

Lognormal shadowing is sometimes treated independently from Rayleigh fading, but the treatment of multiple lognormal interferers is complicated by the fact there is no closed-form expression for the sum of lognormal variables. Attempts to calculate outage probability therefore rely on numerical or statistical approximations. A simple approximation was used by Chan [40] to deal with the interference-limited case when all interferers have the same mean power and the same variability. The median power of the equivalent interferer was calculated as the sum of the median powers of the actual interfering signals, and its variability was assumed to be the same as the actual interfering signals, i.e. σ was not altered.

A more accurate approach has been attributed [41] to Wilkinson, who approximated the sum of several uncorrelated lognormal distributions by another lognormal distribution with a mean equal to the sum of the means and a variance equal to the sum of the variances of the component distributions. The method has been applied to the study of multiple lognormal interferers in interference-limited systems [42–44]. Schwartz and Yeh [45] also derived an equivalent lognormal distribution and used it in a study of cellular systems [46,47]. Their approach was later extended by Safak [48] to include correlated lognormal components.

The problem of multiple interferers having Suzuki statistics has not received much attention in terms of a single equivalent interferer. Lundquist and Peritsky [38] studied the effect of fading and shadowing by generating lognormally distributed random numbers that were used to select the mean values of Rayleigh fading signals; these were then treated using their approximate method for multiple Rayleigh interferers. Chan [40] used a multiple Rayleigh interferer approach as a basis for tackling fading plus shadowing. He assumed the mean power of each of the Rayleigh fading signals to be lognormally distributed and that the mean powers of the Rayleigh fading components were equal. The equality assumption is difficult to justify; nevertheless, Chan integrated over two lognormal distributions to obtain an expression for outage probability in the fading-plus-shadowing case, obtaining a result which somewhat underestimates actual outage probabilities.

Several researchers have based outage probability calculations for multiple Suzuki interferers on results from corresponding multiple lognormal interferer situations. Some have calculated required signal levels by first assuming that only lognormal fading occurs and then adding a protective margin to allow for Rayleigh fading [49]. The Wilkinson method has been used as the basis of two approaches [39,44], but Muammar and Gupta again summed signal voltages rather than powers, hence they obtained pessimistic results.

A few approaches to the single-interferer problem have been made using descriptions in terms of the Nakagami-m distribution [50,51] and the Rice distribution [52]. Chapter 5 showed that under certain conditions the Suzuki distribution can be approximated by a lognormal distribution and this provides the basis of an alternative procedure.

An original and elegant approach to the outage probability problem has been taken by Sowerby and Williamson [36,53]. They have principally used the 'multiple integral' approach to the problem, although the equivalent interferer technique was also employed. Exact outage probability expressions for several cases of practical interest are given, in particular multiple Rayleigh [54], two lognormal [55], and up to four Suzuki interferers. Approximations are presented for several other cases, and the effect of system techniques such as diversity reception is included [56]. Because descriptions in terms of Nakagami-m, Rice and Weibull statistics have been used by several authors, they are also considered.

As an example, they consider an idealised cellular mobile radio system based on a 7-cell repeat pattern with a nominal cell radius of 3 km [57]. The wanted base station is surrounded by six co-channel interfering stations each at a distance of 13.75 km (4.58 × cell radius). The base stations are all assumed identical and transmit just enough power to give a 10% outage probability at their particular cell perimeter. It is assumed that the signals are subjected to fading plus shadowing, i.e. Suzuki statistics are appropriate. Figure 9.26 shows how the 10% outage probability contour recedes from its original 3 km position when the interfering base stations transmit.

The extent of the recession depends on the propagation conditions, the effect of interference being greatest in conditions when the standard deviation σ is high and the exponent n in the propagation law is low. (Note that the trend with n is opposite to the trend for interference-free systems.) These conditions are most likely to exist in urban areas where the cell radii are small. A protection ratio of 6 dB has been assumed in the calculation; a higher value would result in the effect of interference being more severe. For comparison, Figure 9.27 shows a more restricted case of a system in which outage is caused by interference alone and a value of $n = 4$ is assumed. Outage probability contours are shown for fading and shadowing (Suzuki) and shadowing alone (lognormal). Notice that a reduction in outage probability can be achieved by using a technique such as diversity, which alters the signal statistics so that 'shadowing only' is more representative of the conditions experienced.

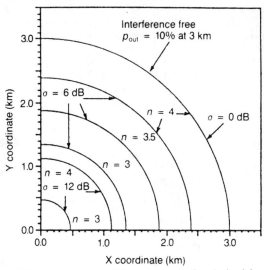

Figure 9.26 The 10% outage probability contours of a desired base station located at the origin, with six co-channel interferers, in a cellular radio system with a 7-cell repeat pattern and a cell radius of 3 km. All base station transmitter powers are sufficient to give 10% outage probability at 3 km in the absence of interference; the protection ratio is 6 dB.

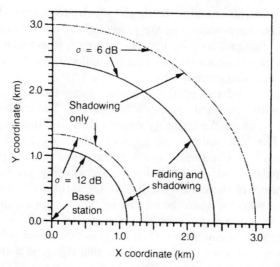

Figure 9.27 Interference-limited 10% outage probability contours of a desired base station at the origin, with six co-channel interferers, in a cellular radio system with a 7-cell repeat pattern and a cell radius of 3 km. All base station transmitter powers are assumed to be equal. A protection ratio of 6 dB and a propagation law exponent of 4 have been used.

REFERENCES

1. Skomal E.N. (1978) *Man-made Radio Noise*. Van Nostrand Reinhold, New York.
2. Herman J.R. (1971) Survey of man-made noise. *Progress in Radio Sci.*, **1**, 315–48.
3. Horner F. (1971) Techniques for the measurement of atmospheric and man-made noise. *Progress in Radio Sci.*, **2**, 177–82.
4. Shepherd R.A. *et al.* (1976) New techniques for suppression of automobile ignition noise. *IEEE Trans.*, **VT25**(1), 2–12.
5. Shepherd R.A. *et al.* (1975) *Improved suppression of radiation from automobiles used by the general public*. Final Report to Federal Communications Commission. Contract FCC-0072, Stanford Research Institute, Menlo Park CA.
6. Shepherd R.A., Gaddie J.C. and Shohara A. (1975) *Measurement parameters for automobile ignition noise*. Final Report SRI 7502-C-2-10, Stanford Research Institute.
7. Andrews J.R. and Arthur M.G. (1977) *Spectrum amplitude – definition, generation and measurement*. NBS Technical Note 699.
8. *IEEE standard for the measurement of impulse strength and impulse bandwidth*. IEEE STD 376-1975.
9. *World distribution and characteristics of atmospheric radio noise*. CCIR Report 322-1.
10. CISPR (1972) *Specifications for CISPR radio interference measuring apparatus for the frequency range 0.15 Mc/s to 30 Mc/s*. CISPR Publication 1. IEC, Geneva.
11. CISPR (1975) *Specifications for CISPR radio interference measuring apparatus for the frequency range 25 Mc/s and 300 Mc/s*. CISPR Publication 2. IEC, Geneva.
12. CISPR (1975) *Specifications for CISPR radio interference measuring apparatus for the frequency range 10 kHz to 150 kHz*. CISPR Publication 3. IEC, Geneva.
13. Shepherd R.A. *et al.* (1975) *Measurement parameters for automobile ignition noise*. Stanford Research Institute, NTIS PB 247766.
14. Parsons J.D. and Turkmani A.M.D. (1986) A new method of assessing receiver performance in the presence of impulsive noise. *IEEE Trans.*, **COM34**(11), 1156–61.
15. Deitz J. *et al.* (1967) *Man-made Noise: Report of the FCC Advisory Committee for Land Mobile Radio Services (ACLMRS)*. Appendix A, Vol. 2, Part 2 (Working Group B-3). Available from NTIS, Springfield VA, PB 174-278).

16. Shepherd N.H. (1978) *Impulsive noise measurements (man-made and atmospheric)*. Working Group 1, International Electrotechnical Commission.

17. IEEE (1968) *Spectrum Engineering – The Key to Progress: Unintended Radiation*. Supplement 9 to Report of the Joint Advisory Committee.

18. Esposito R. (1972) *Research in the characterisation and measurement of man-made electromagnetic noise*. Final Report on DOT-TSC-73, Raytheon Research Division, Waltham MA.

19. CCIR (1974) *Man-made radio noise*. CCIR Report 258-3.

20. Shepherd R.A. (1974) Measurement of amplitude probability distribution and power of automobile ignition noise at hf. *IEEE Trans.*, **VT23**(3), 72–83.

21. Esposito R. and Buck R.E. (1973) A mobile wide-band measurement system for urban man-made noise. *IEEE Trans.*, **COM21**(11), 1224–32.

22. Parsons J.D. and Sheikh A.U.H. (1978) A receiving system for the measurement of noise amplitude distribution. *Proc. IERE Conf. on Radio Receivers and Associated Systems (IERE Conference Publication 40)*, pp. 281–7.

23. Turkmani A.M.D. and Parsons J.D. (1986) A wide band, high dynamic range receiving system for measuring the parameters of impulsive noise. *Proc. 4th Int. Conf. on Radio Receivers and Associated Systems (IERE Conference Publication 68)*, pp. 25–33.

24. Sheikh A.U.H. and Parsons J.D. (1983) Frequency dependence of urban man-made radio noise. *Radio and Electronic Engineer*, **53**(3), 99–106.

25. Spaulding A.D. (1964) Determination of error rates for narrow band communication of binary-coded messages in atmospheric radio noise. *Proc. IEEE*, **52**, 220–1.

26. Conda A.M. (1965) The effect of atmospheric noise on the probability of error for an NCFSK system. *IEEE Trans.*, **COM13**, 281–4.

27. Parsons J.D. and Reyhan T. (1985) Prediction of bit error rate in the presence of impulsive noise: a numerical approach using measured noise data. *Proc. IEE Part F*, **132**(5), 334–42.

28. Parsons J.D. and Turkmani A.M.D. (1986) A new method of assessing receiver performance in the presence of impulsive noise. *IEEE Trans.*, **COM34**(11), 1156–61.

29. Bello P.A. and Esposito R.A. (1969) New method of calculating error probabilities due to impulsive noise. *IEEE Trans.*, **COM17**, 367–79.

30. Engel J.S. (1965) Digital transmission in the presence of impulsive noise. *Bell Syst. Tech. J.*, **44**, 1699–743.

31. Shepherd N.H. (1978) *Impulsive radio noise measurements (man-made and atmospheric)*. International Electrotechnical Commission, Technical Committee No. 12, 12/F WG1 Part 9, Section 1.

32. Parsons J.D. (1984) Propagation and interference in cellular radio systems. *Proc. Conf. on Mobile Radio Systems and Techniques (IEE Conference Publication 238)*, pp. 71–5.

33. Jakes W.C. (1974) *Microwave Mobile Communications*. John Wiley, New York.

34. French R.C. (1979) The effect of fading and shadowing on channel reuse in mobile radio. *IEEE Trans.*, **VT28**(3), pp. 171–81.

35. Sachs H.M. (1971) A realistic approach to defining the probability of meeting acceptable receiver performance criteria. *IEEE Trans.*, **EMC13**(4), 3–6; see also **EMC14**(2), 74–8.

36. Sowerby K.W. (1989) Outage probability in mobile radio systems. PhD thesis, University of Auckland, New Zealand.

37. Engel J.S. (1969) The effects of co-channel interference on the parameters of a small-cell mobile telephone system. *IEEE Trans.*, **VT18**(3), 110–16.

38. Lundquist L. and Peritsky M.M. (1971) Cochannel interference rejection in a mobile radio space diversity system. *IEEE Trans.*, **VT20**(3), 68–75.

39. Muammar R. and Gupta S.C. (1982) Cochannel interference in high capacity mobile radio systems. *IEEE Trans.*, **COM30**(8), 1973–8.

40. Chan G.K. (1984) Design and analysis of a land mobile radio system under the effects of interference. PhD thesis, Carleton University, Ottawa, Canada.

41. Fenton L.F. (1960) The sum of lognormal probability distributions in scatter transmission systems. *IRE Trans.*, **CS8**(1), 57–67.

42. Daikoku K. and Ohdate H. (1983) Optimal channel reuse in cellular land mobile radio systems. *IEEE Trans.*, **VT32**(3), 217–24.

43. Stjernvall J.E. (1985) Calculation of capacity and cochannel interference in a cellular system. *Proc. Nordic Seminar on Digital Land Mobile Radio Communications*, pp. 209–17.
44. Nagata Y. and Akaiwa Y. (1987) Analysis of spectrum efficiency in single cell trunked and cellular mobile radio. *IEEE Trans.*, **VT35**(3), 100–13.
45. Schwartz S.C. and Yeh Y.S. (1982) On the distribution function and moments of power sums with lognormal components. *Bell Syst. Tech. J.*, **61**(7), 1441–63.
46. Yeh Y.S. and Schwartz S.C. (1984) Outage probability in mobile radio telephony due to multiple log-normal interferers. *IEEE Trans.*, **COM32**(4), 380–7.
47. Yeh Y.S., Wilson J.C. and Schwartz S.C. (1984) Outage probability in mobile telephony with directive antennas and macrodiversity. *IEEE J.*, **SAC2**(4), 507–11.
48. Safak A. (1993) Statistical analysis of the power sum of multiple correlated log-normal components. *IEEE Trans.*, **VT42**(1), 58–61.
49. Hughes C.J. and Appleby M.S. (1985) Definition of a cellular mobile radio system. *Proc. IEE Part F*, **132**(5), 416–24.
50. Wojnar A.H. (1986) Unknown bounds on performance in Nakagami channels. *IEEE Trans.*, **COM34**(1), 22–4.
51. Al-Hussaini E.K. (1988) Effects of Nakagami fading on antijam performance requirements. *Electron. Lett.*, **24**(4), 208–9.
52. Oetting J.D. (1987) The effects of fading on antijam performance requirements. *IEEE J.*, **SAC5**(2), 155–61.
53. Sowerby K.W. and Williamson A.G. (1988) Outage probability calculations for multiple cochannel interferers in cellular mobile radio systems. *Proc. IEE Part F*, **135**(3), 208–15.
54. Sowerby K.W. and Williamson A.G. (1987) Outage probability calculations for a mobile radio system having multiple Rayleigh interferers. *Electron. Lett.*, **23**(11), 600–1.
55. Sowerby K.W. and Williamson A.G. (1987) Outage probability calculations for a mobile radio system having two lognormal interferers. *Electron Lett.*, **23**(25), 1345–6.
56. Sowerby K.W. and Williamson A.G. (1988) Selection diversity in multiple interferer mobile radio systems. *Electron. Lett.*, **24**(24), 1511–13.
57. Sowerby K.W. and Williamson A.G. (1989) Estimating service reliability in multiple interferer mobile radio systems. *Proc. Int. Conf. on Mobile Radio and Personal Communication Systems*, Warwick, pp. 48–52.

Chapter 10

Mitigation of Multipath Effects

10.1 INTRODUCTION

We have seen in Chapters 4 and 5 that buildings and other obstacles in built-up areas act as scatterers of the signal, and because of the interaction between the various incoming component waves, the resultant signal at the mobile antenna is subject to rapid and deep fading. The fading is most severe in heavily built-up areas such as city centres, and the signal envelope often follows a Rayleigh distribution over short distances in these heavily cluttered regions. As the degree of urbanisation decreases, the fading becomes less severe; in rural areas it is often only serious when there are obstacles such as trees close to the vehicle.

A receiver moving through this spatially varying field experiences a fading rate which is proportional to its speed and the frequency of transmission, and because the various component waves arrive from different directions there is a Doppler spread in the received spectrum. It has been pointed out that the fading and the Doppler spread are not separable, since they are both manifestations (one in the time domain and the other in the frequency domain) of the same phenomenon. In addition there is the delay spread which leads to frequency-selective fading. This causes distortion in wideband analogue signals and intersymbol interference (ISI) in digital signals. These multipath effects can cause severe problems and, particularly in urban areas, multipath is probably the single most destructive influence on mobile radio systems. Much attention has been devoted to techniques aimed at mitigating the deleterious effects it causes and this chapter reviews some of the available approaches to the problem.

10.2 DIVERSITY RECEPTION

One well-known method of reducing the effects of fading is to use diversity reception techniques. In principle they can be applied either at the base station or at the mobile, although different problems have to be solved. The basic idea underlying diversity reception has been outlined in Section 5.12 and relies on obtaining two or more samples (versions) of the incoming signal which have low, ideally zero, cross-correlation. It follows from elementary statistics that the probability of M independent samples of a random process all being simultaneously below a certain level is p^M where p is the probability that a single sample is below the level. It can be seen therefore that a signal composed of a suitable combination of the various versions will have fading properties much less severe than those of any individual version alone.

Two questions must be answered. How can these independent samples (or versions) be obtained? Then how can they be processed to obtain the best results? Potentially there are several ways to obtain the samples; for example, we could use the fact that the electrical lengths of the scattered paths are a function of the carrier frequency to obtain independent versions of the signal from transmissions at different frequencies. However, *frequency diversity*, as it is called, is not a viable proposition for most mobile radio systems because the coherence bandwidth is quite large (from several tens of kilohertz to a few megahertz depending on the circumstances) and in any case the pressures on spectrum utilisation are such that multifrequency allocations cannot be made.

Two other possibilities are *polarisation diversity* and *field diversity*; polarisation diversity relies on the scatterers to depolarise the transmitted signal, and field diversity uses the fact that the electric and magnetic components of the field at any receiving point are uncorrelated, as shown in Chapter 5. Both these methods have their difficulties, however, since there is not always sufficient depolarisation along the transmission path for polarisation diversity to be successful, and there are difficulties with the design of antennas suitable for field diversity. *Time diversity*, i.e. repeating the message after a suitable time interval, has its attractions in digital systems where storage facilities are available at the receiver (see later). Automatic repeat request (ARQ) systems which use the same underlying principle have been available in conventional mobile radio systems for some years.

It is *space diversity* (obtaining signals from two or more antennas physically separated from each other) that seems by far the most attractive and convenient method of diversity reception for mobile radio. The necessary antenna separation can easily be obtained at base stations and, assuming isotropic scattering at the mobile end of the link, the autocorrelation coefficient of the envelope of the electric field falls to a low value at distances greater than about a quarter-wavelength (Chapter 5). Almost independent samples can therefore be obtained from antennas sited this far apart. At VHF and above, the distance involved is less than a metre and this is easily obtained within the dimensions of a normal vehicle. At UHF it may be feasible even using hand-portable equipment; this will be discussed later.

10.3   BASIC DIVERSITY METHODS

Having obtained the necessary versions of the signal, we must process them to obtain the best results. There are various possibilities, but what is 'best' really amounts to deciding what method gives the optimum improvement, taking into account the complexity and cost involved.

For most communication systems the possibilities reduce to methods which can be broadly classified as *linear combiners*. In linear diversity combining, the various signal inputs are individually weighted and then added together. If addition takes place after detection the system is called a *post-detection combiner*; if it takes place before detection the system is called a *predetection combiner*. In the predetection combiner it is necessary to provide a method of cophasing the signals before addition.

Assuming that any necessary processing of this kind has been done, we can express the output of a linear combiner consisting of M branches as

$$s(t) = a_1 s_1(t) + a_2 s_2(t) + \ldots + a_M s_M(t)$$

$$s(t) = \sum_{k=1}^{M} a_k s_k(t) \tag{10.1}$$

where $s_k(t)$ is the envelope of the kth signal to which a weight a_k is applied.

The analysis of combiners is usually carried out in terms of CNR or SNR, with the following assumptions [1]:

(a) The noise in each branch is independent of the signal and is additive.
(b) The signals are locally coherent, implying that although their amplitudes change due to fading, the fading rate is much slower than the lowest modulation frequency present in the signal.
(c) The noise components are locally incoherent and have zero means, with a constant local mean square (constant noise power).
(d) The local mean square values of the signals are statistically independent.

Different realisations and performances are obtained depending on the choice of a_k, and this leads to three distinct types of combiners: *scanning and selection combiners*, *equal-gain combiners and maximal ratio combiners*. They are illustrated in Figure 10.1. In the scanning and selection combiners only one a_k is equal to unity at any time; all others are zero. The method of choosing which a_k is set to unity provides a distinction between scanning and selection diversity. In scanning diversity the system scans through the possible input signals until one greater than a preset threshold is found. The system then uses that signal until it drops below the threshold, when the scanning procedure restarts. In selection diversity the branch with the best short-term CNR is always selected. Equal-gain and maximal ratio combiners accept contributions from all branches simultaneously. In equal-gain combiners all a_k are unity; in maximal ratio combiners a_k is proportional to the root mean square signal and inversely proportional to the mean square noise in the kth branch.

Scanning and selection diversity do not use assumptions (b) and (c), but equal-gain and maximal ratio combining rely on the coherent addition of the signals against the incoherent addition of noise. This means that both equal-gain and maximal ratio combining show a better performance than scanning or selection combining, provided the four assumptions hold. It can also be shown that in this case maximal ratio combiners give the maximum possible improvement in CNR; the output CNR being equal to the sum of the CNRs from the branches [2]. However, this is not true when either assumptions (b) or (c), or both, do not hold (as might be the case with ignition noise, which tends to be coherent in all branches), in which case selection or scanning can outperform maximal ratio and equal-gain combining, especially when the noises in the branches are highly correlated.

In the remainder of this section we briefly review some of the fundamental results for different diversity schemes. The subject is fully treated by Jakes [3], so the detailed mathematical treatment is not reproduced here.

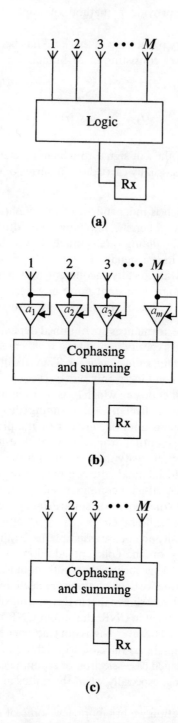

Figure 10.1   Diversity reception systems: (a) selection diversity, (b) maximal ratio combining
$(a_k = r_k/N)$, (c) equal-gain combining.

10.3.1 Selection diversity

Conceptually, and sometimes analytically, selection diversity is the simplest of all the diversity systems. In an ideal system of this kind the signal with the highest instantaneous CNR is used, so the output CNR is equal to that of the best incoming signal. In practice the system cannot function on a truly instantaneous basis, so to be successful it is essential that the internal time constants of a selection system are substantially shorter than the reciprocal of the signal fading rate. Whether this can be achieved depends on the bandwidth available in the receiving system. Practical systems of this type usually select the branch with the highest carrier-plus-noise, or utilise the scanning technique mentioned in the previous section.

For the moment we examine the ideal selector (Figure 10.1(a)) and state the properties of the output signal. We assume that the signals in each diversity branch are uncorrelated narrow-band Gaussian processes of equal mean power; this means their envelopes are Rayleigh distributed and, following the analysis in Appendix B, the PDF of the CNR can be written as

$$p(\gamma) = \frac{1}{\gamma_0} \exp(-\gamma/\gamma_0)$$

The probability of the CNR on any one branch being less than or equal to any specific value γ_s is

$$P[\gamma_k \leqslant \gamma_s] = \int_0^{\gamma_s} p(\gamma_k)\, \mathrm{d}\gamma_k = 1 - \exp(-\gamma_s/\gamma_0) \qquad (10.2)$$

and hence the probability that the CNRs in all branches are simultaneously less than or equal to γ_s is given by

$$P_M(\gamma_s) = P[\gamma_1 \ldots \gamma_M \leqslant \gamma_s] = [1 - \exp(-\gamma_s/\gamma_0)]^M \qquad (10.3)$$

This expression gives the cumulative probability distribution of the best signal taken from M branches.

The mean CNR at the output of the selector is also of interest and can be obtained from the probability density function of γ_s:

$$\bar{\gamma}_S = \int_0^{\infty} \gamma_S p(\gamma_S)\, \mathrm{d}\gamma_S \qquad (10.4)$$

where

$$p(\gamma_S) = \frac{\mathrm{d}}{\mathrm{d}\gamma_S} P(\gamma_S) = \frac{M}{\gamma_0} [1 - \exp(-\gamma_S/\gamma_0)]^{M-1} \exp(-\gamma_S/\gamma_0) \qquad (10.5)$$

and the upper case subscript S is used to denote selection.

Substituting this into (10.4) gives

$$\bar{\gamma}_S = \int_0^{\infty} \frac{\gamma_S M}{\gamma_0} [1 - \exp(-\gamma_S/\gamma_0)]^{M-1} \exp(-\gamma_S/\gamma_0)\, \mathrm{d}\gamma$$

$$= \gamma_0 \sum_{k=1}^{M} \frac{1}{k} \qquad (10.6)$$

The cumulative probability distribution of the output SNR is plotted in Figure 10.2 for different orders of diversity. It is immediately apparent that there is a law of diminishing returns in the sense that the greatest gain is achieved by increasing the number of branches from 1 (no diversity) to 2. Moreover, the improvement is greatest where it is most needed, i.e. at low values of CNR. Increasing the number of branches from 2 to 3 produces some further improvement, and so on, but the increased gain becomes less for larger numbers of branches. Figure 10.2 shows a gain of 10 dB at the 99% reliability level for two-branch diversity and about 14 dB for three branches.

10.3.2   Maximal ratio combining

In this method, each branch signal is weighted in proportion to its own signal voltage/noise power ratio before summation (Figure 10.1(b)). When this takes place before demodulation it is necessary to co-phase the signals before combining; various cophasing techniques are available [4, Ch. 6]. Assuming this has been done, the envelope of the combined signal is

$$r_R = \sum_{k=1}^{M} a_k r_k \tag{10.7}$$

where a_k is the appropriate branch weighting and the subscript R indicates maximal ratio. In a similar way we can write the sum of the branch noise powers as

$$N_{\text{tot}} = N \sum_{k=1}^{M} a_k^2$$

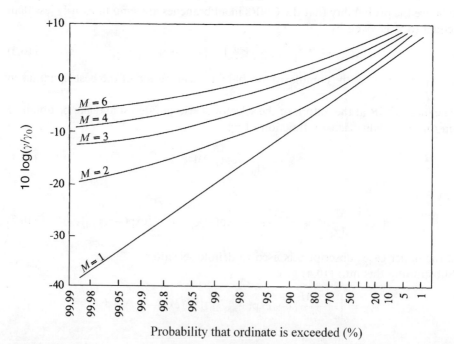

Figure 10.2 Cumulative probability distribution of output CNR for selection diversity systems.

so that the resulting SNR is

$$\gamma_R = \frac{r_R^2}{2N_{tot}}$$

Maximal ratio combining was first proposed by Kahn [2], who showed that if the various branches are weighted in the ratio signal voltage/noise power (i.e. $a_k = r_k/N$) then γ_R will be maximised and will have a value

$$\gamma_R = \frac{\left(\sum r_k^2/N\right)^2}{2N \sum (r_k^2/N)^2} = \sum_{k=1}^{M} \frac{r_k^2}{2N} = \sum_{k=1}^{M} \gamma_k \qquad (10.8)$$

This shows that the output CNR is equal to the sum of the CNRs of the various branch signals, and this is the best that can be achieved by any linear combiner.

The probability density function of γ_R is

$$\gamma_R = \frac{\gamma_R^{M-1}\exp(-\gamma_R/\gamma_0)}{\gamma_0^M(M-1)!} \qquad (\gamma_R \geqslant 0) \qquad (10.9)$$

and the cumulative probability distribution function is given by

$$P_M(\gamma_R) = 1 - \exp(-\gamma_R/\gamma_0) \sum_{k=1}^{M} \frac{(\gamma_R/\gamma_0)^{k-1}}{(k-1)!} \qquad (10.10)$$

It is a simple matter to obtain the mean output CNR from (10.8) by writing

$$\bar{\gamma}_R = \sum_{k=1}^{M} \bar{\gamma}_k = \sum_{k=1}^{M} \gamma_0 = M\gamma_0 \qquad (10.11)$$

thus $\bar{\gamma}_R$ varies linearly with M, the number of branches. Figure 10.3 shows the cumulative distributions for various orders of maximal ratio diversity, plotted from eqn. (10.10).

10.3.3 Equal-gain combining

Equal-gain combining (Figure 10.1(c)) is similar to maximal ratio combining but there is no attempt at weighting the signals before addition. The envelope of the output signal is given by eqn. (10.7) with all $a_k = 1$; the subscript E indicates equal gain. We have

$$r_E = \sum_{k=1}^{M} r_k$$

and the output SNR is therefore

$$\gamma_E = \frac{r_E^2}{2NM}$$

Of the diversity systems so far considered, equal-gain combining is analytically the most difficult to handle because the output r_E is the sum of M Rayleigh-distributed variables. The probability density function of γ_E cannot be expressed in terms of tabulated functions for $M > 2$, but values have been obtained by numerical integration techniques. The curves lie in between the corresponding ones for

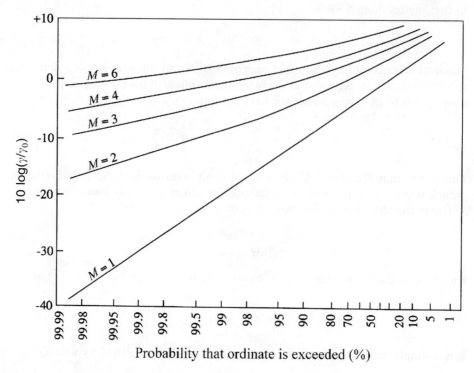

Figure 10.3 Cumulative probability distribution of output CNR for maximal ratio combining.

maximal ratio and selection systems, and in general are only marginally below the maximal ratio curves.

The mean value of the output SNR, $\bar{\gamma}_E$, can be obtained fairly easily as

$$\bar{\gamma}_E = \frac{1}{2NM} \overline{\left(\sum_{k=1}^{M} r_k\right)^2} = \frac{1}{2NM} \sum_{j,k=1}^{M} \overline{(r_j r_k)} \tag{10.12}$$

We have seen in Chapter 5 that $\overline{r_k^2} = E\{r_k^2\} = 2\sigma^2$ and $\overline{r_k} = E\{r_k\} = \sigma\sqrt{\pi/2}$. Also, since we have assumed the various branch signals to be uncorrelated, $\overline{r_j r_k} = \overline{r_j}\,\overline{r_k}$ if $j \neq k$ and in this case (10.12) becomes

$$\bar{\gamma}_E = \frac{1}{2NM}\left(2M\sigma^2 + M(M-1)\frac{\pi\sigma^2}{2}\right)$$

$$= \gamma_0\left(1 + (M-1)\frac{\pi}{4}\right) \tag{10.13}$$

10.4   IMPROVEMENTS FROM DIVERSITY

There are various ways of expressing the improvements obtainable from diversity techniques. Most of the theoretical results have been obtained for the case when the branches have signals with independent Rayleigh fading envelopes and equal mean CNR.

One useful way of obtaining an overall ideal of the relative merits of the various diversity methods is to evaluate the improvement in average output CNR relative to the single-branch CNR. For Rayleigh fading conditions this quantity, $\bar{D}$, is easily obtained in terms of M, the number of branches, using eqns (10.6), (10.11) and (10.13). The results are:

Selection (SC): $$\bar{D}(M) = \sum_{k=1}^{M} \frac{1}{k} \qquad (10.14)$$

Maximal ratio (MRC): $$\bar{D}(M) = M \qquad (10.15)$$

Equal gain (EGC): $$\bar{D}(M) = 1 + \frac{\pi}{4}(M-1) \qquad (10.16)$$

These functions have been plotted in the literature [3, Ch. 5] and show that selection has the poorest performance and maximal ratio the best. The performance of equal-gain combining is only marginally inferior to maximal ratio; the difference between the two is always less than 1.05 dB (this is the difference when $M \to \infty$). The incremental improvement also decreases as the number of branches is increased; it is a maximum when going from a single branch to dual diversity.

Equations (10.14) to (10.16) show that the average improvements in CNR obtainable from the three techniques do not differ greatly, especially in systems using low orders of diversity, and the extra cost and complexity of the combining methods cannot be justified on this basis alone. Looking back at Section 10.3, we see that with selection diversity the output CNR is always equal to the best of the incoming CNRs, whereas with the combining methods, an output with an acceptable CNR can be produced even if none of the inputs on the individual branches are themselves acceptable. This is a major factor in favour of the combining methods.

10.4.1   Envelope probability distributions

The few decibels increase in average CNR (or output SNR) which diversity provides is relatively unimportant as far as mobile radio is concerned. If this were all it did, the same effect could be achieved by increasing the transmitter power. Of far greater significance is the ability of diversity to reduce the number of deep fades in the output signal. In statistical terms, diversity changes the distribution of the output CNR – it no longer has an exponential distribution. This cannot be achieved just by increasing the transmitter power.

To show this effect, we examine the first-order envelope statistics of the signal, i.e. the way the signal behaves as a function of time. Cumulative probability distributions of the composite signal have been calculated for Rayleigh-distributed individual branches with equal mean CNR in the previous paragraphs. For two-branch selection and maximal ratio systems the appropriate cumulative distributions can be obtained from (10.3) and (10.10), and for $M = 2$ an expression for equal-gain combining can be written in terms of tabulated functions. The normalised results have the form:

Selection (SC): $p(\gamma_n) = [1 - \exp(-\gamma_n)]^2$ (10.17)

Maximal ratio (MRC): $p(\gamma_n) = 1 - (1 + \gamma_n) \exp(-\gamma_n)$ (10.18)

Equal gain (EGC): $p(\gamma_n) = 1 - \exp(-2\gamma_n) - \sqrt{\pi\gamma_n} \, \exp(-\gamma_n) \, \mathrm{erf}\sqrt{\gamma_n}$ (10.19)

where γ_n is the chosen output CNR relative to the single-branch mean and $\mathrm{erf}(\cdot)$ is the error function.

Figure 10.4 shows these functions plotted on Rayleigh graph paper with the single-branch median CNR taken as reference; the single-branch distribution is shown for comparison. It is immediately obvious that the diversity curves are much flatter than the single-branch curve, indicating the lower probability of fading. To gain a quantitative measure of the improvement, we note that the predicted reliability for two-branch selection is 99% in circumstances where a single-branch system would be only about 88% reliable. This means that the coverage area of the transmitter is far more 'solid' and there are fewer areas in which signal flutter causes problems. This may be a very significant improvement, especially when data transmissions are being considered. To achieve a comparable result by altering the transmitter power would involve an increase of about 12 dB. Apart from the cost involved, such a step would be undesirable since it would approximately double the range of the transmitter and hence make interference problems much worse. Nor would it change the statistical characteristics of the signal, which would remain Rayleigh.

We have already seen that there is a law of diminishing returns when increasing the number of diversity branches. In equal-gain combiners the use of two-branch diversity increases reliability at the −8 dB level from 88% to 99%; three-branch

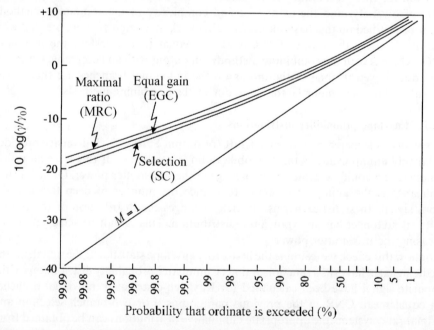

Figure 10.4 Cumulative probability distributions of output CNR for two-branch diversity systems.

increases it further to 99.95%; and four-branch increases it to $> 99.99\%$. At the mobile it would be difficult economically to justify the use of anything more complicated than a two-branch system, but the base station is another matter.

The theoretical results in this chapter have been derived for uncorrelated Rayleigh signals (exponentially distributed CNRs) with equal mean square values, but some attention has been given in the literature to non-Rayleigh fading, correlated signals and unequal mean branch powers. Most of the theoretical results available have been obtained for selection and/or maximal ratio systems since these are mathematically tractable, but they are believed to hold, in general terms, for equal-gain combiners.

Maximal ratio combining still gives the best performance with non-Rayleigh fading. The performance of selection and equal-gain systems depends on the signal distribution; the less disperse the distribution (e.g. Rician with large signal-to-random-component ratio), the nearer equal-gain combining approaches maximal ratio combining. In these conditions selection becomes relatively poorer. For more disperse distributions, selection diversity can perform marginally better than equal-gain combining, although the average improvement $\bar{D}(M)$ of equal-gain systems is not substantially degraded.

The performance of all systems deteriorates in the case of correlated fading, especially if the correlation coefficient exceeds 0.3. Maximal ratio combining continues to show the best performance; equal-gain combining approaches maximal ratio as the correlation coefficient increases, and its performance relative to selection diversity also improves. However, some improvement is still apparent even with correlation coefficients as high as 0.8 and it is interesting to speculate on the reasons for this.

Fundamentally, as we have already seen, diversity is useful in removing the very deep fades which cause the greatest system degradation. However, in statistical terms, these deep fades are comparatively rare events; a Rayleigh signal is more than 20 dB below its median level for only 1% of the time. We can anticipate therefore that even with two signals which have a fairly high overall correlation, there remains a low probability that both will be suffering a rare event (i.e. a deep fade) at the same time. It is likely that much of the diversity advantage will be retained even when significant correlation exists, and this can be seen from Figure 10.5 which shows the cumulative probability distribution function for a two-branch selection diversity system when the inputs have various degrees of correlation.

If the signals in the various branches have different mean square values, a diversity improvement based on the geometric mean (i.e. average of the dB values) of the signal powers is to be expected, at least in the low-probability region of the curves.

10.4.2 LCR and AFD

The previous section has illustrated the effects of diversity on the first-order statistics of the signal envelope. However, some theoretical predictions can also be made about higher-order statistics such as the level crossing rate (LCR) and the average fade duration (AFD).

An early analysis of this problem was due to Lee [5], who investigated equal-gain combining. Assuming that the envelope of the combiner output signal and its time derivative are both independent random processes, it was shown that the level

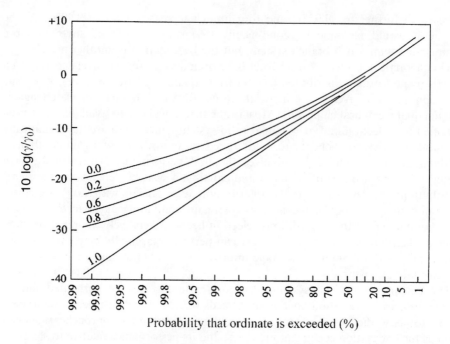

Figure 10.5 Cumulative probability distributions of output CNR for a two-branch selection diversity system with various branch correlations.

crossing rate at a mobile depends on the antenna spacing d and the angle α between the antenna axis and the direction of vehicle motion (Figure 10.6). This can be extended to a unified analysis [6] for other two-branch predetection systems assuming Rayleigh fading signals; the effects of correlation can also be included.

In the nomenclature used previously (Chapter 5) the level crossing rate N_R and the average fade duration $E\{\tau_R\}$ at a given level R are given by

$$N_R = \int_0^\infty \dot{r} p(R, \dot{r}) \, \mathrm{d}\dot{r}$$

$$E\{\tau_R\} = \frac{P(R)}{N_R}$$

If we assume equal noise power N in each branch and we take this into account so that $r^2/2N$ represents the combiner output CNR, then we can exactly compare the effects of the different diversity systems on the LCR and AFD of the combiner output r. It is shown in Appendix D that the effective signal envelopes can be expressed as

$$r(t) = \begin{cases} \max\{r_1(t), r_2(t)\} & \text{SC} \\[2ex] \dfrac{r_1(t) + r_2(t)}{\sqrt{2}} & \text{EGC} \\[2ex] \sqrt{r_1^2(t) + r_2^2(t)} & \text{MRC} \end{cases} \qquad (10.20)$$

hence we obtain

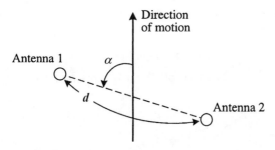

Figure 10.6 Antenna configuration at the mobile.

$$\dot{r}(t) = \begin{cases} \dot{r}_1(t), \ r_1(t) \geqslant r_2(t) \\ \dot{r}_2(t), \ r_1(t) < r_2(t) \end{cases} \quad \text{SC} \\ \\ \dfrac{\dot{r}_1(t) + \dot{r}_2(t)}{\sqrt{2}} \quad \text{EGC} \\ \\ \dfrac{r_1(t)\dot{r}_2(t) + r_2(t)\dot{r}_1(t)}{\sqrt{r_1^2(t) + r_2^2(t)}} \quad \text{MRC} \end{cases} \quad (10.21)$$

It can also be shown that $\dot{r}(t)$ is a Gaussian random variable, hence the mean value $m_{\dot{r}}$ and the variance $\sigma_{\dot{r}}^2$ can be found. Hence, for independent fading signals, the normalised level crossing rate (i.e. the number of crossings per wavelength) at the level R is given by

$$N_{R,2} = \begin{cases} 2\sqrt{2\pi}\rho \, \exp(-\rho^2)[1 - \exp(-\rho^2)] & \text{SC} \\ \\ \sqrt{2\pi}\rho \exp(-\rho^2)\left[\exp(-\rho^2) + \dfrac{\sqrt{\pi}}{2\rho}(2\rho^2 - 1)\,\text{erf}\,\rho\right] & \text{EGC} \\ \\ \sqrt{2\pi}\rho^3 \exp(-\rho^2) & \text{MRC} \end{cases} \quad (10.22)$$

From eqn. (5.44) the normalised rate for a single branch is $\sqrt{2\pi}\rho \exp(-\rho^2)$.

The level crossing rates given by eqn. (10.22) show that, as expected, diversity substantially reduces the LCR at low levels, but the rate at higher levels is increased.

The effect of correlation between the signals on the two branches is to increase the LCR at low levels. For two mobile antennas with omnidirectional radiation patterns, the received signal envelopes fade independently when the antenna spacing is very large but the correlation increases as d is reduced, until for very small spacings the single-branch LCR (Figure 5.13) is approached. The angle α is more important for large antenna spacings than for small spacings.

Equation (5.46) shows that the average fade duration depends on the ratio between the cumulative distribution function $P(R)$ and the level crossing rate N_R. Closed-form expressions for the CDF of selection and maximal ratio systems are available in the literature [3]. Equal-gain combining can be approximated by using the CDF for maximal ratio combining and replacing the average signal power σ^2 of a single branch with $\sigma^2\sqrt{3}/2$ [3]. For independently fading signals, eqns (10.17) to

(10.19) apply to two-branch systems; again using the nomenclature of Chapter 5, the normalised AFD is then given by

$$L_{R,2} = \frac{[1 - \exp(-\rho^2)]^2}{2\sqrt{2\pi}\,\rho\,\exp(-\rho^2)[1 - \exp(-\rho^2)]} \qquad (10.23)$$

$$= \frac{1}{2\sqrt{2\pi}}\left(\frac{\exp\rho^2 - 1}{\rho}\right) \qquad \text{SC}$$

Similarly,

$$L_{R,2} = \begin{cases} \dfrac{1}{\sqrt{2\pi}}\, \dfrac{\exp\rho^2 - \exp(-\rho^2) - \sqrt{\pi}\rho\,\text{erf}\,\rho}{\rho\exp(-\rho^2) + (2\rho^2 - 1)(\sqrt{\pi}/2)\,\text{erf}\,\rho} & \text{EGC} \\[4mm] \dfrac{1}{\sqrt{2\pi}}\, \dfrac{\exp\rho^2 - (1 + \rho^2)}{\rho^3} & \text{MRC} \end{cases} \qquad (10.24)$$

Again we recall from eqn. (5.49) that for a single branch

$$L_R = \frac{1}{\sqrt{2\pi}}\left(\frac{\exp\rho^2 - 1}{\rho}\right)$$

The normalised average fade durations corresponding to eqns. (10.23) and (10.24) are shown in Figure 10.7, with the single-branch values included for comparison. Equations (10.23) and (5.49) indicate that two-branch selection diversity halves the AFD for independent signals, and indeed the result can be generalised to conclude that the average duration of fades is reduced by a factor equal to the number of branches, i.e. $L_{R,M} = L_R/M$. We can infer that a similar result holds for equal-gain and maximal ratio combiners.

The effect of envelope correlation is carried through into the results for AFD since they are simply related to those for LCR. Again, there are considerable differences for $\alpha = 0$ and $\alpha = \pi/2$. When $\alpha = \pi/2$ (i.e. the antennas are perpendicular to the direction of vehicle motion) the antenna spacing is of far less importance than when $\alpha = 0$.

10.4.3 Random FM

Diversity techniques can also be effective in reducing the random FM present in the signal, but the effectiveness depends upon the manner in which the system is realised. For a single branch, the probability density function of the random FM experienced by a mobile receiver moving through an isotropically scattered field was described in Chapter 5 and for the electric field it is given by eqn. (5.32). The analysis leading to an expression for the random FM in a selection diversity system amounts to determining the random FM on the branch which, at any particular time, has the largest envelope; it is a rather complicated procedure.

No closed-form expression for the power spectrum is obtainable, but since the baseband frequencies in a narrowband speech system (300–3000 Hz) are much greater than the spread of the Doppler spectrum, an asymptotic solution as $f \to \infty$

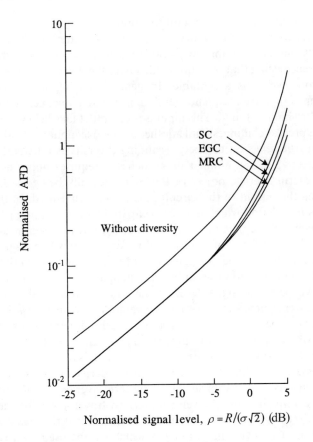

Figure 10.7 Normalised average fade duration (AFD) in wavelengths for two-branch diversity systems.

is sufficient. To give some idea of the magnitude of the quantities involved, a two-branch selection diversity system has an output random FM about 13 dB lower than that of a single-branch system. The use of three-branch diversity further improves this to about 16 dB. Selection diversity therefore provides a significant reduction provided the highest baseband modulation frequency is much larger than the Doppler frequency.

The effectiveness of the combining methods in reducing random FM is highly dependent on the method of realisation. If, during the cophasing process necessary in predetection combiners, the signals are all cophased to one of them, then the output random FM is the same as that of the reference branch. If the sum of all the signals is used as the reference, the output random FM is reduced. In some systems [7] it is possible to completely eliminate random FM and even a single-branch receiver using this kind of demodulation process would have its random FM completely eliminated.

10.5 SWITCHED DIVERSITY

A major disadvantage of implementing true selection diversity as described in Section 10.3.1 is the expense of continuously monitoring the signals on all the branches. In some circumstances it is useful to employ a derivative system known as scanning diversity. Both selection and scanning diversity are switched systems in the sense that only one of a number of possible inputs is allowed into the receiver, the essential difference being that in scanning diversity there is no attempt to find the best input, just one which is acceptable. In general, the inputs on the various branches are scanned in a fixed sequence until an acceptable one, i.e. an input above a predetermined threshold, is found. This input is used until it falls below the threshold, when the scanning process continues until another acceptable input is found.

Compared with true selection diversity, scanning diversity is inherently cheap to build, since irrespective of the number of branches it requires only one circuit to measure the short-term average power of the signal actually being used. Scanning recommences when the output of this circuit falls below a threshold. In this context 'short-term' refers to a period which is short compared with the fading period or, in the mobile radio context, the time taken by the vehicle to travel a significant fraction of a wavelength. A basic form of scanning diversity is shown in Figure 10.8(a), although it is not essential for the averaging circuit to be connected to the front-end of the receiver. The simplest form uses only two antennas, and switching from one to the other occurs whenever the signal level on the antenna in use falls sufficiently to activate the changeover switch. In this form it is commonly known as *switched diversity*.

Some advantage can be gained from a variable threshold, because a setting which is satisfactory in one area may cause unnecessary switching when the vehicle has moved to another location where the mean signal strength is different. Figure 10.8(b) shows a modified system in which the threshold level is derived from the mean signal-plus-noise in the vicinity of the vehicle. The long-term average is computed over a period comparable with the time the vehicle takes to travel about 10 wavelengths, and the attenuator setting determines the threshold in terms of the mean input level.

Basically, there are two switching strategies which can be used, and these cause different behaviour when the signals on both antennas are in simultaneous fades. The *switch-and-examine* strategy causes the system to switch rapidly between the antennas until the input from one of them rises above the threshold. In the *switch-and-stay* strategy the receiver is switched to, and stays on, one antenna as soon as the input on the other falls below the threshold, irrespective of whether the new input is acceptable or not. Selection diversity is subject to deep fading only when the signals on both branches fade simultaneously, but in addition to this, deep fades can be caused in switched systems by a changeover to an input which is already below the threshold and with the signal entering a deep fade. Although in this case, use of the switch-and-examine strategy allows a marginally quicker return to an acceptable input, it causes rapid switching with an associated noise burst, and for this reason the switch-and-stay strategy is preferable in normal circumstances.

Although the ability of switched systems to remove deep fades is inferior to that of selection, the difference can be made small at low signal levels (where diversity has most to offer) and its inherent simplicity therefore makes switched diversity an attractive proposition for mobile use.

10.6 THE EFFECT OF DIVERSITY ON DATA SYSTEMS

Earlier in this chapter we used CNR as the criterion by which to judge the effectiveness of a diversity system. This is an important parameter in analogue (particularly

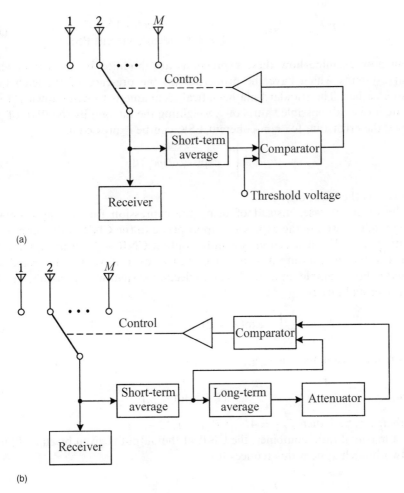

Figure 10.8 Scanning diversity: (a) simple system, (b) system with variable threshold.

speech) transmissions since it is related to the fidelity with which the original modulating signal is reproduced at the system output. However, the techniques of selection or combining diversity can equally be applied to all data transmission formats, and in these systems fidelity as such is unimportant provided the correct decision is made. In other words, to assess the effectiveness of diversity on data transmission systems, we should determine the reduction in error rate which can be achieved from their use. As an example we consider binary FSK and PSK systems which produce fairly simple results and are useful to illustrate the principle.

The form of the error probability expressions for FSK and PSK when the signals are subject to additive Gaussian noise are well known, and can be written as follows [8]:

$$P_e(\gamma) = \tfrac{1}{2} \exp(-\alpha\gamma) \quad \begin{cases} \alpha = \tfrac{1}{2} & \text{noncoherent FSK} \\ \alpha = 1 & \text{differentially coherent PSK} \end{cases} \qquad (10.25)$$

$$P_e(\gamma) = \tfrac{1}{2}\,\mathrm{erfc}(\alpha\gamma) \quad \begin{cases} \alpha = \tfrac{1}{2} & \text{coherent FSK} \\ \alpha = 1 & \text{ideal coherent PSK} \end{cases} \tag{10.26}$$

We can now examine how these expressions are modified by the use of various diversity systems which have the properties (in the presence of Rayleigh fading) discussed earlier. The standard mathematical technique is to write down $P_e(\gamma)$ and integrate it over all possible values of γ, weighting the integral by the PDF of γ. For example, the error rate for non-coherent FSK can be expressed as

$$P_e = \tfrac{1}{2} \int_0^\infty \exp(-\gamma/2) p(\gamma)\, d\gamma \tag{10.27}$$

where $p(\gamma)$ is the PDF of γ.

In the diversity case, instead of using the expression for $p(\gamma)$ appropriate to Rayleigh fading, we use the expression appropriate to the CNR at the output of the diversity system. For a selection system the output CNR is given by eqn. (10.5), so the BER at the system output is the integral of P_e over all values of γ, weighted by this factor. For example, in a two-branch selection system with non-coherent FSK, the error probability is

$$P_{e,2} = \frac{1}{2} \int_0^\infty \exp(-\gamma_s/2)\frac{2}{\gamma_0}[1 - \exp(-\gamma_s/\gamma_0)]\, d\gamma_s$$

This is readily evaluated, yielding

$$P_{e,2} = \frac{4}{(2 + \gamma_0)(4 + \gamma_0)} \tag{10.28}$$

Note that if $\gamma_0 \gg 1$ then $P_{e,2} = 4P_{e,1}^2$; $P_{e,1} = 1/(2 + \gamma_0)$.

For a maximal ratio combiner, the CNR at the output is given by eqn. (10.9) and for a two-branch system this reduces to

$$P_{M,2}(\gamma_R) = \frac{\gamma_R \exp(-\gamma_R/\gamma_0)}{\gamma_0^2}$$

So, for non-coherent FSK transmissions, we have

$$P_{e,2} = \frac{1}{2} \int_0^\infty \exp(-\gamma_R/2)\frac{\gamma_R}{\gamma_0^2} \exp(-\gamma_R/\gamma_0)\, d\gamma_R$$

Again this is readily integrable:

$$P_{e,2} = \frac{2}{(2 + \gamma_0)^2} = 2P_{e,1}^2 \tag{10.29}$$

As a simple numerical example, consider a non-coherent FSK system with a BER of 1 in 10^3 in Rayleigh fading. Using two-branch selection diversity the BER is

$$4 \times (1 \times 10^{-3})^2 = 4 \times 10^{-6}$$

and with two-branch maximal ratio combining we get

$$2 \times (1 \times 10^{-3})^2 = 2 \times 10^{-6}$$

Coherent detection systems produce similar substantial reductions in error rate.

The ability of diversity systems to reduce the duration of fades implies that another very important advantage to be gained from the use of diversity is a significant reduction in the lengths of error bursts. Rayleigh fading tends to cause a burst of errors when the signal enters a deep fade, and since diversity tends to smooth out these deep fades, it not only reduces the error rate but also affects the error pattern by causing the errors to be distributed more randomly throughout the data stream. This in turn makes the errors easier to cope with, and if error-correcting codes are used to improve error rate, much shorter codes can be used in conjunction with diversity than would be necessary without it.

10.7 PRACTICAL DIVERSITY SYSTEMS

Of the three basic schemes, equal-gain combining seems to be an optimum compromise between the complexity of having to provide branch weighting in maximal ratio combining, and the smaller improvement yielded by selection diversity. In situations very often encountered in the mobile ratio environment, equal-gain combining also tends to come closer to maximal ratio combining and it departs from the performance of selection diversity; this is true, for example, when there are correlated signal envelopes or one predominant wave. However, selection can perform better than the two combining systems where coherent noise is present, and this is sometimes the case at VHF in urban environments, polluted with man-made noise. Since selection may introduce its own switching noise, it is difficult to assess its true superiority with respect to the combining methods. No practical comparative data between the various systems is readily available, and it does not seem that there is one 'ideal' system that will always outperform all others in the mobile radio environment.

Let us return briefly to the question of predetection and post-detection systems. The distinction between them was made at the beginning of Section 10.3 but it has not been apparent in the discussion above. Leaving aside selection and switched systems for the moment, in many cases there are very sound reasons to implement a predetection system if a combiner is to be used.

In principle it is irrelevant whether the signals are combined before or after demodulation when the demodulation process is linear, but of vital importance in any system where the detector has threshold properties (e.g. FM discriminators). This is because combining methods can produce an output CNR which is better than any of the input CNRs. If there are a number of branch signals, all of which are individually below the detector threshold, they should be combined before detection in order to produce a CNR which is above the threshold. In this way we not only gain the diversity advantage, but also fully exploit the characteristics of the detector in further improving the output SNR. This is obviously not the case when post-detection combining is used.

10.8 POST-DETECTION DIVERSITY

Postdetection diversity is probably the most straightforward if not the most economical technique among the well-known diversity systems. The cophasing function is no longer needed since after demodulation only baseband signals are present. The earliest diversity systems were of the post-detection type where an operator manually selected the receiver that sounded best; in effect, this was a form of selection diversity.

In post-detection combining diversity, the equal-gain method is the simplest. Two or more separately received signals are added together to produce the combined output with equal gain in all the diversity branches. However, in an angle modulation system, the output SNR will be reduced drastically when the signal in one of the diversity branches falls below the threshold, because the faded branch then contributes mainly noise to the combined output. As in predetection systems, the best performance comes from maximal ratio combining, with each branch gain weighted according to the particular branch SNR. Post-detection maximal ratio combiners therefore require a gain-control stage following the detector, and the required weighting factor for each branch can be obtained by using a measure of the amplitude of the received signal envelope before detection or a measure of the out-of-band noise from the detector output. The first method provides an indication of the receiver input SNR only if the receiver noise is constant. The second method will provide a good indication of the receiver SNR even if the receiver input noise changes.

In an analogue system using angle modulation, the demodulated output signal level from the discriminator is a function of the frequency deviation only if the receiver input signal level is above threshold. The output noise level will vary inversely with the input down to the threshold and it will increase non-linearly below it. Brennan [1] has shown that it makes little difference to the performance of a post-detection combining receiver that utilises angle modulation whether the weighting factors follow the output SNR exactly or whether the receiver merely 'squelches', i.e. discards the output of a particular branch when its input falls below the threshold. This is because if all the branches are already above threshold there is little to be gained by further weighting. Below threshold the noise increases rapidly, thereby reducing the output SNR by a significant amount; this means that the branch gain has to be reduced accordingly. Since the reduction in gain is so large, it makes little difference if the branch is discarded altogether.

Selection and switched diversity can both be implemented in the post-detection format, with some advantages. With selection diversity there are no amplitude transients, since the switchover takes place when the two signals are (nominally) of equal value. Abrupt changes in amplitude are still possible with switched diversity, but phase transients have no meaning in the post-detection context. As a result it is likely that in data communication systems the errors caused by the switching process will be much fewer with post-detection systems than with predetection systems.

An interesting implementation of post-detection diversity is possible for QDPSK, which is the modulation scheme used in the TETRA system. Figure 10.9 shows receiver structures suitable for selection, maximal ratio and equal-gain systems [9]. For selection diversity the estimate of signal power is obtained using a window having a width equal to the symbol period. For good performance this has to be much shorter than the average fade duration in the channel, but this is not normally a problem. For maximal ratio combining, the appropriate weightings have to be determined and $|x_k(t) \| x_k^*(t - T_{sym})|$ results in the structure of Figure 10.9(b). This effectively merges the differential decoder and the weighting circuitry, thus minimising the hardware. The output signal is [10]:

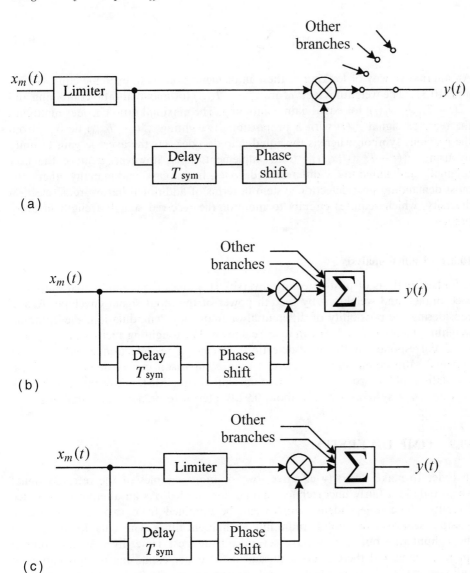

Figure 10.9 Post-detection combiners incorporating a differential detector: (a) selection diversity, (b) maximal ratio combiner, (c) equal-gain combiner.

$$y(t) = \sum_{k=1}^{M} x_k x_k^*(t - T_{\text{sym}}) \tag{10.30}$$

For equal-gain combining, the limiter effectively ensures that the weighting in each branch is $|x_k^*(t - T_{\text{sym}})|$, so the output is

$$y(t) = \sum_{k=1}^{M} x_k \frac{x_k^*(t - T_{\text{sym}})}{|x_k(t)|} \tag{10.31}$$

An alternative way of looking at these implementations is to examine eqns. (10.30) and (10.31). The weighting factors are $x_k^*(t - T_{\text{sym}})$ for maximal ratio combining and $x_k^*(t - T_{\text{sym}})/|x_k(t)|$ for equal-gain combining. The maximal ratio decoder multiplies the received signal $x_k(t)$ with a proportional weighting $x_k^*(t - T_{\text{sym}})$ derived from the previous symbol, whereas the equal-gain decoder sets the average gain to unity by using $x_k^*(t - T_{\text{sym}})/|x_k(t)|$. Both implementations, however, achieve the task originally designated to the differential decoder. In this case, and in many others, the most demanding post-detection system in terms of additional hardware is selection diversity, which requires circuits to monitor the received signal strength in every branch.

10.8.1   Unified analysis

A unified analysis of post-detection diversity [11] takes the demodulated output of each branch and weights it by the vth power of the input signal envelope. Again, considering the possibility of differential or frequency demodulation, the optimum weighting factor is $v = 2$. It can also be shown that weighting factors of $v = 1$ and $v = 2$ correspond, in the post-detection system, to predetection equal-gain and maximal ratio combiners respectively, so a comparison can be made. Numerical calculations of bit error rate with minimum shift keying (MSK) show that two-branch post-detection systems are only about 0.9 dB inferior to predetection combiners.

10.9   TIME DIVERSITY

In order to make diversity effective, two or more samples of the received signal which fade in a fairly uncorrelated manner are needed. As an alternative to space diversity, these independent samples can be obtained from two or more transmissions sent over the mobile radio link at different times. This cuts down the data throughput rate but it does have several advantages. Time diversity uses only a single antenna and there is no requirement for either cophasing or duplication of radio equipment. In principle it is simple to implement, although it is only applicable to the transmission of digital data, where the message can be stored and transmitted at suitable times.

The principal consideration in time diversity is how far apart in time the two messages should be, in order to provide the necessary decorrelation. In practice the time interval needs to be of the order of the reciprocal of the maximum baseband fade rate $2f_{\text{m}}$, i.e.

$$T > \frac{1}{2f_{\text{m}}} = \frac{\lambda}{2v} \tag{10.32}$$

For a mobile speed of 48 kph and a carrier frequency of 900 MHz the required time separation is 12.5 ms; this increases as the fade rate decreases and it becomes infinite when $v = 0$, i.e. when the mobile is stationary. Theoretically the advantages are then lost, but at UHF the wavelength is so small that minor movements of people and objects ensure the standing wave pattern is never truly stationary.

Nevertheless, it is worth examining the potential for time diversity in the mobile radio environment; we take as an example the case when the same data is transmitted twice with a repetition period T. A single antenna is used at the receiver. The relationship between the received signal envelope $r(t)$ and the data sequence $\ldots a_{-1}, a_0, a_1 \ldots$ is depicted in Figure 10.10(a). At the receiver the nth data element a_n ($n = \ldots -1, 0, 1, \ldots$) is received twice and the original and repeated data are demodulated from two samples of the fading signal received at different times. Hence the number of diversity branches is 2, and this type of diversity is equivalent to a two-branch system with the signal envelopes $r(t)$ and $r(t - T)$.

One simple method of using the received data is to output the data element a_n associated with the larger signal envelope. In this case the system is directly analogous to selection diversity, with the resultant signal envelope after selection represented as

$$r_0(t) = \max\{r(t), r(t - T)\}$$

as shown in Figure 10.10(b).

Analysis has shown that the average fade duration and level crossing rates are substantially reduced by the use of time diversity, provided certain criteria are met [12]. In appropriate circumstances, therefore, time diversity can be effective in reducing the rate at which error bursts occur. To obtain some diversity advantage, $f_m T$ should exceed about 0.5.

An alternative method which avoids the need to monitor the signal strength associated with the reception of each data symbol, is to transmit the sequence not twice but three or more times and to form an output by a majority decision (symbol by symbol) on the various versions received. This is simpler, but eats seriously into the data throughput rate. Nevertheless, it is used to protect the various data messages sent over the forward and reverse channels in the TACS system. Eleven repeats are used in base-to-mobile transmissions on the forward voice channel (FVC); the remaining links use five repeats.

Significant advantages accrue from this simple 'majority voting' technique. By simulating a communication system using Manchester-encoded data at 8 kbit/s, PSK modulation, ideal coherent demodulation, and a mobile speed of 40 kph, it has been shown that the BER in a Rayleigh fading channel is reduced from about 2×10^{-2} to about 2×10^{-4} [13]. Improved benefits are obtainable with slightly more sophisticated processing; for example, repeating several times and using the symbol received at the time of highest signal strength (analogous to selection diversity), or using majority voting after weighting each received symbol by a factor which is a function of the signal strength at the time it was received.

Many mobile transceivers provide, as one of their outputs, a signal strength indication in decibels (the RSSI), and the latter technique, which is similar to maximal ratio diversity, could use this to advantage. Linear combining (unity weighting factor) produces a greater improvement in BER than majority voting for a given number of repeats; alternatively it is possible to reduce the number of repeats

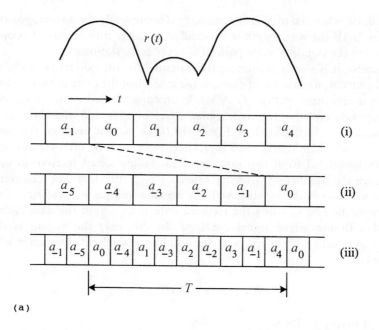

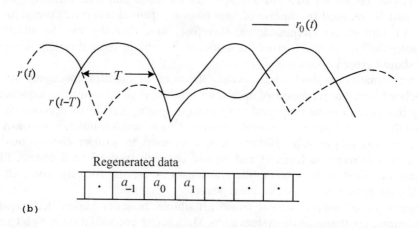

Figure 10.10 Time diversity. (a) Signal envelope and data sequence: (i) original data, (ii) delayed data, (iii) transmitted data. (b) Relationship between $r(t)$, $r(t - T)$, $r_0(t)$ and the regenerated data.

while maintaining the same BER performance. Linear combining using three repeats offers the same BER performance as a five-repeat simple majority voting scheme, and it has the potential to improve channel utilisation considerably.

10.10 DIVERSITY ON HAND-PORTABLE EQUIPMENT

Space diversity is implemented in a number of operational cellular radio systems. In most cases the diversity system exists at the base station where antenna separations of tens of wavelengths are readily available. The correlation between the field components at spatially separated points is covered in Section 5.12 but only the smallest field-probing antennas, which are too inefficient for normal transceiver applications, detect distinct components of the field at a single location. Practical receiving antennas produce an output which is a function of the total electromagnetic field over an extended region of space. Nevertheless, the correlation between the signals obtained from real antennas separated by several wavelengths (as at a base station) is reasonably well approximated by the correlation between the electric fields at points corresponding to the antenna locations; the approximation is certainly good enough to be used for estimates of the separation required for a space diversity system.

Space (antenna) diversity can also be used on vehicles and, conceptually, on hand-portable equipment. The Clarke and Aulin models, however, predict that in an isotropically scattered field the correlation between the electric field components at small spatial separations is high enough to reduce the diversity advantages signi-ficantly, and it seems to have been a tacit assumption for many years that this would make it pointless to implement a diversity system on hand-portable equipment. However, the relatively small antenna separation that can be accommodated on hand-portable equipment means that the output from a given antenna in a certain electromagnetic field is also influenced by the mutual impedance between it and other antennas which form part of the diversity system.

In these circumstances the Clarke and Aulin models are clearly inadequate tools for calculating the correlation between the signals, since the correspondence between signal and field component correlation breaks down. Indeed, although these theoretical models predict that the correlation increases rapidly for points less than 0.4λ apart, there is experimental evidence [14] showing that the correlation between signals obtained from real antennas with fairly small (i.e. subwavelength) spacings is still low enough to offer considerable diversity benefit.

Figure 10.11 shows some measured results obtained under a variety of different circumstances, compared with Clarke's theoretical prediction for an isotropically scattered field. They lead to the conclusion that diversity reception on hand-portable equipment is a realistic aim in the context of current and future systems operating at UHF.

Theoretical studies and simulation techniques [15,16] have been used to provide an explanation for the observed effects. Clearly the nature of the field in which the antennas are located is important – we have seen this earlier in the context of correlation at the mobile and base station ends of the radio link – as is the far-field radiation pattern of the antenna configuration. The far-field pattern contains, implicitly, the effects of mutual impedance between elements.

The antenna correlation between two antenna configurations can be determined as follows. Suppose that, in terms of an $\{r, \theta, \phi\} = \{r, \Omega\}$ spherical coordinate system, the far-field patterns of the two configurations are given by

$$\mathbf{E}_1(\Omega) = E_{1\theta}(\Omega)a_\theta(\Omega) + E_{1\phi}(\Omega)a_\phi(\Omega)$$
$$\mathbf{E}_2(\Omega) = E_{2\theta}(\Omega)a_\theta(\Omega) + E_{2\phi}(\Omega)a_\phi(\Omega)$$

$$(10.33)$$

where a_θ and a_ϕ are unit vectors associated with the Ω direction; $E_{1\theta}, E_{1\phi}, E_{2\theta}$ and $E_{2\phi}$ are the complex envelopes of the θ and ϕ components of the field patterns of

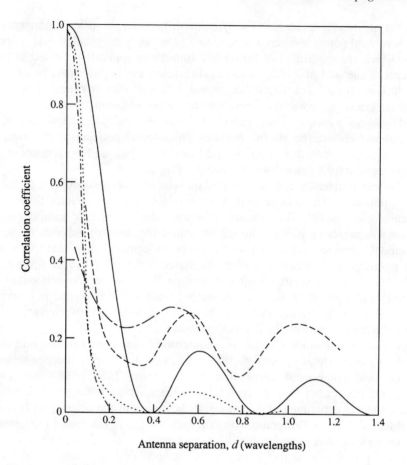

Figure 10.11 Correlation coefficient as a function of antenna spacing: (——) field autocorrelation (after Clarke); other curves are for measurements reported by Japanese researchers.

configurations 1 and 2 respectively and each pattern is measured with respect to the origin of that particular configuration.

If we now assume that configuration 1 is at the true origin of the coordinate system and the position of configuration 2 is defined by a vector d in the coordinate system, then the pattern of configuration 1 is as above, but the pattern of configuration 2 becomes

$$\tilde{E}_2(\Omega) = \tilde{E}_{2\theta}(\Omega)a_\theta(\Omega) + \tilde{E}_{2\phi}(\Omega)a_\phi(\Omega) \qquad (10.34)$$

where

$$\tilde{E}_{2\theta}(\Omega) = E_{2\theta}(\Omega) \exp[-jk\boldsymbol{d} \bullet a_r(\Omega)] \qquad (10.35)$$

and similarly for $\tilde{E}_{2\phi}$.

Then, if $P_\theta(\Omega)$ and $P_\phi(\Omega)$ are the distributions of the a_θ-polarised and a_ϕ-polarised waves respectively, the antenna correlation can be defined as

$$\rho_a^2 = \frac{\left| \int\int_\Omega (E_{1\theta}\tilde{E}_{2\theta}^* P_\theta + E_{1\phi}\tilde{E}_{2\phi}^* P_\phi) \, d\Omega \right|^2}{\int\int_\Omega (E_{1\theta}E_{1\theta}^* P_\theta + E_{1\phi}E_{1\phi}^* P_\phi) \, d\Omega \int\int_\Omega (\tilde{E}_{2\theta}\tilde{E}_{2\theta}^* P_\theta + \tilde{E}_{2\phi}\tilde{E}_{2\phi}^* P_\phi) \, d\Omega}$$

$$= \frac{\left| \int\int_\Omega (E_{1\theta}E_{2\theta}^* P_\theta + E_{1\phi}E_{2\phi}^* P_\phi) \exp[\,jk\mathbf{d} \bullet a_r(\Omega)] \, d\Omega \right|^2}{\int\int_\Omega (E_{1\theta}E_{1\theta}^* P_\theta + E_{1\phi}E_{1\phi}^* P_\phi) \, d\Omega \int\int_\Omega (\tilde{E}_{2\theta}\tilde{E}_{2\theta}^* P_\theta + \tilde{E}_{2\phi}\tilde{E}_{2\phi}^* P_\phi) \, d\Omega} \quad (10.36)$$

where * represents complex conjugate and $\int\int_\Omega \ldots d\Omega$ denotes integration over all angles.

We now consider a practical system consisting of two parallel dipoles. The geometry is defined in Figure 10.12 and we are interested in the correlation between the two configurations shown in Figure 10.13(a) and (b). With this geometry $E_\phi = 0$ and

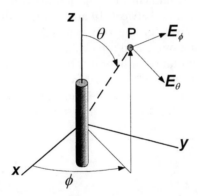

Figure 10.12 Coordinate system and dipole orientation.

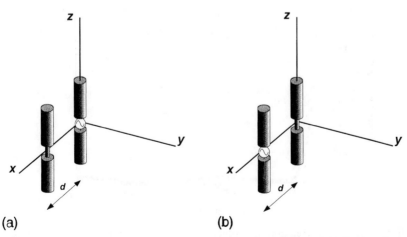

(a)   **(b)**

Figure 10.13 Antenna configuration showing driven and terminated $\lambda/2$ dipoles; the roles are reversed in (a) and (b).

since the origins of the two configurations are identical, $d \equiv 0$. In these circumstances eqn. (10.36) reduces to

$$\frac{\left| \int \int_{\Omega} E_{1\theta} E_{2\theta}^* P_{\theta} \, d\Omega \right|^2}{\int \int_{\Omega} E_{1\theta} E_{1\theta}^* P_{\theta} \, d\Omega \int \int_{\Omega} E_{2\theta} E_{2\theta}^* P_{\theta} \, d\Omega} \tag{10.37}$$

The far-field radiation patterns of these two configurations depend on factors which include the spatial separation and the impedance used to terminate the undriven antenna. As part of an extensive study [17], it has been shown that a terminating impedance of $71\,\Omega$ (which would match an isolated dipole) is a good choice as far as correlation and efficiency are concerned.

Figure 10.14 shows how the antenna correlation varies with dipole separation when a $71\,\Omega$ termination is used. The correlation falls rapidly as the separation increases and is negligible if $d > 0.2\lambda$. The diversity gain depends on the form in which the diversity system is implemented but lies between 5 and 7 dB at the 90% cumulative probability level and between 10 and 12 dB at the 99% level.

In practice there is no need for the two antennas to be identical; on a hand-portable, for example, a monopole and a patch antenna might well be an attractive proposition. In these circumstances a hybrid form of spatial (antenna), pattern and polarisation diversity exists. Indeed, in the foregoing discussion of closely spaced dipoles, the diversity effect is more accurately identified as a combination of both pattern and spatial effects. Regardless of how the (antenna) diversity is actually

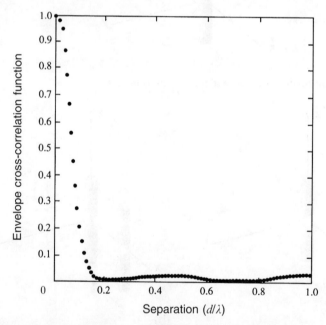

Figure 10.14 Antenna correlation as a function of separation for a resistive termination of $71\,\Omega$. The incoming waves are assumed to arrive with equal probability from all directions in three-dimensional space (Courtesy P.S.H. Leather).

achieved, however, the correlation of the signals produced by any pair of antennas can always be calculated using eqn. (10.36).

10.11   DISCUSSION AND CONCLUSIONS

Switched diversity is potentially an economical diversity method since it is simple in concept, and with careful design it could be an effective way of improving performance at very low cost. Predetection combining can increase the CNR before detection, which produces an effect similar to threshold extension in analogue FM systems; in the presence of uncorrelated Gaussian noise its performance is very good. Post-detection combining may not be economical, since it involves duplication of the predetection parts of the receiver. However, it is effective in reducing BER and can be easily implemented.

All the diversity schemes produce a substantial improvement in signal quality or a reduction in the BER over what is obtainable from a single receiver. Direct comparison of the results with the theoretical predictions is difficult for several reasons. First, the theory gives the BER as a function of the CNR, and this is difficult to measure in practice. Secondly, the detection process in the practical receiver is often different from the process assumed in the theory. Nevertheless, the improvements obtained by using diversity as opposed to a single receiver are substantial and are of the same order as theory predicts; in practice the improvement is the factor of greatest importance.

With the exception of time diversity most diversity techniques do not eat into the information bandwidth, and a two-branch diversity system produces an improvement at the 99% reliability level comparable with the improvement from a 12 dB increase in transmitter power. It removes the vast majority of signal dropouts making speech clearer, and it reduces error rate by more than one order of magnitude. There are also other advantages such as reduction of random FM; with some kinds of predetection combiner the random FM can be completely eliminated, and this can be an important consideration at higher carrier frequencies.

As far as implementation is concerned, although predetection combining can outperform other systems in the presence of uncorrelated Gaussian noise, this may not be representative of the conditions prevailing at VHF, where there are often many noise sources such as the ignition systems of other vehicles in close proximity to the radio installation. The total inputs (signal plus noise) to the various branches may then be sufficiently correlated to impair the performance of combining systems, and selection diversity becomes the optimum technique in these circumstances.

10.12   INTERLEAVING

Interleaving is a relatively simple technique which is extensively used, often in association with other techniques, to mitigate the effects of fading. If a stream of digits (a data stream) from a single source is sent via a Rayleigh fading channel then there will be two observable effects. Firstly, the overall error rate will be higher in the fading channel than it would be in a channel having a constant CNR equal to the mean CNR in the fading channel. Secondly, whereas errors in the non-fading channel occur randomly throughout the bitstream, the fading causes errors to occur in bursts coinciding with the rapid reductions in short-term CNR resulting from that fading.

Interleaving is used to introduce some time diversity into a digital communication system so that data bits which are generated consecutively are not transmitted consecutively. When the data stream is reconstructed at the receiver, the errors have been effectively randomised. Interleavers take two basic forms, a block structure or a convolutional structure. In a block interleaver, the source data, which may have been encoded by a speech coder, is read into a two-dimensional store. To explain the action, assume that the first m bits are read into the first column, the second m bits into the second column, etc. The block interleaver in Figure 10.15 can thus accommodate mn bits. For transmission purposes the stored bits are read out in rows. This has the effect of separating consecutive source bits by m-bit periods because the input sequence is 1, 2, 3, 4, ... and the output sequence is 1, $(m+1)$, $(2m+1), (3m+1), \ldots$. Provided the separation between consecutively generated bits is long enough, a time diversity effect is achieved. At the receiver end of the link, the deinterleaver performs the opposite function by storing the received bits in rows and reading them out in columns.

Interleavers are extensively used in second-generation cellular radio systems, e.g. GSM, in association with speech coders. Because of the structure of these coders, some of the source bits are far more important than others in ensuring successful transmission of a message. It is therefore vital to protect these bits from error, and spreading them throughout the data stream is a step in this process. In practice this is not the only step that is taken, but it would obviously be disastrous if the important bits were transmitted consecutively and were subject to an error burst.

A major problem with interleaving is the delay associated with the process, since the received message cannot be fully decoded until the complete transmitted block containing mn bits has been received and deinterleaved. Fortunately, in the case of digitised speech, intelligibility is readily maintained, and there is little subjective annoyance to the listener provided the delays do not exceed ~ 35 ms, Interleavers in existing cellular radio-telephone systems have delays which do not exceed this amount. For digitised speech, block interleavers are well matched to block codes and

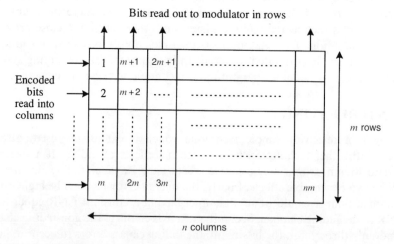

Figure 10.15 Structure of a block interleaver.

convolutional interleavers, which have a different structure, are well matched to convolutional codes [18].

10.13 CHANNEL EQUALISATION

Equalisation is a very important anti-multipath technique in wideband systems and has received much attention in recent years. Generally, fading in mobile radio channels is space (or time) selective and frequency selective; both have been discussed earlier. Frequency-selective fading arises whenever the bandwidth of the transmitted signal is comparable to the spread in delayed multipath propagation echoes; it causes deep dynamic fades to occur in the channel transfer function, and in the absence of any suitable signal processing in the receiver this leads to significant distortion of the signal and hence to intersymbol interference (ISI).

ISI is the major barrier to high-speed digital transmissions over mobile radio channels but it is possible to exploit the diversity implicit in the various echo paths if the radio receiver can constructively add the incoming multipath components. Adaptive signal processing with special receiver features (i.e. adaptive equalisation) offers such a possibility and this can be explained with reference to the time-variant transfer function $T(f, t)$ described in Chapter 6.

We know that the channel is dynamic and therefore the components of $T(f, t)$ will decorrelate in time and frequency as the channel characteristics change. Moreover, time and frequency variations are clearly related, so the decorrelation time is related to the fading rate or Doppler spread as eqn. (10.32) shows. Likewise, in the frequency domain, the signal spectral components also decorrelate as the coherence between the echo paths decreases. The multipath delay spread is related to the frequency coherence (correlation) bandwidth as shown in Chapter 6.

10.13.1   Adaptive equalisers

The decorrelation time (or Doppler spread) in the channel determines the rate at which the receiver must adapt in time. In other words, it defines the *receiver learning* or *training time*. The spread in time delays provides a measure of the diversity implicit in the channel, and to benefit from the multipath diversity effectively, the receiver must learn the channel characteristics accurately and quickly and it must be able to track changes at an appropriate rate. There is, however, a limit to the rate at which the receiver can learn, because increasing the transmission rate to help distinguish multipath echoes leads in itself to ISI.

In practice the operating modes of an adaptive equaliser include *training* and *tracking*. For training purposes the transmitter sends a fixed-length sequence which is either a known pseudo-random binary sequence or a predetermined pattern of bits. The equaliser uses this *training sequence* to establish the short-term channel characteristics and to optimise its settings. Following the training sequence the user data is sent. The adaptive algorithm in the equaliser tracks the changing channel and the equaliser continually updates its settings to follow changes in the channel characteristics. Of course, equalisers need periodic retraining depending on the rate at which the channel characteristics change, and the rate at which any equaliser converges to new optimum settings depends on its structure and the algorithm used. Equalisers are well suited to TDMA systems such as GSM where the data is sent in short time slots, and the training sequence can be sent at the beginning of a slot.

The most explicit form of equaliser is the tapped delay line (TDL) filter (Figure 10.16). The similarity to Figure 6.3 is obvious and the TDL is therefore, in many senses, a basic or generic equaliser. If the tap delays are judiciously chosen to correspond to the major delayed paths encountered in the multipath environment and the tap weights $a_0, a_1, \ldots, a_N$ are adjusted to maximise the signal output in the presence of noise, then the TDL equaliser is essentially a matched filter [19]. The structure in Figure 10.16 has N delay elements, $N+1$ taps and $N+1$ complex weighting elements; the complex weighting elements are updated continuously by the adaptive algorithm. This in turn has, as its input, an error signal derived from the difference between the output and a reference, which could be an exact scaled replica of the transmitted training sequence or a known property of that sequence. In noisy fading channels the equaliser performs three distinct functions:

- It removes ISI
- It derives implicit diversity
- It performs noise filtering.

The TDL equaliser is an example of a linear filter arranged to separate the super-imposed components caused by multipath echoes, weight them suitably and then add them in a constructive manner. Implementation in the form of a lattice filter is also possible [20]. To perform effectively, this type of equaliser must behave as an inverse filter of the channel, hence in a frequency-selective channel the equaliser amplifies the weak spectral components and attenuates the strong ones in order to provide an overall flat frequency response and a linear phase response.

10.14 NON-LINEAR EQUALISERS

In the TDL filter described above and in the lattice filter, the reconstructed message is not used directly in the feedback path to the adaptive algorithm. Equalisers of this kind are known as *linear* equalisers and can provide very effective performance. However, linear equalisers do not perform well where there are deep spectral nulls

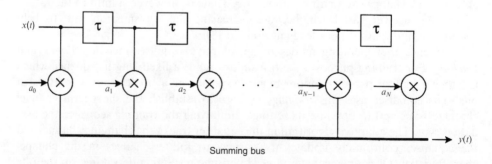

Figure 10.16 Structure of a tapped delay line (TDL) filter (generic equaliser).

in the channel response and in cases where there is severe distortion. Improved performance can be obtained if the output stream is used directly to assist in adjusting the equaliser; equalisers with this feature are known as *non-linear equalisers*. Different types have been developed and the two most common are briefly described.

10.14.1 Decision feedback equalisers

To completely remove ISI, the effects of both past and future ISI pulses must be cancelled. Figure 10.17 illustrates these effects for an alternating sequence 1010, . . . ; the ISI due to the previous pulse at $t = kT$ is I_p and the ISI due to a pulse in the future is I_f. To remove these ISI components the shapes of these pulses must be known, so the equaliser begins by estimating the expected multipath echo structure. In the decision feedback equaliser (DFE) the future ISI is removed directly by a TDL filter known as the forward filter. The past ISI is estimated from the hard decisions made on the previously detected bits; if the past bits were correctly detected then the backward filter, which is also a TDL filter, removes the ISI caused.

This decision-directed action aims to minimise the error between the detector input and output. The DFE structure is shown in Figure 10.18. It suffers from an error propagation effect if the detected symbols are in error; the extent of this effect depends on the length of the backward filter and the signal-to-noise ratio at the demodulator input. Results reported in the literature [21] indicate a typical performance loss of 2 dB due to incorrect decisions. In general, the performance of a DFE degrades significantly under severe ISI in a noisy channel.

10.14.2 MLSE Viterbi equaliser

The DFE makes hard decisions on a symbol-by-symbol basis. An optimum equaliser, however, will make decisions on a sequence of symbols, choosing the sequence which has the maximum likelihood of having being sent at the transmitter. Maximum likelihood sequence estimation (MLSE) is known to be the optimum approach to equalisation in additive white Gaussian noise(AWGN) channels [20]; the essential task is to select a sequence of symbols from a set of candidate sequences

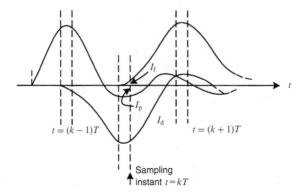

Figure 10.17 The effects of intersymbol interference ($I_f = I_p$).

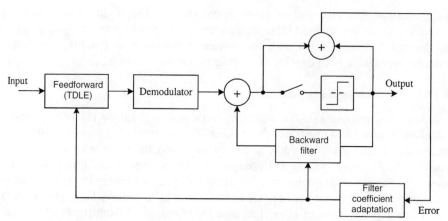

Figure 10.18   Structure of the decision feedback equaliser (DFE).

available at the MLSE output. A particular realisation of the equaliser (Figure 10.19), helps to present a simplified view of the equaliser action.

A preamble or training sequence is sent by the transmitter and used by the channel estimator to obtain an estimate of the multipath structure. A candidate sequence, of sufficient length to cover the maximum multipath delay expected in the channel, is generated and convolved with the estimated multipath channel response. This ISI-corrupted candidate sequence is then compared with the actual demodulated sequence and a *metric*, i.e. a measure of similarity, is generated. Ideally this metric should directly relate to the likelihood (probability) of the candidate sequence being the actual transmitted sequence. In practice a metric based on the Euclidean distance on a cumulative symbol-by-symbol basis will be devised, the aim being to select the sequence closest in 'distance' to the demodulated sequence. This sequence is expected to be the maximum likelihood sequence with respect to the transmitted sequence.

Viterbi algorithm

The equaliser algorithm can assign a state for each ISI bit (symbol) combination permissible, and the number of symbols affected by the delay due to multipath

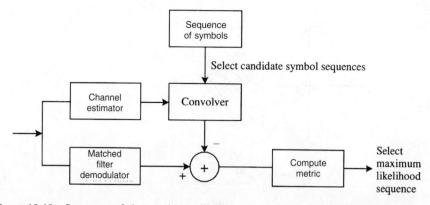

Figure 10.19   Structure of the maximum likelihood sequence estimator (MLSE).

propagation will determine the number of allowable states. The maximum likelihood candidate can be found by determining the trajectory of the symbols through a trellis with the maximum accumulated metric. As an example, consider the four-state trellis diagram in Figure 10.20. Here two sequences originating from a known state (say 00) converge into the same state after 3 symbols. This means that the two symbol sequences 000 and 011 relate to the same ISI state if the multipath structure affects more than one symbol. To determine which of the two sequences should be selected, the accumulated path metrics for the two sequences are computed as

$$d_1 = \sum_{i=1}^{N} d_{i1} \quad d_2 = \sum_{i=1}^{N} d_{i2}$$

If the accumulated path metrics are such that $d_1 > d_2$ then the sequence d_1 is selected as a 'survivor' for the state 00. It is quite possible that all four states have a survivor at the same depth in the trellis. The reason for selecting a survivor is based on the observation that if symbol sequences re-emerge after a merge then the sequence with the largest distance will continue to have the largest distance on the arrival of the next symbol. This observation by Viterbi [22] results in reduced computation and led to the discovery of the well-known *Viterbi algorithm* (VA).

10.15 CHANNEL CODING

Redundant symbols selectively introduced into a transmitted data stream can be used to form a coding scheme which gives some protection against errors in reception caused by imperfections in the channel. *Error-detecting codes* can recognise the presence of errors; *error-correcting codes* can also correct a limited number of errors within a data block of a given length. The effectiveness of channel coding depends on how the assembled codewords spanning the fades counteract the errors in transmission. The fades distribute the errors as a combination of bursts and, of course, random errors. One approach already discussed in Section 10.12 is to

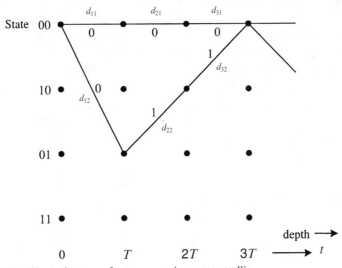

Figure 10.20   The trajectory of a sequence in a state trellis.

interleave the transmitted symbols and hence to disperse the error bursts so that a form of error coding suited to random errors may be employed effectively.

Channel codes are arranged first to detect errors and then to correct some or all of the detected errors at the receiver. This process is known as *forward error correction* (FEC) since the receiver attempts, or accomplishes, error correction on a one-way link. Some errors, however, may remain undetected depending on the properties of the specific code employed. Another strategy, also mentioned earlier, is to use a two-way link to acknowledge errors detected in the receiver. Automatic repeat request (ARQ) messages are sent by the transceiver requesting a repeat of the previous message. ARQ may be used alone in data transmission systems to cope with the effects of shadowing, but in a poor channel the repeat requests will decrease the information throughput and add delays in transmission. FEC is therefore much preferred for speech services to keep delay to a minimum; it is usually supplemented with a moderate span of interleaving to assist the channel decoder. In practical applications, two types of FEC codes are normally employed: linear block codes and convolutional codes.

10.15.1   Linear block codes

A block code is a fixed-length vector with n elements (symbols), of which k elements are information-bearing. In all cases $n > k$, hence there are $(n - k)$ redundant elements, known as parity symbols. Block codes are usually described as (n, k, d) or in short form as (n, k) codes; d is the so-called *Hamming distance* of the code. The Hamming distance is a measure of the difference between two codewords, reflecting the number of positions in which they differ. The probability that one codeword is confused with another codeword due to transmission errors therefore decreases as d increases.

Encoding

Redundancy is the crucial factor in error correction. If we consider a codeword of n binary symbols, the number of possible binary sequences of length n symbols is 2^n. If all these sequences were legitimate codewords, there would be no basis on which to distinguish between codewords. By increasing the redundancy, however, the codewords become unique. To illustrate this let us assume that $k = 2$ and $n = 5$. The ratio k/n is known as the *rate* of the code. Table 10.1 illustrates the case when 3 parity bits

Table 10.1   Block encoding lookup table

Tx codewords	00 | 110	01 | 101	10 | 011	11 | 000
	|	|	|	|
	|	|	|	|
	|	|	|	|
	|	|	|	|
	00010	01100	10001	11001
Single-error codewords at receiver	00111	01111	10010	11010
	01110	01001	10111	11100
	10110	00101	11011	10000
	00010	11101	00011	01000
Double-error codewords at receiver	00000	10101	11111	
	10100	00001	01011	

are added in a rate 2/5 code to make the 4 transmitter codewords (alphabet) unique so that all single errors at the receiver can be detected. In the lookup table the errors are listed under each valid codeword to make the task of a simple decoder straightforward.

Algebraic encoding

Practical encoders and decoders usually exploit the algebraic structure of the code; the basic principle can be illustrated as follows. Let a k-bit vector describe the information sequence as a row matrix:

$$d = [d_1 d_2 d_3 \ldots d_k] \tag{10.38}$$

After the inclusion of $r = (n-k)$ redundant bits, the coded vector is

$$c = [c_1 c_2 c_3 \ldots c_k c_{k+1} \ldots c_n] \tag{10.39}$$

In a systematic code the first k elements are identical to the information bits. The remaining $(n-k)$ parity bits are generated by a linear operation as follows:

$$c_{k+1} = p_{11}d_1 \oplus p_{21}d_2 \oplus \ldots \oplus p_{k1}d_k$$
$$c_{k+2} = p_{12}d_1 \oplus p_{22}d_2 \oplus \ldots \oplus p_{k2}d_k$$
$$\vdots$$
$$c_n \quad = p_{1r}d_1 \oplus p_{2r}d_2 \oplus \ldots \oplus p_{kr}d_k$$

where $\oplus$ indicates modulo 2 addition.

To illustrate the coding process, we consider a certain (7,3) code. The parity check matrix is

$$p = \begin{bmatrix} p_{11} & p_{21} & \cdots & p_{k1} \\ p_{12} & p_{22} & \cdots & p_{k2} \\ \vdots & \vdots & & \vdots \\ p_{1r} & p_{2r} & \cdots & p_{kr} \end{bmatrix} = \begin{bmatrix} 1 & 1 & 0 & 0 \\ 0 & 1 & 1 & 0 \\ 1 & 1 & 1 & 1 \end{bmatrix}$$

The parity check symbols are obtained for an information vector [1 0 1] by the multiplication

$$[c_4 c_5 c_6 c_7] = [1 \ 0 \ 1] \begin{bmatrix} 1 & 1 & 0 & 0 \\ 0 & 1 & 1 & 0 \\ 1 & 1 & 1 & 1 \end{bmatrix}$$

which gives the parity symbols

$$c_4 = 1.1 \oplus 0.0 \oplus 1.1 = 0$$
$$c_5 = 1.1 \oplus 0.1 \oplus 1.1 = 0$$
$$c_6 = 1.0 \oplus 0.1 \oplus 1.1 = 1$$
$$c_7 = 1.0 \oplus 0.0 \oplus 1.1 = 1$$

and hence the code word [1 0 1 0 0 1 1].

Algebraic decoding

At the receiver the information and parity check bits are separately identified. The information bits detected are then used with the parity check matrix p to generate the parity check bits at the receiver, $c'_{k+1}, c'_{k+2}, \ldots, c'_n$, and these parity check bits are compared with the received parity check bits. If the bits have been received error-free then $c'_{k+1} = c_{k+1}, c'_{k+2} = c_{k+2}, \ldots, c'_n = c_n$. In addition, if the columns of the parity check matrix are all unique then it is possible to locate errors in transmission. The detailed consideration of decoders is beyond the scope of this book but extensive literature exists [23].

10.15.2   Convolutional codes

Convolutional encoders are implemented using a serial shift register and modulo 2 adders; Figure 10.21 shows the encoder structure for a rate $R = \frac{1}{2}$ code with a constraint length $k = 3$. The constraint length corresponds to the encoder memory, i.e. no more than $(k - 1)$ previous information bits influence the new coded bit. The switch selects the modulo 2 bits from each encoder arm and multiplexes the coded bits. Consideration of a simple binary sequence reveals that the output of each modulo-2 arm is a binary convolution of the input sequence and the binary shift register coefficients selected, i.e. the bit positions selected in the modulo 2 addition.

   The decoding of convolutional codes is best performed by using the Viterbi algorithm. The number of states in the decoder trellis is 2^{k-1}, i.e. this is the number of bits that influence the trellis trajectory.

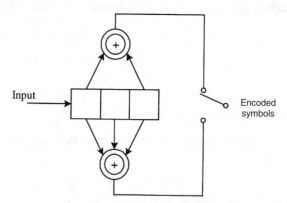

Figure 10.21   A convolutional encoder for a code with $R = \frac{1}{2}$ and $K = 3$.

10.16   CODES FOR FADING CHANNELS

Because of the error propagation effect caused by the encoder memory, convolutional codes are not able to cope with error bursts, especially when the Viterbi decoder is used. On the other hand, linear block codes, especially the non-binary Reed–Solomon (RS) codes, are particularly suited to channels where errors appear in bursts compatible with its block length. These codes may be adapted to suit a specific channel by introducing, for example, bit interleaving to disperse errors.

10.16.1 Performance of codes in fading channels

The BER performance of convolutional and block codes may be obtained using the standard mathematical technique of averaging the 'static' BER over the fading envelope of the channel or by equivalently describing the error characteristics of the channel. A simple model sometimes used to characterise the channel errors depicts the channel as being in a 'good' state when the envelope exceeds a prescribed threshold and in a 'bad' state when it remains below this threshold. This two-state description is known as the Gilbert–Elliott model [24] and is widely used in simplified analyses. The probability of error in the good state, knowing the static BER $P_e(\gamma)$ and the PDF $p(\gamma)$ of the fading, is then given by

$$P_G = \int_{r_T}^{\infty} P_e(\gamma) p(\gamma) \, d\gamma \qquad (10.40)$$

and similarly in the bad state

$$P_B = \int_{0}^{r_T} P_e(\gamma) p(\gamma) \, d\gamma \qquad (10.41)$$

where r_T is an appropriate threshold level. The overall BER can then be estimated using these equations, weighting each by a factor equal to the fraction of the total time spent in the good and bad states.

BER for convolutional codes

The upper bound on the BER for convolutional codes after Viterbi decoding is expressed by the inequality

$$P_{e_{VA}} \leqslant \frac{1}{k} \sum_{d=d_f}^{\infty} c_d P_d \qquad (10.42)$$

where k is the constraint length, d_f is the free distance of the code and c_d is the number of information symbols that lead to trellis paths with distance d. The probability that the decoder selects a wrong path at a distance d is P_d. The derivation of an expression for P_d depends on the decoder metrics d, P_B, P_G and the error distributions in the good and bad states [25]. Such derivations are very cumbersome, however, and the upper bounds on BER can be obtained more easily by simulation.

BER of block codes

Channel measurement information can improve the BER performance of codes because if a measure of the symbol reliability is available at the decoder output then unreliable symbols may be erased and substituted by symbols likely to decrease the block error probability. Various decoder strategies are available in the literature [23], but here we follow a simple procedure [25]. An amplitude estimate is used and a reliability threshold prescribed. The probability of symbol erasure then depends on the fraction of time that the fading signal amplitude spends below the reliability threshold r_T. The probability of erasure is

$$P_E = \int_0^{r_T} p(r) \, dr \tag{10.43}$$

and the average BER in an erased symbol is

$$P_{b/E} = \frac{1}{P_E} \int_0^{r_T} P_b(\gamma) p(\gamma) \, d\gamma \tag{10.44}$$

where the notation $P_{b/E}$ refers to the conditional bit error probability given that an erasure has occurred. We can easily deduce from equation (10.41) that the probability of no erasures is

$$P_{\bar{E}} = 1 - P_E \tag{10.45}$$

and the conditional probability of an error in a non-erased symbol becomes

$$P_{S/E} = \frac{1}{P_{\bar{E}}} \int_{r_T}^{\infty} P_S(\gamma) p(\gamma) \, d\gamma \tag{10.46}$$

where we distinguish a symbol from a bit to allow consideration of non-binary, i.e. m-bit, Reed–Solomon codes. In equation (10.46) $P_S(\gamma)$ is the static symbol error ratio. The average bit error probability given that non-erased bits are received is simply

$$P_{b/\bar{E}} = \frac{1}{P_{\bar{E}}} \int_{r_T}^{\infty} P_b(\gamma) p(\gamma) \, d\gamma \tag{10.47}$$

and the symbol error probability for a symbol consisting of m bits in a random error channel is

$$P_S(\gamma) = 1 - [1 - P_b(\gamma)]^m \tag{10.48}$$

The BER lower bound for an (n, k) block code can be derived in the form

$$P_b \geq \frac{1}{n} \sum_{e=0}^{d-1} \binom{n}{e} P_E^e (1 - P_E)^{n-e} \sum_{t=\lceil \frac{d-e}{2} \rceil} \binom{n-e}{t} P_{S/\bar{E}}^t (1 - P_{S/E})^{n-e-t}$$

$$\times (e P_{b/E} + t P_{b/\bar{E}} / P_{S/\bar{E}}) + \frac{1}{n} \sum_{e=d}^{n} \binom{n}{e} P_E^e (1 - P_E)^{n-e} \{ e P_{b/E} + (n-e) P_{b/E} \}$$

$$\tag{10.49}$$

when e erasures are assumed to occur in the received codeword (block). The notation $\lceil \frac{a}{b} \rceil$ represents the largest integer of the ratio $\frac{a}{b}$.

10.17 SPEECH CODING

Speech coding is an essential part of any digital radio-telephone system. It is not an anti-multipath technique, however, so we will treat it very briefly. Speech coders convert an analogue speech signal into a digital signal through a process of analogue-to-digital conversion. Pulse code modulation (PCM) is the most widely known speech encoding technique, but variants of PCM have arisen from a desire to reduce the rate of information transmission and hence to reduce the transmission bandwidth. Well-known techniques such as differential PCM (DPCM) and delta modulation (DM) encode the difference between the actual signal (speech) sample and a predicted value based on previous samples.

For many years, interest has focused on speech encoders that are able to adapt the quantiser step sizes to achieve a further reduction in transmission rates, e.g. ADPCM. Another class of speech coder exploits the intrinsic characteristics of speech to derive low residual error by effective linear predictive coding (LPC). Pulse excitation is used to minimise the residual error by varying the position and amplitude of these pulses before the quantisation of this error. In mobile telephony the key aim is to reduce the transmission rate in order to improve spectrum efficiency, and in this context only two speech coder classes have received significant attention: sub-band coders (SBC) and pulse-excited linear predictive coders (LPC)

10.17.1 Sub-band coders

The basic elements in SBC speech coders are shown in Figure 10.22, a simplified block diagram. Typically between 4 and 16 sub-bands are used. The filters are realised as finite impulse response (FIR) filters and implemented in hardware using quadrature mirror filter (QMF) elements [26]. ADPCM with a maximum of 4 bits per sub-band is usually used in the quantiser encoder. Between the sub-bands either a fixed or adaptive bit allocation may be used. A fixed bit allocation tends to give smoother speech quality whereas adaptive bit allocation tends to offer better quality, although not always with enough robustness against errors caused by fading. Speech coders at 16 kbit/s have been developed for digital cellular applications.

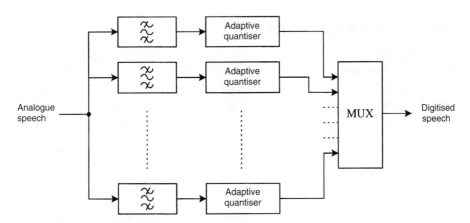

Figure 10.22 Structure of a sub-band speech coder (SBC).

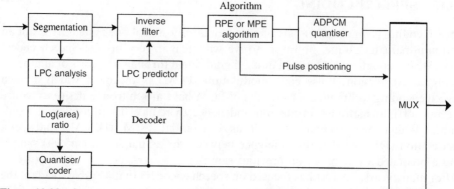

Figure 10.23   Structure of a pulse-excited LPC speech coder.

10.17.2   Pulse-excited coders

A block diagram of a pulse-excited LPC speech coder is shown in Figure 10.23. Segments of the sampled speech are subjected to an LPC analysis which comprises computation of the autocorrelation functions (ACF) and extraction of the predictor coefficients [27]. These coefficients are transformed to reduce quantisation sensitivity and then quantised into coded words. In a feedback loop the coded coefficients are decoded and converted back into predicted speech samples; an inverse filter then generates the residual error.

Two different approaches can be adopted for representing the residual error signal: multipulse LPC (MPE) and regular pulse-excited LPC (RPE-LPC). The MPE algorithm varies the position and amplitude of a small number of pulses in order to minimise the prescribed error criterion, and some MPE algorithms also operate a long-term prediction (LTP) around the quantised residual signal. The RPE algorithm differs from MPE in that it employs a regular pulse train with fewer pulse positions occurring more frequently. More details of speech coders are presented in the literature [28,29]; they include comparisons between speech coders with respect to speech quality, transmission delay and complexity.

10.18   THE RAKE RECEIVER

The so-called RAKE receiver, first proposed by Price and Green [30], is a spread-spectrum receiver that is able to track and demodulate resolvable multipath components. It allows a number of independently fading echoes to be isolated, and the effects of multipath fading can be combatted or even exploited to advantage. A RAKE receiver can combine the delayed replicas of the transmitted signal to improve reception quality. Essentially this is a sophisticated time diversity technique which has enormous potential for future wideband cellular radio systems which use CDMA techniques.

Particularly in outdoor environments, the delays between multipath components can be quite large, certainly greater than one chip period of the CDMA sequence, and as we have seen in Chapter 8 (Figures 8.21 and 8.22) these components have low average correlation. The RAKE receiver, which has the architecture in Figure 10.24, provides a number of correlation receivers for the M strongest components. Weighting amplifiers provide a linear combination of the correlator outputs for bit detection. Correlator 1 is synchronised to the first multipath component, which is often but not always the strongest. Correlators 2 to M are synchronised to later components which, as discussed, have low mutual cross-correlation. The obvious advantages of this receiver include the fact that if the output of one of the correlators is corrupted by fading, the others may not be, and the corrupted path may be effectively discarded by the weighting process. This is in contrast to the case of a simple one-path receiver which cannot recover synchronisation and may, in these circumstances, produce a high BER.

The outputs of the M correlators are weighted to form an overall decision as shown in Figure 10.24; the output is given by

$$R = \sum_{m=1}^{M} W_m R_m$$

The weighting coefficients can be chosen to produce the equivalent of maximal ratio combining (MRC) or equal-gain combining (EGC). The same compromises and trade-offs exist; maximal ratio combining provides a better output SNR but at the expense and complexity of providing the weighting amplifiers.

The design of a practical RAKE receiver is influenced by a number of factors. Clearly a receiver comprising a large number of branches (or 'fingers') is expensive both financially and in terms of power consumption. There is likely to be little

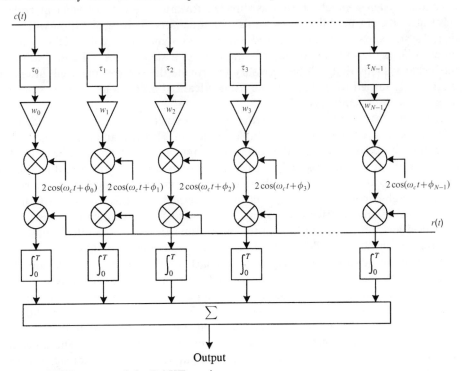

Output

Figure 10.24 Structure of the RAKE receiver.

further increase in output SNR after the few strongest multipath echoes have been captured. Experimental evidence indicates that the number of relatively strong, distinguishable multipaths is fairly small, and a 3-finger or 4-finger receiver is therefore likely to be a good compromise, technically and economically. Experimental results confirm this: output SNR has been computed for single-path (equivalent to SC), all-path EGC and all-path MRC, from available channel-sounding information [31].

All-path MRC produced an increase, at the 50% probability level, of up to about 8 dB over a single-path receiver, but all-path EGC often produced a deterioration. The reasons for this are as discussed in Section 10.4: weak paths add little signal power but contribute the same noise power as the strong paths. All-path receivers are unrealistic and further analysis has been performed using receivers with 5, 4, 3 and 2 paths. This has confirmed that EGC has very little to offer and has shown that the difference in performance between 4-path and all-path receivers is negligible. In IS-95 – a system designed for wide area coverage and using a much smaller band-width than the channel sounder mentioned above – 3-path or 4-path RAKE receivers are used at the base station; this probably represents an optimum choice taking all factors into account.

10.19   SMART ANTENNAS

During the development and implementation of second-generation cellular radio-telephone systems, much effort was directed towards the area of efficient coders, spectrally efficient modulation methods and equalisers. Third-generation wideband digital systems, soon to be introduced, will provide significantly enhanced services to an even larger, international user community. As a step towards the realisation of intelligent systems that will be necessary to provide the services required, attention has recently turned to spatial filtering using advanced antenna techniques, so-called *adaptive* or *smart* antennas [32].

Spatial filtering using antenna arrays is not in itself a new concept. It has been known for many years that a suitable choice of element amplitude and phase weighting can be used to steer the beams of an antenna array, pointing beams in the direction of wanted signals and/or steering nulls to coincide with the direction of interferers. Recently, however, multipath suppression has become a more overt aim as far as high-capacity cellular systems are concerned, and this together with main-beam steering and interference nulling has allowed a smart antenna system to be envisaged as a promising technique for maximising the carrier-to-interference ratio at base stations and mobiles, and for improving system capacity [33].

Basically an adaptive antenna array, a smart antenna, is a system whose time, frequency and spatial response can be tailored to suit a specific purpose by amplitude and phase weighting of the various elements in the array and by feedback control. A generic adaptive array is shown in Figure 10.25. The beam-forming network takes as its inputs the signals from the various array elements; it applies amplitude and phase weighting to each and subsequently sums them to form an output. The difference between the output and a reference (the desired output) forms an error signal which, together with the element signals, is fed to an adaptive controller. This provides

Antenna array

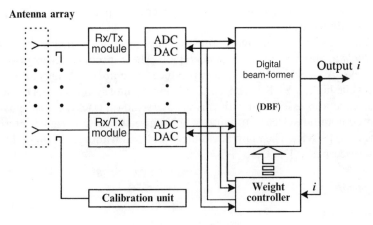

Figure 10.25 A generic adaptive array (smart antenna).

control signals to the weighting networks, driving the output towards its desired value.

This generic system is easily envisaged in analogue form; indeed the origins of this type of processing date back many years to systems operating at RF or IF [34] where signals from the array were amplified, phase-shifted and summed, all in analogue form, and then downconverted to baseband. Nowadays fast ADC circuits allow the signals from each array element to be converted to complex digital numbers at high sampling rates so that the whole process, beam-forming and adaptive control, can be implemented at baseband. This provides high accuracy and a number of other advantages [35].

Current first- and second-generation cellular networks use omnidirectional or sectored base station antenna systems to provide intensive coverage over wide areas in urban and rural environments, and base station diversity reception (a simple form of smart antenna system) is also widely employed. There are two obvious ways to increase capacity: build additional cell sites, i.e. indulge in further 'cell splitting', or implement further sectorisation. Neither is really attractive because cell sizes in many areas are already approaching the limit set by practical hand-off delays, and sectorisation has also been utilised to its practical limit. Smart antennas therefore represent an innovative and realistic advance in the technology, particularly if they can be used as part of a truly integrated system design.

10.19.1 Considerations and possibilities

Current and future cellular systems operate in a variety of scenarios outdoors and indoors using macrocells, microcells and in some cases picocells, and they all present different problems. In macrocells using relatively high base station antennas, for example, signals are received at the base station with a fairly narrow spread in angle of arrival (Section 5.13). In this case an adaptive antenna cannot be used to discriminate against multipath, and the major benefits are gained from main-beam steering or interference nulling. This is not the case in micro- or pico-cellular environments nor at the mobile end of the link. However, there are no current proposals for adaptive antennas on mobiles; the concept presents enormous practical and financial difficulties.

Three main categories of smart antenna have been identified: *switched beam*, *direction-finding* and *optimum combining* [32]. These produce their responses in different ways. A switched-beam system produces a number of beams and the beam which gives the highest SNR is chosen. Direction-finding systems focus on acquiring the spatial directions associated with various users and on tracking their movements. Optimum combining systems attempt to maximise the output signal/interference-plus-noise ratio [SINR]. The main advantages and disadvantages of these three types are summarised below [32,33,36–39].

Switched beam

✓ Easily deployed
✓ Tracking at beam switching rate

× Low gain between beams
× Limited interference suppression
× False locking possibility

Direction-finding

✓ Tracking at angular change rate
✓ No reference signal required
✓ Easier downlink beam-forming

× Lower overall CIR gain
× Susceptible to signal inaccuracies, needs calibration
× Concept not applicable to small-cell non-LOS situations

Optimum combining

✓ Optimum SINR gain
✓ Accurate calibration unnecessary
✓ Good performance when number of elements is less than number of signals

× Difficult downlink beam-forming
× Good reference signal needed
× High update rates required

When the smart antenna is envisaged as a spatial filter, i.e. pointing a beam in the direction of a wanted signal or a null in the direction of an interferer, there is a further property that can be identified: spatial filtering for interference reduction (SFIR) and space division multiple access (SDMA). SFIR is illustrated in Figure 10.26(a), which shows a number of co-channel cells within a given service area. Each cell supports one user on the frequency f_k, but if that user is served via a beam from a smart antenna at a central base station (and power control is also used) then the

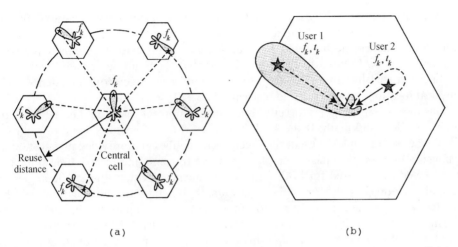

(a)   (b)

Figure 10.26 (a) Spatial filtering for interference reduction (SFIR); (b) space division multiple access (SDMA) (Courtesy G.V. Tsoulos).

overall level of interference will be reduced. Alternatively, a lower reuse distance could be employed in the system. With SDMA, illustrated in Figure 10.26(b), two or more users within the same cell can use the frequency f_k at the same time t_k, provided they are located in different spatial directions with respect to the base station. This can be termed *dynamic sectorisation* to distinguish it from the static or fixed sectorisation mentioned earlier. There is a need for monitoring, however, and an intracell handover to another frequency becomes necessary if the angular separation between two users becomes too small.

In practice there are different operating environments which place different constraints on the potential for improvement [40]. Some scatterers are local to the mobile and some are local to the base station; these scatterers produce the effects discussed in Chapter 5. There are also remote (distant) scatterers which can give rise to independently fading paths with large delays and a wide spread of spatial angles of arrival. In macrocells with fairly high base station antennas the change in the angle of arrival at the base station, for the signal from a given mobile, depends on the velocity of that mobile and its distance from the base station (assuming that contributions from remote scatterers are non-existent or can be ignored).

Suppose the mobile is moving at 200 kph on a circle of radius 1 km, centred on the base station, even then the angular velocity is no more than 3° per second. This dictates the rate at which the direction of the relevant beam has to be updated in order to track the mobile and it is clear that the greater the beamwidth, the lower the required update rate. Other factors also have to be taken into account. TDMA systems often use discontinuous transmission (DTX) to reduce the transmission rate when there is little or no speech actually passing [41, Ch. 8], and because this reduces the rate at which speech bursts are sent, the angular change between bursts is increased, possibly by a factor greater than 20. Compensatory measures may be necessary to deal with this.

In microcells the situation is quite different. There are often scatterers in close proximity to the base station and the signal arriving at the base station from a given mobile can no longer be associated with a narrow sector (spatial angle). In the extreme, waves may arrive from a wide variety of directions, much as they do at the mobile. In practice, although the dominant direction only exists where there is a very strong or maybe LOS path, there is still a spatial 'signature' associated with each mobile and this has to suffice for user location. The direct correspondence with physical location, however, no longer exists.

This discussion brings us directly to another important question, i.e. the base-to-mobile or downlink path. If the same frequency is used for uplinks and downlinks, i.e. time division duplex is used, then reciprocity applies and provided no substantial change in user location has taken place between transmit and receive time slots, the same beam can be used for both paths. In frequency division duplex the angles of arrival remain sensibly the same – they are a function of the locations of the scatterers rather than the transmission frequency – but the fading characteristics on the uplink and downlink paths are uncorrelated, especially if the duplex separation is 40 MHz as it is in GSM. Although this presents a challenge in the context of smart antennas, approaches to the problem have been suggested. Some of these [42–44] are essentially oriented towards signal processing while others exploit characteristics of the air interface, for example by adaptive resource allocation, i.e. allocating the downlink more radio channels in TDMA systems or more bandwidth in CDMA systems in an attempt to balance the benefits. Sophisticated diversity techniques have also been suggested [45].

The most difficult problem related to smart antennas, however, is practical implementation. Clearly there would be increased system complexity and the requirement for an integrated approach within an 'intelligent' system. Nevertheless, there are positive indications firstly because there are methods that can be used to reduce complexity [46,47] and secondly because technological advances may well be able to support smart antennas by the time that third-generation systems reach maturity. In any case, although the cost may seem high, the capacity gains and other benefits that will accrue make this a very attractive technology.

Provided a smart antenna can be incorporated into a cellular system, a number of benefits become available:

- The diversity gain offered by the adaptive system reduces fading, as seen earlier, and this means that RF power can often be reduced. Power control requirements are also eased.
- The information inherent in the system about mobile location and speed is a factor in deciding which adjacent cell is best placed to take over any mobile when handover is required. Moreover, the information available should permit the handover process to be optimised, i.e. made 'smart' as opposed to 'hard' or 'soft'.
- The spatial filtering properties can be used to provide transparent handover and simultaneously to combat the near far effect in direct sequence CDMA systems [48].

A number of other operational benefits such as increased capacity, coverage extension and the ability to support value-added services also become available. How many are exploited will depend on the maturity and characteristics of the system concerned.

10.20 WIDEBAND MODULATION: THE ALTERNATIVE

In earlier chapters we discussed the variety of transmission impairments that arise in radio communication systems as a result of the channel characteristics. The three-stage model (Figure 5.3) gives a good insight into the principal propagation phenomena that need to be considered. In narrowband systems, where the signal bandwidth is small compared with the coherence bandwidth of the channel, the fading is said to be 'flat'; this implies that all spectral components of the signal are affected in a similar way. In this context it is worth remembering that the small-scale multipath effects which give rise to Rayleigh fading are far more destructive in terms of performance degradation than the variations in median signal strength caused by shadowing and range from the transmitter. Moreover, multipath effects are far more difficult to combat. If no other steps were taken, the increase in transmitter power necessary to maintain a given performance threshold (SNR or BER) in the presence of multipath would be far greater than the increase required to deal with shadowing or variations in range.

In wideband systems, particularly in dense urban areas, the time delays associated with the various propagation paths are sufficiently different to cause signal components at different frequencies to be affected in different ways. The magnitude of the frequency transfer function then exhibits large variations over the band of interest. The fading is now 'frequency-selective' and this can cause severe distortion in the spectrum of the received signal. If the delay spread of the channel is comparable with, or larger than, the period of a data symbol then the received data symbols overlap causing errors due to intersymbol interference.

In addition to these effects, apparent in narrow- and wideband systems, there is always the Doppler shift in the received signal caused by motion. The spectrum of the transmitted signal is not merely displaced in frequency, it is actually spread out (Section 5.4) and this causes further distortion. Generally, in any system, for a given combination of carrier frequency, mobile speed and data rate, there is a certain irreducible error rate in a dense scatterer environment. In a well-designed system this IBER, as it is called, is well below the acceptable BER for the system and remains unnoticed. However, it sets a limit on system performance, and since it arises from the channel characteristics such as delay spread, rather than CNR, it cannot be reduced by improving the CNR, for example by increasing the transmitter power. The multipath mitigation techniques described earlier in this chapter all have a place in reducing the deleterious effects caused by these channel characteristics.

In the discussion above it is implicitly implied that the bandwidth of the transmitted signal is the minimum realistic bandwidth consistent with the information to be conveyed over the channel. This can range from a few tens of kilohertz in analogue voice systems to a few hundred kilohertz in a digital TDMA system such as GSM. However, there exists the possibility of intentionally spreading the transmitted signal over a bandwidth much larger that the required minimum – indeed much greater than the coherence bandwidth of the channel – to gain the necessary improvements in a completely different way. A detailed discussion of these spread spectrum-systems is well beyond the scope of this book but there are two basic possibilities.

First there is frequency hopping (FH), in which the relatively narrowband signal is transmitted using a carrier which 'hops' sequentially from one frequency to another;

the range is often determined using a PRBS and exceeds the channel coherence bandwidth. There is then only a very small probability that the signal spectrum will coincide with a minimum in the frequency transfer function and even then the coincidence will only exist for a short time before the carrier hops to another frequency.

Alternatively there is direct-sequence spread-spectrum (DS/SS) modulation, in which the information modulation is spread using a PRBS keyed at a high rate, as in the channel sounder of Chapter 8. Note that both FH and DS/SS have a built-in multiple access strategy since users can be assigned different spreading codes (PRBSs), coupled with which there is the natural security that comes from using wideband digital systems.

How wide must be the spectrum in FH or DS/SS systems in order to mitigate multipath effects? Intuitively it should be large; indeed the fact that the signal spectrum spans many lobes (maxima and minima) of the frequency transfer function is a key requirement in ensuring that the major part of the signal energy gets to the receiver and is not subject to the nulling that might destroy a narrowband signal.

10.20.1   Mitigation bandwidth

Recent studies [49–52] have focused on defining a more systematic measure (metric) of the ability of DS/SS signalling to reduce the variability of the received signal in a scattering environment. In these studies the received power, which is much less variable with DS/SS signalling, is described and displayed in a quite different way from the method used for narrowband signals (see Figure 5.2). Figure 10.27 shows normalised received power on a linear scale as a function of distance moved in wavelengths, under specified conditions.

The variability is now expressed in terms of the statistical variance about the mean, as indicated in the figure. The dependence of this normalised variance on the bandwidth of the DS/SS signal in a dense scatterer environment can be expressed in terms of the transmitted power spectral density $|U(f)|^2$ and the delay spread S [49], and simple analytical expressions are available for the normalised variance $\sigma^2[\,|s(t)|^2\,]$ for several different types of modulation [52]. These are plotted in Figure 10.28; the signal variable is T_c/S, i.e. the chip duration T_c, in units of RMS delay spread.

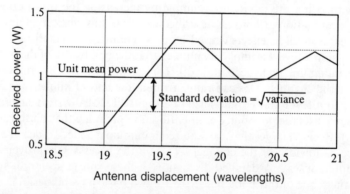

Figure 10.27   Normalised received power (linear scale) as a function of distance moved, for a DS/SS signal with chip duration equal to delay spread (Courtesy F.A. Amoroso).

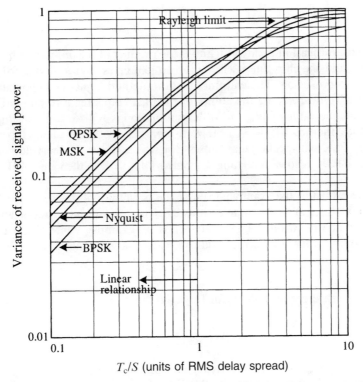

Figure 10.28 Received signal power as a function of T_c/S for different types of modulation (Courtesy F.A. Amoroso).

Amoroso [53] has pointed out that the curves in Figure 10.28 are reasonably linear for small values of T_c/S, indicating direct proportionality between $\sigma^2[|s(t)|^2]$ and T_c/S. In other words, the ability of DS/SS signalling to pass energy over a fading channel is proportional to the width of the DS/SS spectrum; the spectrum width is expressed in multiples of the coherence bandwidth of the channel. This, in turn, is proportional to $1/S$, while the width of the DS/SS spectrum is proportional to $1/T_c$. Thus the ability of DS/SS signalling to mitigate fading is proportional to S/T_c. The reason why Figure 10.28 shows that the variance of the received signal power is proportional to T_c/S is precisely because the variance itself is inversely proportional to the ability of DS/SS signalling to mitigate fading.

The constant of proportionality linking $\sigma^2[|s(t)|^2]$ to T_c/S is an important parameter of the chip modulation scheme. In fact, it enables us to define a quantity known as the *mitigation bandwidth*. The proportionality relationship may be written as

$$\sigma^2[|s(t)|^2] = \frac{1}{W_m^0} \frac{T_c}{S} \tag{10.50}$$

In this expression W_m^0 is termed the *mitigation bandwidth per chip rate* $(= W_m^0 T_c)$ of the chip modulation type. The concept becomes clearer if eqn (10.50) is rewritten as

$$\sigma^2[|s(t)|^2] = \frac{T_c}{W_m^0}\frac{1}{S} = \frac{1}{W_m S} \tag{10.51}$$

Here W_m is the mitigation bandwidth and is the product of the mitigation bandwidth per chip rate and the chip rate itself. Amoroso [53] gives values of W_m^0 for various types of chip modulation. The chip modulation method is just as important as the chip rate itself in determining the ability of the final chip stream to mitigate fading. As an example the value of W_m^0 for MSK modulation is 1.7046, so if the chip rate is 20 Mbit/s then the net W_m is 34.1 MHz. In contrast, with the same chip rate but using binary PSK with rectangular pulses (for which $W_m^0 = 3$), the value of W_m^0 is 60 MHz; for QPSK, also with rectangular pulses ($W_m^0 = 1.5$), the value of W_m is only 30 MHz.

Finally we return to the example given at the beginning of this section, where we commented on the loss of communications efficiency as a result of Rayleigh fading. Taking differentially encoded PSK (DPSK) as an example, eqn. (10.25) gives the error probability in the presence of additive white Gaussian noise (AWGN) as

$$P_e(\gamma) = \tfrac{1}{2}\exp(-\gamma)$$

whereas the mean error probability in Rayleigh fading is

$$\overline{P_e}(\gamma) = \frac{1}{2(1+\gamma_0)}$$

where γ_0 is the mean value in the Rayleigh fading channel.

To maintain a BER of 10^{-3} in the AWGN channel, the necessary value of γ is 7.93 dB whereas in Rayleigh fading this increases to 26.98 dB – a loss of 19.05 dB. Fortunately, the use of DS/SS signalling can overcome a large part of this loss. Simulation [52] has been used to investigate the relationship between $\overline{P_e}(\gamma)$ and γ_0 over a wide range of values of T_c/S for DPSK detection on bits with rectangular binary chip pulses, and for DPSK detection on bits with MSK modulation.

A number of simplifying approximations were made, but in general terms the BER curves migrated from the Rayleigh fading case towards the AWGN case as the value of T_c/S was increased. Moreover, the shape of the BER curve tended to resemble that of a narrowband Rician fading channel. Normally, the Rician factor K indicates the ratio of the power in the steady (LOS) component of the signal to the power in the multipath components (Section 5.11) and although the simulation did not include an actual LOS path, it was possible to produce an approximate formula linking the hypothetical K factor to the parameters of the DS/SS system:

$$K = 2W_m S$$

This clearly reflects that if $W_m = 0$ then $K = 0$, i.e. a narrowband Rayleigh fading channel, and it also reflects that as $W_m \to \infty$ so $K \to \infty$. Infinity corresponds to perfect mitigation of the effects of fading, so that in effect we have an AWGN channel. This can never be achieved in practice but the reported simulation and the approximate relationship derived from it indicate very clearly that the greater the bandwidth able to be accommodated in a given system, the greater the mitigation of the Rayleigh fading effects.

The major considerations as summarised by Amoroso are:

- For effective mitigation of fading effects, the chip interval T_c should be, at most, equal to the RMS delay spread S; preferably it should be much smaller.
- For minimal intersymbol interference the symbol duration should be much longer than S.
- The chip stream should be as random as possible, i.e. the chip values should be statistically uncorrelated.
- Even in a fast-moving vehicle, the distance travelled by the antenna during one symbol period should be less than 0.38λ. Otherwise coherent symbol correlation becomes difficult and a high IBER may result.
- The mean BER is the average over the full distance travelled by the receiving antenna. If the antenna moves very slowly (perhaps it is being carried by a pedestrian) or remains stationary, it may be more meaningful to calculate the worst-case BER over a specified large percentage of all the antenna locations.

A chip-matched filter is essential to symbol detection at those frequently occurring antenna locations where the received power is a minimum, dispersion is a maximum and chip pulses arrive with essentially zero mean value. These situations occur about once per wavelength of antenna travel. With DPSK the matched filter could be as simple as a one-symbol time delay.

REFERENCES

1. Brennan D.G. (1959) Linear diversity combining techniques. *Proc. IRE*, **47**, 1075–102.
2. Kahn L.R. (1954) Ratio squarer. *Proc. IRE*, **42**, 1704.
3. Jakes W.C. (ed.) (1974) *Microwave Mobile Communications*. John Wiley, New York.
4. Parsons J.D. and Gardiner J.G. (1988) *Mobile Communication Systems*. Blackie, Glasgow.
5. Lee W. C.-Y. (1978) Mobile radio performance for a two-branch equal-gain combining receiver with correlated signals at the land site. *IEEE Trans.*, **VT27**, 239–43.
6. Adachi F., Feeney M.T. and Parsons J.D. (1988) Effects of correlated fading on level crossing rates and average fade durations with predetection diversity reception. *Proc. IEE Part F*, **135**(1), 11–17.
7. Granlund J. (1956) *Topics in the design of antennas for scatter*. Technical Report 135, MIT Lincoln Lab, Lexington MA, pp. 105–13.
8. Schwartz M., Bennett W.R. and Stein S. (1966) *Communication Systems and Techniques*. McGraw-Hill, New York.
9. Adachi F. and Ohno K. (1991) BER performance of QDPSK with post-detection diversity reception in mobile radio channels. *IEEE Trans.*, **VT40**(1), 237–49.
10. Dernikas D. (1999) Performance evaluation of the TETRA radio interface employing diversity reception in adverse conditions. PhD thesis, University of Bradford.
11. Adachi F. and Parsons J.D. (1988) Unified analysis of post-detection diversity for binary digital FM mobile radio. *IEEE Trans.*, **VT37**(4), 189–98.
12. Adachi F., Feeney M.T. and Parsons J.D. (1988) Level crossing rate and average fade duration for time-diversity reception in Rayleigh-fading conditions. *Proc. IEE Part F*, **135**(1), 11–17.
13. deToledo A.F. (1989) Investigation of time diversity techniques for digital mobile radio. MSc thesis, University of Liverpool.
14. Leather P.S.H. and Parsons J.D. (1996) Handheld antenna diversity experiments at 450 MHz. *Proc. IEE Colloquium on Multipath Countermeasures*, 1996/20, pp. 4/1 to 4/6.

15. Vaughan R.G. and Andersen J.B. (1987) Antenna diversity in mobile communications. *IEEE Trans.*, **VT36**(4), 149–72.
16. Leather P.S.H. and Massey P. (1996) *Antenna diversity from two closely spaced dipoles.* Philips Research Report RP3492.
17. Leather P.S.H. (1996) Antenna diversity for handportable radio at 450 MHz. PhD thesis, University of Liverpool.
18. Rappaport T.S. (1996) *Wireless Communications.* Prentice Hall, Englewood Cliffs NJ.
19. Lucky R.W., Salz J. and Weldon E.J. Jr (1968) *Principles of Data Communication.* McGraw-Hill, New York.
20. Proakis J.G. (1989) *Digital Communications.* McGraw-Hill, New York.
21. Monson P. (1977) Theoretical and measured performance of a DFE modem on a fading multipath channel. *IEEE Trans.*, **COM25**, 1144–53.
22. Viterbi A.J. (1967) Error bounds for convolutional codes and an asymptotically optimum decoding algorithm. *IEEE Trans.*, **IT13**, 260–9.
23. Blahut R.E. (1983) *Theory and Practice of Error Control Codes.* Addison-Wesley, Reading MA.
24. Gilbert E.N. (1960) Capacity of a burst error channel. *Bell Syst. Tech. J.*, **39**, 1253–65.
25. Hagenauer J. and Lutz E. (1987) Forward error correction coding for fading compensation in mobile satellite channels. *IEEE Trans.*, **SAC5**(2), 215–25.
26. Hanes R.B., Goody C. and Attkins, P. (1986) An efficient 16 kb/s speech codec for land mobile radio applications. *Proc. Second Nordic Seminar DMR II.*
27. Vary P., Sluyter R.J., Galand C. and Rosso M. (1987) RPE-LPC codec – the candidate for the GSM radio communication system. *Proc. Int. Conf. on Digital Land Mobile Radio Communications*, Venice, pp. 507–16.
28. Un C.K. and Magill D.T. (1975) The residual-excited linear prediction vocoder with transmission rates below 9.6 kb/s. *IEEE Trans.*, **COM23**, 1466–74.
29. Atal B.S. and Remde J.R. (1982) A new mode of LPC excitation for reproducing natural-sounding speech at low bit rates. *Proc. ICASSP*, pp. 614–17.
30. Price R. and Green P.E. (1958) A communication technique for multipath channels. *Proc IRE*, **46**(3), 555–70.
31. Nche C. (1995) UHF propagation measurements for future CDMA systems. PhD thesis, University of Liverpool.
32. Tsoulos G.V. (1999) Smart antennas for mobile communication systems: benefits and challenges. *IEE Electron. Commun. J.*, pp. 84–94.
33. Ho M.-J., Stuber G.L. and Austin M.D. (1998) Performance of switched-beam smart antennas for cellular radio systems. *IEEE Trans.*, **VT47**(1), 10–19.
34. Hudson J. (1981) *Adaptive Array Principles.* Peter Peregrinus, London.
35. Tsoulos G.V., Beach M. and McGeehan J.P. (1997) Wireless personal communications for the 21st century: European technological advances in adaptive antennas. *IEEE Commun. Mag.*, **35**(9), pp. 102–9.
36. TSUNAMI Project Final Report R2108/ERA/WP1.3/MR/P/096/b2 (1996).
37. Tangemann M. (1995) Near-far effects in adaptive SDMA system. *Proc. 6th International Symposium on Personal, Indoor and Mobile Radio Communications*, Toronto, Canada, pp. 1293–7.
38. Fuhl J. and Molisch A. (1996) Capacity enhancement and BER in a combined SDMA/TDMA system. *Proc. 46th IEEE Vehicular Technology Conference*, Atlanta AA, pp. 1481–5.
39. Xu G. *et al.* (1994) Experimental studies of space division multiple spectrally efficient wireless communications. *Proc ICC'94* New Orleans LA, pp. 800–4.
40. Ward C. *et al.* (1996) Characterising the radio propagation channel for smart antenna systems. *IEE Electron. Commun. Engng J.*, **8**(4), 191–200.
41. Steele R. (ed.) (1992) *Mobile Radio Communications.* Pentech Press, London.
42. Winters, J. (1993) Two signalling schemes for improving the error performance of FDD transmission systems using transmitter antenna diversity. *Proc. VTC'93*, Secaucus, NJ, pp. 85–8.
43. Gerlach D. and Paulraj A. (1994) Adaptive transmitting antenna arrays with feedback. *IEEE Signal Process. Lett.*, **1**(10), 150–3.

44. Rayleigh G., Oigarri S., Jones V. and Paulraj A. (1995) A blind adaptive transmit antenna algorithm for wireless communication. *Proc ICC'95*), pp. 1495–99.
45. Paulraj A. (1997) Smart antennas in wireless communications: technology overview. *Proc. 4th Workshop on Smart Antennas in Wireless Mobile Communications*, Stanford University, Stanford CA.
46. Ward C., Hargrave P. and McWhirter J. (1986) A novel algorithm and architecture for adaptive digital beamforming. *IEEE Trans.*, **AP34**(3), 338–46.
47. Gockler H. and Schenermann H. (1982) A modular approach to a 60-channel trans-multiplexer using directional filters. *IEEE Trans. Commun.*, **30**(7), 1588–613.
48. Tsoulos G., Athanasiadou G., Beach M. and Swales S. (1998) Adaptive antennas for microcellular and mixed cell environments with DS-CDMA. *Kluwer Wireless Pers. Commun. J.*, **7**(2/3), 147–69.
49. Holtzman J. and Jalloul L.M.A. (1994) Rayleigh fading effect reduction with wideband DS/CDMA signals. *IEEE Trans. Commun.*, **42**(3), 1012–16.
50. Amoroso F. (1993) Effective bandwidth of DSPN signaling for mitigation of fading in dense scatterers. *Electron. Lett.*, **29**(8), 661–2.
51. Amoroso F. (1993) Improved method for calculating mitigation bandwidth for DSPN signals. *Electron. Lett.*, **29**(20), 1743–5.
52. Amoroso F. (1994) Investigation of signal variance, bit error rates and pulse dispersion for DSPN signaling in a mobile dense scatterer ray-tracing model. *Int. J. Satellite Commun.*, **12**(6), 579–88.
53. Amoroso F. (1996) Use of DS/SS signaling to mitigate Rayleigh fading in a dense scatterer environment. *IEEE Personal Commun. Mag.*, **3**(2), 52–61.

Chapter 11

Planning Radio Networks

11.1 INTRODUCTION

In earlier chapters we discussed the characteristics of the radio propagation channel in some detail. We introduced methods for predicting the mean signal level within a small area in rural, suburban and urban environments and it became clear that this is a complicated process involving a knowledge of several factors, including the details of the terrain, the building clutter and the extent of foliage along the radio path. Most importantly perhaps, it became apparent that signal strength prediction is not an exact science; the mean signal in a small area can be predicted using any of the methods discussed in Chapters 3 and 4, but the prediction is only an estimate. Not only is it inexact in itself, there will also be variations about the mean as the mobile moves around within the small area concerned.

The variations have lognormal statistics with a standard deviation which depends on the nature of the local environment. Superimposed on these variations in the local mean signal (which are known as slow fading) are much more rapid and deep variations (known as fast fading), caused by multipath propagation in the immediate vicinity of the mobile. These follow Rayleigh statistics over fairly short distances. We have also discussed other important characteristics of the channel such as noise, mentioned the interference that can affect a given user in a multi-user environment, and considered additional parameters that are important in so-called wideband channels where the signal bandwidth is such that frequency-selective fading and intersymbol interference arise.

We can now take a look, albeit brief, at how this information can be brought together in order to plan a radio network for a specific purpose. We will find that some factors are more important than others and that radio system planning involves far more than merely estimating the signal strength and its variability. Cellular radio systems are very important in the modern world and they will be used as examples throughout this chapter. Cellular systems also require a well-designed frequency assignment plan based, among other things, on an assessment of the amount of teletraffic offered to the system in certain locations and at certain times. These aspects of system planning have not been mentioned so far and will only be treated very briefly here.

11.2 CELLULAR SYSTEMS

Cellular systems were introduced in Chapter 1 when we were considering area coverage techniques. Many excellent explanations of the general strategy exist in the literature, so a very short account will be sufficient to set the scene.

If a fixed amount of radio spectrum is available to provide a given service, then the traditional problem faced by system designers is how to balance the apparently conflicting requirements of area coverage and system capacity. We discussed in Chapter 1 the question of using a powerful transmitter on a high site and concluded that while this was ideal for public service broadcasting it was completely contrary to the requirements of a mobile radio communication service. Recognising this, the regulatory authorities in many countries have, from the very early days, set limits on base station transmitter powers in order to improve frequency reuse opportunities, thereby obliging system designers to invent other strategies to achieve area coverage. Here again there are different considerations, and a technique which suits a private mobile radio system operating in a single town or city is unlikely to be optimum for implementing a national network.

Nevertheless, the provision of wide area coverage will always involve the development of an infrastructure of radio and/or line links to connect together a number of base stations via one or more control points, so that the nearest base station to any mobile can be used to relay messages to and from that mobile. Creating a national network using only radio links is clearly very complicated and costly; in any case a ready-made alternative, the public telephone network, is already available. If this is used as the backbone infrastructure to connect the base stations together, then provided there are many connection points between the base stations and the fixed network, each base station only has to cover a small area. This in itself is a major step towards achieving much greater frequency reuse. Moreover, in principle, a mobile within the coverage area of any base station has available to it the full facilities of the national and international telephone network.

The potential of this strategy was realised many years ago but before any systems could be implemented some major issues had to be solved:

- Much higher carrier frequencies had to be used so that the radio coverage from any base station could be defined and constrained (more or less) to a desired area or cell. This had technological and regulatory implications.
- It was necessary to develop methods of addressing individual mobiles, and locating and continuously monitoring the position of all active mobiles in the system. Calls directed to any mobile could then be routed via the base station which offered the best radio path, and mobiles wishing to initiate a call could gain access to the network via the appropriate base station. This required a new generation of electronic exchanges (switches) and low-cost processing power at both base stations and mobiles to handle the overhead associated with setting up and monitoring the progress of telephone calls.

Cellular schemes [1] represent the most technologically advanced method of area coverage and they are now highly developed and well documented. They are specifically engineered so that overall system performance is limited by interference rather than by noise and they operate at frequencies of 900 MHz and above where, in any case, receiver noise is likely to dominate over external, man-made noise.

Frequency reuse is a fundamental concept in cellular systems, but careful planning is necessary to avoid performance degradation by co-channel interference, i.e. interference with calls in one cell caused by a transmitter in another cell where the same set of frequencies are used. If a fixed number of radio channels are available for a given cellular system, they can be divided into several sets, each set being allocated for use in a given small area (a cell) served by a single base station. The greater the number of channels available in any cell, the more simultaneous telephone calls that can be handled but the smaller the total number of cells that make up a cluster which uses all the channels.

Suppose there are 56 channels in total: they can be split into four groups of 14 or seven groups of 8 after which the channels have to be reused. Capacity is maximised by a design which uses a small cluster size repeated often, but this increases the potential for interference since co-channel cells are geographically closer together. However, not only is it necessary to reuse channel sets in a number of different cells, it is also necessary that every mobile transceiver can be tuned, on command from the central control, to any of the available channels, including those designated as 'control' channels.

This is necessary firstly because a mobile can be located anywhere in the total coverage area of the system and can therefore be required to operate on a channel associated with any cell; and secondly because it can cross a cell boundary during the progress of a call. When this is detected, the central control instructs the mobile to retune to a different channel – one associated with the new cell – and at the same time it initiates a handover of the call to the new base station.

The principle is that if a set of channels (a subset of the total) is available in a given cell, a mobile is allocated exclusive use of a channel (go-and-return) on demand, but only for the duration of the call. When the call is complete, or if the mobile crosses a cell boundary, the channel is returned to the pool and can be reallocated to another mobile. This is known as *dynamic channel assignment*, or by analogy with fixed telephone networks, *trunking*. It requires agile, low-cost frequency synthesisers at base stations and in mobiles; it also implies that quiescent mobiles, i.e. those that are active but not engaged on a call, must automatically tune to a predesignated control channel associated with the cell in which they are located, so that instructions can be sent and received.

The word 'channel' has been used to describe the resource allocated to a mobile in order to make a call. In first-generation analogue systems using FDMA, the available spectrum is divided into narrow channels, typically 25 kHz apart in systems such as TACS. These channels are allocated to the cells that make up a cluster in a manner that will be discussed later. A mobile initiating or receiving a call is allowed exclusive use of one of the channels allocated to the cell where it is located at the time the call is set up, and it retains exclusive use of that channel until the call ends, or it experiences a handover as a result of crossing a cell boundary. In second-generation digital systems such as GSM, the available spectrum is split into much wider channels, 200 kHz apart, and these are allocated to cells in a similar way.

In GSM, however, TDMA is used and a mobile initiating or receiving a call is allocated exclusive use of one of the time slots associated with a carrier. In other words, a mobile is allocated use of the whole bandwidth, but for only part of the time. If a handover is necessary, the mobile will have to tune to a new carrier frequency and use a new time-slot within the TDMA frame. In third-generation systems, soon to be implemented, it is likely that CDMA will be used as the multiple access technique. A mobile initiating or receiving a call will then be allocated a code which will enable it to use the whole of the bandwidth for the whole of the time, interference being limited by the fact that the codes allocated to various mobiles are different and mutually orthogonal.

In planning a cellular radio-telephone system it is necessary to use a cluster size such that all the clusters fit together to cover the desired service area without leaving any gaps. Although there are a number of cell shapes that could be used, and would satisfy this criterion (e.g. squares and triangles), a hexagon is the ideal model for radio systems since it approximates the circular coverage that would be obtained from a centrally located base station and it offers a wide range of cluster sizes determined by the relationship

$$N = i^2 + ij + j^2 \qquad (11.1)$$

Here i and j are positive integers, or zero, and $i \geqslant j$ Any value of N given by this relationship produces clusters which tessellate, and the planned overall coverage area has the appearance of a mosaic. Table 11.1 shows various allowable cluster sizes which satisfy eqn. (11.1) and the 7-cell cluster (for which $i = 2$ and $j = 1$) proved a good choice in early analogue systems.

The layout of a basic cellular system proceeds from a knowledge of the two shift parameters i and j as follows. Starting from any cell as a reference, move i cells along any of the chains of hexagons (6 in number) that emanate from that cell; turn anticlockwise by 60°; move j cells along the chain that lies in this new direction. The cell so located should use the same set of channels as the original reference cell. Other co-channel cells can be found by returning to the reference cell and moving along a different chain of hexagons using the same procedure. Figure 11.1 shows how this procedure can be used to build up a system comprising 7-cell clusters. Once the location of all the cells using channel set A has been determined, it is not necessary to work through the procedure again for other cells, e.g. cells marked B; the pattern of cells around all cells marked A is the same as that around the reference cell.

How far apart are cells which use the same channel set? This is a major factor in determining the probability of co-channel interference. The distance D between the centres of cells which use the same set is often called the *repeat distance* or *reuse distance*. It can be determined in terms of the cell radius R and is given by

Table 11.1 Some possible values of cluster size N

i	j	N
1	0	1
1	1	3
2	0	4
2	1	7
2	2	12
3	2	19
4	1	21

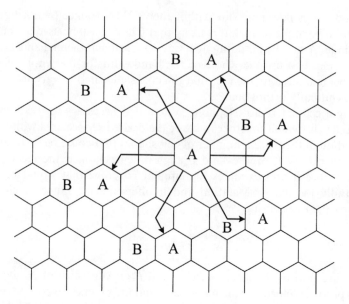

Figure 11.1 Determining co-channel cells; here $i = 2$ and $j = 1$, realising 7-cell clusters.

$$\frac{D}{R} = \sqrt{3N} \qquad\qquad (11.2)$$

11.2.1 Interference considerations

The design of any cellular radio-telephone system must include ways of limiting adjacent channel as well as co-channel interference. Receivers normally contain IF filters which significantly attenuate signals on those channels adjacent to the wanted channel, but it is highly desirable to avoid circumstances in which a strong adjacent channel signal is present, as this will inevitably degrade performance. The first step towards this is to adopt a frequency allocation strategy in which adjacent carrier frequencies are not used in the same cell. In practice this is relatively straightforward and the largest possible difference is maintained between the frequencies used to make up a given set. For example, suppose that the available channels are numbered sequentially from 1 upwards and the frequency difference between channels is proportional to the difference between their channel numbers. If N disjoint channel sets are required in a given system then the nth would contain channels n, $(n + N)$, $(n + 2N)$, $(n + 3N)$, etc. Thus in a 7-cell system the 4th set would contain channels 4, 11, 18 and 25.

In addition to this, a mobile located near the edge of its serving cell is approximately equidistant from its wanted base station transmitter and one adjacent cell base station (maybe more). Propagation factors and fading can combine to make the adjacent channel signal up to 30 dB stronger than the wanted signal, causing severe problems. It is also desirable, therefore, that the adopted strategy should avoid the use of adjacent channels in any pair of adjacent cells. With cluster sizes of $N = 3$ or 4, excellent for overall system capacity, this is impossible since in a 3-cell cluster each cell is adjacent to the other two, and in a 4-cell cluster there are two cases in which one of the cells is adjacent to the other three.

The 12-cell cluster permits the adjacent channel criterion to be satisfied completely but at the expense of an increased D/R ratio and a reduced capacity per cell. In consequence the 7-cell cluster is usually preferred; it allows the adjacent channel criterion to be more closely approached because, although the centre cell is adjacent to all the other 6 cells, each cell on the outer ring is adjacent to only the centre cell and two others.

MacDonald's paper [1] contains an appendix which summarises the fundamentals of hexagonal cellular geometry and presents a simple algebraic method for using the coordinates of the cell centre to determine which channel set should be used in that cell. It was developed with first-generation systems in mind, but the principles remain generally applicable. The method is illustrated in Figure 11.2, which shows a convenient coordinate system. The positive halves of the two axes intersect at an angle of 60° and the unit distance along each axis is $\sqrt{3}$ times the cell radius; the radius being defined as the distance from the cell centre to any vertex. This geometry allows the centre of every cell to fall on a point specified by a pair of integer coordinates.

In this coordinate system the distance d_{12} between two points having coordinates (u_1, v_1) and (u_2, v_2) is

$$d_{12} = \sqrt{(u_2 - u_1)^2 + (u_2 - u_1)(v_2 - v_1) + (v_2 - v_1)^2} \qquad (11.3)$$

Thus the distance between the centres of adjacent cells is unity and the cell radius is

$$R = \frac{1}{\sqrt{3}} \qquad (11.4)$$

The number of cells per cluster, N, can be calculated fairly easily. We have already described the way in which co-channel cells can be located and Figure 11.1 gives an illustration. Equation (11.3) shows that the distance between the centres of these cells is

$$D = \sqrt{i^2 + ij + j^2} \qquad (11.5)$$

Figure 11.1 further illustrates the universal fact that any cell has exactly six equidistant neighbouring co-channel cells and that the vectors from the centre of any cell to these co-channel cells are separated in angle from one another by multiples of 60°. The next step is to visualise each cluster as a large hexagon (Figure 11.3). In reality a cluster is composed of a group of contiguous hexagonal cells and cannot itself be hexagonal; nevertheless, the large hexagon can have the same area as a cluster. The seven cells labelled A in Figure 11.3 are reproduced from Figure 11.1 and the centre of each of these cells is also the centre of a large hexagon representing a cluster.

Each A cell is embedded in precisely one large hexagon, just as it is contained in precisely one cluster. All large hexagons have the same area, just as all clusters have the same area, and the area of the large hexagon equals the area of the cluster. We know that the distance between the centres of adjacent cells is unity, so the distance

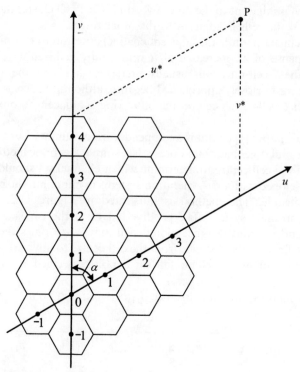

Figure 11.2   Coordinate system for hexagonal cell geometry.

between the centres of large hexagons is $\sqrt{i^2 + ij + j^2}$. The pattern of the large hexagons is clearly an exact replica of the cell pattern, scaled by a factor of $\sqrt{i^2 + ij + j^2}$, so N, the total number of cell areas contained in the area of the large hexagon, is the square of this scaling factor, i.e.

$$N = i^2 + ij + j^2$$

indicated by eqn. (11.1). Using equations (11.4), (11.5) and (11.1) we can obtain the relationship quoted earlier:

$$\frac{D}{R} = \sqrt{3N}$$

In certain cases of practical interest, specifically when the smaller of the shift parameters j equals unity, a simple algebraic algorithm exists to determine the frequency set to be allocated to any cell. In these cases it is convenient to label each cell in a cluster with the integers 0 to $N - 1$. The correct label for the cell that lies at (u, v) is then given by

$$L = [(i + 1)u + v] \quad \mathrm{mod}\ N \tag{11.6}$$

Application of this simple formula causes all cells which should use the same frequency set to have the same numerical label.

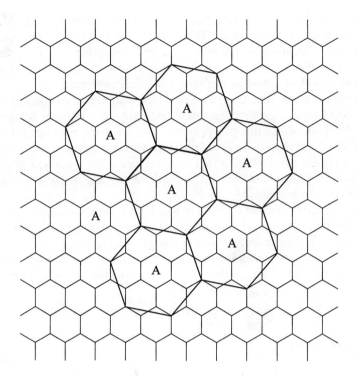

Figure 11.3 Determining the number of cells per cluster; this example is related to Figure 11.1 and is for a 7-cell repeat pattern.

11.3 RADIO COVERAGE

The quality of service experienced by an individual subscriber to a radio-telephone network depends on a number of factors. Among the more important is the strength of the wanted signal at the subscriber terminal. Coverage is the generic term used to describe this; it also embraces the assumption that sound engineering design has been used to obtain a balanced link so that the subscriber terminal produces an adequate signal at the base station receiver.

Other factors include the probability of interference and the availability of the necessary resources within the radio and fixed network segments to accommodate calls, to hand them over as necessary and to avoid dropped calls. We will return to these topics later. None of these factors will remain constant throughout a large network. They will depend on parameters such as the morphological characteristics of the area, the number of subscribers and the extent of frequency reuse.

11.3.1 Coverage of a small area

The term 'coverage' is used in a generic sense to mean the area that is served by a base station, or a number of base stations which form a network. However, to say an individual base station covers a given area does not mean that an adequate signal strength exists at all (100%) of locations within that area. It means that an adequate signal exists at a very high percentage of locations within the cell (the exact percentage remains to be defined); this is a compromise between the impossible task of covering every location while providing an acceptable level of service to subscribers within the cell and not causing interference to subscribers in adjacent cells. The calculations of coverage can be approached as follows.

We assume that the coverage area of a given base station is approximately circular and that the local mean signal strength in a small area at a radius r is lognormally distributed. We understand this to imply that the local mean (averaged over the Rayleigh fading) in decibels is a normal random variable x with mean value $\bar{x}$ and standard deviation σ. We recognise that x and $\bar{x}$ are often expressed in dBm. To avoid confusion, $\bar{x}$ is the value that can be predicted by any of the available signal strength prediction techniques.

Let x_0 be the receiver threshold level for which an acceptable output is obtained. Again we realise that the value of x_0 is not necessarily the receiver noise threshold but can take into account interference and fading margins (see later). We wish to know the percentage of locations (incremental areas) at the given radius $r = R$, where the signal x is above the threshold level. The probability density function of x is given by

$$p(x) = \frac{1}{\sigma\sqrt{2\pi}} \exp\left[\frac{-(x - \bar{x})^2}{2\sigma^2}\right] \tag{11.7}$$

and the probability that $x \geqslant x_0$ is

$$P_{x_0}(R) = P[x \geqslant x_0] = \int_{x_0}^{\infty} p(x)\,\mathrm{d}x$$

$$= \frac{1}{2}\left[1 - \operatorname{erf}\left(\frac{x_0 - \bar{x}}{\sigma\sqrt{2}}\right)\right] \tag{11.8}$$

If we have predicted values for $\bar{x}$ and σ for the small area concerned, then we can use eqn. (11.8) to estimate the percentage of locations at a given radius R where the average signal exceeds the value x_0. Table 11.2 shows the location probability for various values of $(x_0 - \bar{x})$ and σ. As an example, at a radius where the receiver threshold level is 10 dB below the mean value of the lognormal distribution and $\sigma = 10$ dB, we have

$$P_{x_0}(R) = \frac{1}{2}\left[1 - \operatorname{erf}\left(\frac{-1}{\sqrt{2}}\right)\right] = 0.84$$

Table 11.2   Location probability (% area coverage)

$x_0 - \bar{x}$ (dB)	Location probability (%)			
	$\sigma = 4\,\mathrm{dB}$	$\sigma = 6\,\mathrm{dB}$	$\sigma = 8\,\mathrm{dB}$	$\sigma = 10\,\mathrm{dB}$
-15	> 99	> 99	97	93.3
-10	> 99	99	89.5	84
-5	89	79.5	73.5	69
-2	69	63	60	58
0	50	50	50	50

In other words, 84% of locations at a radius R from the given base station have a signal strength above the threshold.

11.3.2 Coverage area of a base station

It is vital for radio system planners to be able to estimate the coverage area of a base station. This can be done by extending the analysis in the previous section to estimate the percentage of locations within a circle of radius R (which in this case represents the cell boundary) where the signal exceeds the given threshold level x_0. This gives a measure of the base station coverage and hence the quality of service. An analysis presented by Jakes [2] proceeds as follows. We define the fraction of useful service area F_u within a circle of radius R as that area where the received signal exceeds x_0. If P_{x_0} is the probability that x exceeds x_0 in a given incremental area dA, then

$$F_u = \frac{1}{\pi R^2} \int P_{x_0} \, dA \qquad (11.9)$$

Jakes points out that in a practical situation it would be necessary to break the integration down into small areas for which P_{x_0} can be estimated and then sum over all such areas. However, a useful indication can be obtained by assuming that the mean received signal strength follows an inverse power law with distance from the base station, i.e. it varies as r^{-n}. Then $\bar{x}$ (dB or dBm) can be written as

$$\bar{x} = \alpha - 10 \log_{10}\left(\frac{r}{R}\right) \qquad (11.10)$$

where α is a constant determined from the transmitter power, the height and gain of the base station antenna, etc. Using eqn. (11.8) we obtain

$$P_{x_0} = \frac{1}{2}\left[1 - \mathrm{erf}\left(\frac{x_0 - \alpha + 10\,n\,\log_{10}(r/R)}{\sigma\sqrt{2}}\right)\right] \qquad (11.11)$$

Making the substitutions

$$a = \frac{x_0 - \alpha}{\sigma\sqrt{2}} \quad \text{and} \quad b = \frac{10n\log_{10}\mathrm{e}}{\sigma\sqrt{2}}$$

and noting the general relationship

$$\log_b N = \frac{\log_a N}{\log_a b}$$

we obtain

$$P_{x_0} = \tfrac{1}{2}\left[1 - \mathrm{erf}(a + b\log_\mathrm{e}(r/R))\right] \qquad (11.12)$$

Again, we can write eqn. (11.9) as

$$F_u = \frac{2}{R^2} \int_0^R r P_{x_0} \, dr$$

and thus reach the expression

$$F_u = \frac{1}{2} - \frac{1}{R^2} \int_0^R r \, \text{erf}(a + b \log_e(r/R)) \, dr \tag{11.13}$$

This can be evaluated by making the substitution $t = a - b \log_e(r/R)$, which leads to the equation

$$F_u = \frac{1}{2} - \frac{2 \exp(2a/b)}{b} \int_a^\infty \exp(-2t/b) \text{erf}(t) \, dt \tag{11.14}$$

This is a standard integral listed in tables [3]; the solution is

$$F_u = \frac{1}{2}\left\{ 1 + \text{erf}(a) + \exp\left(\frac{2ab + 1}{b^2}\right)\left(1 - \text{erf}\frac{ab + 1}{b}\right)\right\} \tag{11.15}$$

This is a rather complicated equation, but it simplifies considerably for the special case when $\bar{x} = x_0$ at $r = R$, i.e. when at the cell edge, the predicted mean is equal to the threshold level. In this case

$$F_u = \frac{1}{2} + \frac{1}{2}\exp\left(\frac{1}{b^2}\right)\left[1 - \text{erf}\left(\frac{1}{b}\right)\right] \tag{11.16}$$

When an inverse power law is assumed, the important parameter for coverage is σ/n. Figure 11.4 shows a plot of F_u as a function of σ/n for various values of $P_{x_0}(R)$ (the percentage of locations on the cell boundary where the signal level exceeds the threshold). It shows, for example, that if $n = 4$ and $\sigma = 8$ dB, typical values for built-up areas, then the signal level is above the threshold at 94% of locations within the cell if 75% of locations on the boundary are covered. Similarly, 71% location coverage results from 50% boundary coverage. This does not necessarily mean that service is not available at the remaining locations; we are dealing with averages here

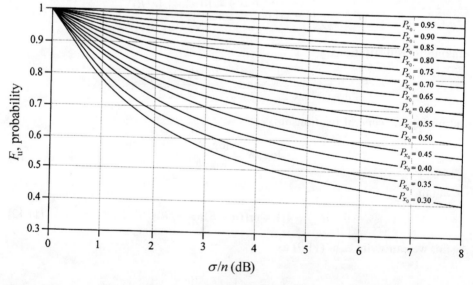

Figure 11.4   Location probability F_u for the various values of edge probability P_{x_0}.

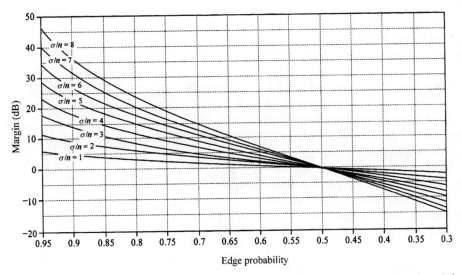

Figure 11.5 The required lognormal margin as a function of edge probability, plotted for various values of σ/n.

and it only means that at these locations a subscriber has a less than 50% probability of establishing a connection to the network.

In practice a service provider may wish to provide 90% coverage of locations within a given cell. For any given value of σ/n, Figure 11.4 gives the percentage of boundary locations that have to be covered in order to achieve this; and for $\sigma/n = 2$ the percentage of boundary locations is 72%. This introduces a further factor, because methods of estimating signal strength produce the median, i.e. the value exceeded at 50% of locations. To guarantee coverage at a greater percentage of locations on the cell boundary, the median signal strength on this boundary will need to exceed the receiver threshold value x_0, not just be equal to it. The necessary margin can be calculated fairly easily because we know that the local mean signal has lognormal statistics. Figure 11.5 shows the required margin as a function of the edge probability. Again it is drawn with σ/n as a parameter. Returning to the example above, if $\sigma/n = 2$ then Figure 11.5 shows that a margin of about 4 dB (increase in signal strength) is needed to move from a probability of 50% to a probability of 72%.

11.4 PLANNING TOOLS

Planning tools are complicated software packages which comprise a number of modules that enable the engineer to plan a mobile radio network. Central among them is a modelling tool which facilitates the automatic assessment and calibration of environment-specific propagation models and the prediction of signal coverage over a wide geographical area. It is unrealistic to go into detail since planning tools vary widely in complexity and capability, but we can consider the requirements and give short descriptions of the most important modules.

We expect that as a minimum a planning tool will produce the following as its outputs:

- A plan of base station site locations
- A frequency assignment plan
- Traffic information
- A coverage map and analysis of the likely service provision.

We expect the last item to include a statement along the lines that service will be available at, say, 95% of locations on roads outside buildings, to 90% of users who are using hand-portable equipment in cars, at 85% of locations inside residential buildings in suburban areas, and at 75% of locations inside office buildings in urban areas.
　　We expect to provide as inputs:

- Terrain data of the proposed service area
- Land usage (clutter) information
- Representative propagation measurements
- Network roll-out plans(if available)

Modules which are typically available include:

- Mapping data import facility
- Profiler module
- Modelling and survey analysis module
- Propagation prediction module
- Coverage analysis module
- Interference analysis module
- Automatic resource planning module
- Real-time sites grouping module
- Configuration database module
- Automatic neighbours list generation module
- Traffic dimensioning module

The *mapping data import facility* permits different types of mapping information to be imported into a planning tool. The most important inputs are a digital terrain model (DTM) of the relevant area and land-cover clutter data. Information which defines the locations of roads, railways, rivers, postcode boundaries and administrative district boundaries can also be included. Street names, city names, motorway codes and province names can be loaded and displayed when needed. The data can be derived from paper maps, digital maps, satellite or aerial photography and from published gazettes of administrative information. Both digital terrain model and land-cover clutter data are commonly loaded in raster form and can use any convenient pixel size. Mapping resolution can be as fine as 10 m or as coarse as 200 m.
　　The *profiler module* enables network planners to investigate terrain profiles for microwave link studies or simply to examine profiles along radials between any two points for modelling analysis. The total number of intervening obstacles along any path can be determined from stored terrain data, and the diffraction loss can be estimated. The distance between points and the propagation mode (whether line-of-sight, partial line-of-sight or non-line-of-sight) can be determined and displayed

graphically. Terrain profiles can be obtained between any two points: from point to point, from a base station site to a point, or from a base station site to another base station site. Several user-selectable knife-edge diffraction techniques such as Bullington, Epstein–Peterson, Japanese, Deygout, Giovaneli, and Edwards and Durkin are normally stored, and user-selectable rounded-hill diffraction techniques are also available, such as the technique due to Hacking. The crest radii for the estimation of loss are deduced directly from the terrain profile.

The *modelling and survey analysis module* is aimed at developing one or more radio propagation models. A fundamental task, it is the foundation of all planning processes. Planning tools always permit the import of survey measurement data and test-mobile measurement data for model calibration and network optimisation as indicated above. They commonly provide an automatic, flexible and empirically based modelling capability to be used for assessment, evaluation and calibration of environment-specific propagation prediction models. Users are provided with the maximum flexibility in selecting model parameters and in modelling specific data, clutter and transmission condition categories. Nowadays model calibration is undertaken automatically; the use of a multivariate optimisation technique enables the planner to derive the most suitable coefficients and parameters so the chosen propagation model accurately describes the propagation characteristics of the imported data.

By comparing the imported data with a generic propagation model and 'tuning' the coefficients to get the best fit, it is possible to produce models with optimum RMS error (typically 6 dB) and zero mean error. The automatic model-tuning feature guarantees the prompt realisation of an accurate and sensible model in a very short time. A typical modelling tool would feature several possible propagation prediction models, including the extended COST231–Hata model, the original Okumura–Hata model, the Walfisch–Ikegami model and one or more microcell models.

The module contains software routines to deal with several other items that are important in the modelling process.

- Terrain and clutter profiles between the transmitter and the receiver can be constructed using a standard technique such as the Edwards–Durkin (row, column and diagonal interpolation) or bilinear interpolation.
- The effective antenna height, applicable to both transmitter and receiver sites, can be calculated using one of several stored algorithms. These normally include terminal height above ground, terminal height plus ground height, height above least mean square fit to terrain, height above average elevation (as in Okumura).
- The effect of clutter can be considered in terms of clutter at receiver location (local clutter), interpolated clutter taking into account the effect of surrounding clutter types nearest to the receiving point using a bilinear interpolation technique, or profile clutter that gives an unweighted average value of clutter factors over a user-selectable distance. The profile clutter model ensures consideration of the clutter effects in the direct path between transmitter and receiver.

Radiation patterns for many types of antenna, both transmitter and receiver, are also stored within the planning tool, and it is possible to calculate the effects of these patterns on measured data. This ensures that the true path loss between isotropic antennas can be estimated for all measurement points.

The survey analysis part of this module facilitates the analysis of radio surveys imported into the planning tool as an aid to developing an accurate prediction model. Radio surveys can be loaded in a suitable format and multiple survey files representing a variety of areas are particularly useful. This data can be examined in terms of parameters such as distance from transmitter (minimum and maximum values), signal level (minimum and maximum values) type of clutter encountered, propagation mode (line-of-sight, partial line-of-sight and non-line-of-sight) and receiver height (minimum and maximum).

The information can be used globally or individually to optimise the prediction model. It is then possible to produce X–Y plots of signal strength, path loss, diffraction loss, effective antenna height, predicted path loss and residual errors versus distance. Contour plots can also be produced and overlaid on terrain, clutter or scanned-map backdrops, enabling the user to identify and examine problematic areas. Having done all this, a comprehensive global and local analysis platform is available which produces an assessment of the accuracy of selected models in terms of the mean error, RMS and standard deviation of error, and the correlation between measured and predicted path loss.

Models can also be assessed in terms of performance with respect to individual measurement data files and as a function of specific parameters such as clutter type for line-of-sight, partial line-of-sight and non-line-of-sight transmission conditions and for regions which are near, intermediate or far from the transmitter site. Assessing prediction performance in the near and intermediate regions is primarily to establish how well a particular model will predict coverage; in distant regions the aim is to assess interference prediction capability. The correlation between the measured path loss and individual predicted losses such as diffraction loss and distance dependence can also be computed. Planners can then select model parameters which offer the highest correlation with the measured data.

A *microcell modelling module* is included in modern planning tools and this makes use of detailed building data. Such modules model the corner loss effect observed by many researchers and system operators. The effects of base station antenna radiation patterns are also included. Models are based on the dual-slope corner loss model in line-of-sight and non-line-of-sight areas in a microcellular environment (section 4.4.2).

The *prediction module* itself takes the model or models which have been optimised within the planning tool and uses them to predict coverage from a large number of chosen or potential base station sites. Predictions can usually be produced for areas that range from diameters of 1 km to over 100 km, using any suitable mapping resolution. Predictions can be carried out on an individual cell or for groups of cells intended to cover, say, a given metropolitan area or a county. It is possible to produce predictions for a given base station site using more than one prediction model, or to produce a number of predictions for any individual site using different antenna heights, radiation patterns and downtilts, or different transmitter powers.

The *coverage analysis module* then allows the production of composite coverage plots from multiple sites. If the composite plot shows gaps in coverage or excessive overlaps from certain base stations then the parameters of individual base stations, i.e. antenna height and pattern, orientation, downtilt and transmitter power, can be altered and the effect displayed graphically. It is possible to optimise coverage interactively in this way. It is also possible to display equal-power boundaries where handover between one cell and another is likely to take place.

Expected system coverage can also be predicted at various predefined thresholds such as might be appropriate for vehicle-borne installations, portables in the streets, portables in buildings, etc., in various areas which are influenced by different amounts of building clutter loss. Coverage analysis modules can usually predict the base station most likely to provide the best service (the best server) for a mobile in a given area. They can also produce predictions for the second-best server. These predictions are very useful because they indicate the amount of traffic that might be handled by specific base stations and this helps in dimensioning the network.

Interference analysis is a vital step in the design of cellular radio telephone systems because they are designed to be limited by interference rather than by noise. Propagation prediction modules need to produce accurate predictions at short and intermediate ranges for coverage calculations, but it is equally important to have predictions at long range for interference estimation. Interference analysis modules can normally calculate, analyse and display composite co-channel and adjacent channel downlink interference in user-specified regions. Analysis can be undertaken for traffic-only carriers, control-only carriers or for all carriers; worst-case, average and total interference can also be examined.

Analysis can be carried out for a given class of mobile using a specified signal level and, for example, urban in-building coverage with 90% location probability. It is possible to examine the percentage of covered area with an interference level above a specified threshold in areas having a given class of clutter.

An *automatic resource planning module* is usually incorporated in planning tools. This can produce both regular and irregular frequency assignment plans. Regular frequency planning has been discussed earlier and has proved flexible and simple to implement. It is particularly useful for network expansion. However, the amount of traffic that is offered to a network is not uniform throughout the whole service area, and the amount of resource that needs to be allocated to certain cells has to reflect this. Non-regular automatic frequency planning algorithms provide good results when they are first applied to a radio network in which all the base station locations and carrier allocations are known.

Optimisation of the available radio resources is accomplished using a heuristic channel and colour code assignment algorithm, which attempts to minimise the area where interference is likely to exist, or the amount of traffic that is likely to experience interference. These algorithms are based on approaches such as genetic algorithms, mathematical programming or simulated annealing [4]. Carriers, or colour codes, can be assigned to the whole network or to specific areas of interest. The introduction of new sites into an existing frequency plan can be addressed fairly easily.

Furthermore, because most of the dropped calls in a network occur in the handover boundary regions, resource planning modules also support the inclusion of the interference calculated in these areas. They include features such as automatic frequency planning of broadcast control channel (BCCH) and traffic channel (TCH) carriers and base station identity codes (BSIC) in GSM systems and the ability to automatically group carriers for a given frequency reuse pattern, e.g. 4-cell or 7-cell repeats.

The interference calculation uses a mutual interference table that can be calculated for a user-defined co-channel interference threshold level, adjacent channel interference threshold and a certain coverage level. The calculation can be undertaken for the entire coverage area or selected sub-areas. The frequency assignment plan can be created for traffic-only carriers, control-only carriers or all available carriers. Colour code planning is available as indicated above, and eight colour codes are planned throughout GSM networks. Colour codes are assigned using the following characteristics:

- Co-channel and adjacent cells are not assigned the same colour codes.
- Co-channel cells which belong to the neighbours list of a given cell are not assigned the same colour code.
- The assignment of co-colour codes is based on maximising separation.

Checking mechanisms are built into the module to create a flag if any of the assignment rules are violated, and modules that support GSM system planning will provide support for planning networks with partial or network-wide cell deployment of frequency hopping, power control and discontinuous transmission (DTX).

The *planning module* recognises that not only is it important to realise a frequency plan with low interference distributed throughout the network, it is also important to generate a plan which will minimise dropped calls. Some system planners simply assume that a plan with minimum interference is sufficient; unfortunately, simple allocation errors such as assigning the same BSIC to cells which are co-channels within the neighbours list of a given cell will result in dropped calls at the boundary of coverage.

The best planning tools allow the user to define several restrictions on frequency plans, such as minimum combiner spacing, minimum co-sited cell carrier spacing and minimum neighbour carrier spacing; they also incorporate analysis features which enable the planning engineer to identify potential locations of dropped calls. In these modules co-channel and adjacent channel neighbours are identified and flagged, and carrier usage statistics are produced; this enables the planner to identify carriers which are being used too often. Any assignments which violate the assignment strategy are identified. Rogue cells, which cause the most significant interference, are also identified. These cells are usually regarded as candidates for antenna downtilt or reorientation to improve interference performance.

The *real-time site grouping module* enables users to perform planning processes such as prediction and coverage analysis on a select group of sites or all the sites within the site database. For example, sites can be grouped on the basis of a city (e.g. Manchester or London) or a county, by operational status (e.g. surveyed or accepted) or by geographical location. All the possible site filtering mechanisms are stored in the site database, and a number of active site lists can be generated as required. Site lists can also be based upon sites which meet a combination of criteria, and individual sites can be added or deleted from site lists.

The *configuration database module* contains a comprehensive configuration database which is a combination of several individual databases. These include:

- *Cell site database*: includes information such as cell identification codes, location, propagation models, antenna types, orientation and downtilt specific to cells.

- *Carrier database*: includes cell control and traffic channels, colour codes, hopping sequence numbers (for GSM systems), number of required carriers, etc.
- *Neighbouring cell database*: contains handover margins for all neighbouring cells.
- *Exceptions database*: contains a list of the forbidden carriers on a per cell basis as well as the separations between cells.
- *Mutual interference database* contains measures of mutual interference in terms of area and traffic between cells.

Other features include the ability to assign different transmitter heights, effective radiated powers, propagation models and antenna types to collocated cell sites; to assign common site parameters to more than one site; and to track by date and time all site configuration changes.

The *traffic dimensioning module* enables the creation of traffic raster information using various methods. It can import data regarding 'live' traffic actually measured on the network and can create Erlang maps of traffic density (erlang/km^2). Tables of the required number of channels per cell can be produced for a given grade of service (GOS) using the Erlang B model, the Erlang C model or dedicated circuit models, and cells not meeting the required grade of service are identified.

11.4.1 Self-regulating networks

In order to maintain a high-quality network while increasing the subscriber base, there is a need to continually improve frequency planning processes, using accurate prediction models and measurement data where possible. Planning tools will support the incorporation of measured data collected from a network for the improvement of the frequency planning process. Data is collected using a standard engineering test-mobile handset and this is used to create a C/I matrix containing the difference in received signal levels (RXLEVs) from a serving cell and all non-serving surrounding cells. The module utilises this data alongside similar mutual interference tables derived from prediction to realise a plan based upon real-world data, wherever this is available. After deployment of new plans, the regulation process is continually repeated with incremental improvements in network performance each time.

Finally, once the network has been planned, the planning tool can download information to a proprietary network management system. To refine the frequency plan further, the planning tool can also retrieve valuable statistics from proprietary software that monitors base station traffic and handovers. Both uploading and downloading processes ensure that correct parameter information is present in all network elements.

11.5 A MODELLING AND SURVEY ANALYSIS MODULE

Following the brief overview in Section 11.4, we can now look at one or two aspects in a little more detail. We have already suggested that the production of one or more propagation models is central to the planning process; models are required to estimate the coverage from base stations and for subsequent optimisation of the network. It is essential to input representative propagation data, and a series of measurements should be conducted from as many base stations as possible covering the area of interest. Typically, sites should be selected with the following criteria in mind:

- The site should be suitable in terms of its surrounding clutter and terrain features as required for accurate representation of the area.
- The heights of the surrounding buildings should be representative of heights to be used for radio network planning.
- There should be full or partial clearance of the rooftop area at the site to ensure the view of the base antenna is not obstructed in any way by rooftop clutter.
- There should be full or partial 360° clearance up to a distance of about 400 m from the base station.

Detailed planning of survey routes should also be conducted, taking into account the need to cover major and minor roads, provincial motorways, expressways and freeways. Measurements conducted within tunnel areas, on bridges, on overpasses and in underpasses should be tagged.

11.5.1   Data preparation

Survey data should be collected in all areas surrounding the base stations to ensure that all clutter types are reasonably represented in the analysis. A good propagation model cannot be obtained if the different clutter types are incorrectly represented. Furthermore, measurements should be conducted at distances between 200 m and about 10 km from the base station.

When the survey data is imported into the planning tool, the header files for each survey route should contain all necessary information such as the location of the base station, the effective radiated power, the antenna type and the base station antenna height. Furthermore, other relevant information such as the noise floor of the measurement system should be noted. Similarly, it is prudent to identify any measurements made at a distance of less than 300 m from the base station as it may sometimes be necessary to exclude them in order to minimise the effect of the base station antenna radiation pattern.

11.5.2   Model calibration

The tuning of propagation models involves determination of the different coefficient values in the propagation equation so that the residual RMS value of the error reaches the lowest possible value (known as the global minimum). The first step in the calibration process is to adopt a suitable model structure that is able to explain most of the propagation effects observed in the measurement data set.

General model

Many planning tools adopt a basic model structure similar to the Okumura–Hata model with the addition of correction factors for knife-edge and rounded-hill diffraction. The basic equation can be expressed in the form:

$$P_{Rx} = P_{Tx} + C_{CT} + C_d \log d + C_{dh} \log d \log h_b + C_h \log h_b + C_{dk} K_{dk} + C_{dr} K_{dr}$$
$$+ C_{Cl} K_{Cl} + G_T(\theta,\phi) + G_R(\theta,\phi) \qquad (11.17)$$

where

P_{Rx} is the received signal strength (dBm)
P_{Tx} is the transmitted power (dBm)
C_{CT} is a fixed correction term
d is the distance (m)
h_b is the base station antenna height
K_{dk} is the knife-edge diffraction loss
K_{dr} is the rounded-hill diffraction loss
K_{Cl} is the clutter factor

$G_T(\theta, \phi)$ and $G_R(\theta, \phi)$ are the transmitter and receiver polar patterns. The correction term C_{CT} accounts for the effects of frequency and other non-specific factors in the model; C_{CT} and K_{Cl} combine to give the signal strength at a distance of 1 m. The initial values of the coefficients associated with the various terms are set to the values in Table 11.3. The distance and height dependence parameters are exactly the same as those used in the Hata model.

Before moving on, we consider the matter of clutter. Clutter is very important since the development of an accurate propagation model requires a detailed and accurate clutter database. By adopting a standard methodology and a sound definition for a clutter classification, areas having the same effect on radio signals would be classified similarly wherever they are situated within the service area. Clutter factors model the losses or gains associated with the different types of environment . The clutter loss or gain depends on street width, building density and vegetation density in the area concerned.

The clutter factor for urban areas is always lower than for suburban areas, and in turn this is lower than the clutter factor for open areas. Table 11.4 shows a typical example of a simple clutter database. Initially, the value of the clutter factor K_{Cl} is set to a default value of zero. However, meaningful values can be obtained by selecting one of the available clutter algorithms. Some of the options are:

- *Local clutter*: clutter loss/gain is computed by considering the clutter type at each specific receiver location.
- *Interpolated clutter*: this uses the weighted sum of clutter factors for the four pixels closest to the receiver location.
- *Profile clutter*: clutter loss/gain is computed as the weighted sum of interpolated clutter factors for the clutter types in the path between the receiver and the transmitter over a user-definable distance.

Table 11.3 Initial values of coefficients

Parameter	Initial value
C_{CT}	Arbitrary
C_d	-44.9
C_{dh}	6.55
C_h	0.0
C_{dk}	-0.5
C_{Cl}	1.0

Table 11.4   Clutter classification

Clutter type	Urbanisation level[a]
Dense urban	Business districts consisting of very tall building structures and office complexes Residential areas with very tall buildings
Urban	Residential areas with a mixture of tall blocks Detached and semi-detached houses
Mixed rangeland	Desert land, open fields and farmlands, sand dunes, open spaces in urban areas
Ocean	Ocean, lakes, streams and canals
Undefined	Unused contents of data file

[a] Urbanisation level is described in terms of building heights and street widths.

The process of developing a model starts with the clutter factors initially set to zero and the clutter algorithm set to *local clutter*. Using the modelling and survey analysis tool, the resultant RMS and mean error values for all the individual data files are then found, together with the overall RMS and mean error values for the entire measurement data set. The clutter factors are then adjusted such that the mean error figures are close to zero. Care is needed to ensure that the values of the different clutter factors are appropriate; for example, the clutter factor for an open area must be higher than the value set for a suburban area. No other parameters are adjusted at this stage. The automatic modelling facility is then used to obtain new coefficients for all the parameters.

The tuning of the model is now nearly complete. It remains to select the profile clutter algorithm and adjust the path clutter distance until no further improvement is obtained in the RMS error. Typically the path clutter distance is set to 1.5 km. The distribution of residual errors can now be examined to identify problematic areas. Furthermore, since the performance of the model can be assessed in terms of distance regions, individual data files and clutter, the user can concentrate on specific erroneous data.

11.5.3   Developing a model

Obtain a benchmark model

A benchmark model should first be obtained using the automatic modelling capability; its performance should then be assessed in terms of the RMS and mean error for all clutter types, individual base station data sets, and different distance regions.

Examine the residual errors

The geographical distribution of the residual errors should then be examined. Where very large residual errors occur, the following questions need to be addressed:

- Was there sufficient clearance at the antenna site in the direction of the area where the largest residual errors occurred?

    If insufficient clearance was obtained, and the measurement cannot be repeated, the obstructed segment of the data set should be excluded.
- Was the data recorded on the border between clutter types?

    Measurements conducted on the border between two clutter types, may be represented by the wrong clutter type, and this may cause residual errors of several decibels. These data points should be excluded.
- Were the roads on which data was recorded elevated, or were they underpasses with the land level higher on either side?

    Where this is observed, the user may choose to exclude these specific measurement points, or edit the clutter database and include a clutter type which represents elevated roads or underpasses.
- Are the measurement points very close to the base station (< 400 m) and do they possibly have line-of-sight?

    This condition is difficult to model. For example, if the base station has true line-of-sight with a particular road with tall buildings on either side, a canyon effect will result and very high signals will be obtained. In a real network, regions close to the base station will always have high signals and will be served by the same base station. It is therefore more important to be able to predict signal level at medium distances to establish coverage fringe areas, and at large distances for interference purposes. Measurement points in close proximity to the base station should be excluded if they are known to have line-of-sight with it.
- Were the measurement conditions (transmitter power, cable losses, antenna positioning and recorded antenna height) correct?

    All recorded heights should be confirmed.
- Is the clutter classification in the area correct?

    A major contributor to the largest residual errors are the inaccuracies in the clutter classification. A standard methodology and sound definition for clutter classifications help to minimise this possibility; however, where a large standard deviation is observed for a particular clutter type, there is every indication that the clutter definition contains a wide-range of characteristics. In this situation it would be better to reclassify clutter types.
- Is the propagation environment from the base station different from all other sites?

    In some cases it may be impossible to combine data sets from several regions to create one generic propagation model. If site visits confirm this, it is better to have a different model for each area.

Realise a better model

The revised measurement data and clutter edits should then be used to realise a better propagation model as described in step 1.

Repeat steps 2 and 3

Steps 2 and 3 should be repeated until the optimum RMS figures are obtained.

Check the RMS errors

Finally the performance of the model should be examined to ensure it provides an overall RMS error of about 6–7 dB for the entire data set, and to ensure it yields RMS errors of this order for each of the individual base stations. It is usually considered essential to have an overall mean error of 0 dB.

11.5.4   Limits on coefficients

It is essential that the model structure reflects expectations of signal behaviour in different environments. For example, signal level is expected to decrease with distance and the presence of more buildings, and increase with base station antenna height and the presence of line-of-sight. It is usual to set limits on the coefficients to be used and some typical values are:

$$C_{CT} < 0 \text{ (always)}$$
$$-25 < C_d < -45$$
$$-12 < C_h < +12$$
$$0.3 < C_{dk} < 1.4$$
$$0 < C_{dh} < 12$$
$$C_h + C_{dh} \log d > 0 \text{ (always)}$$

Planning models almost invariably include a display option which allows the user to examine a selection of X–Y plots, including signal level or residual errors from prediction versus distance from transmitter. This is particularly useful in identifying regions where a tuned propagation model may be failing. Typically a user may wish to view signal strength versus distance, residual error versus distance and predicted path loss versus measured path loss.

11.5.5   Microcell model

Microcell models are usually based upon the dual-slope plus corner loss concept, as indicated earlier. Here the microcell is defined as a cell of radius 0.5–1.0 km in which the base station antenna is mounted at street-light level, well below the average height of the surrounding buildings. Because of this relatively low elevation, the influence of the propagation environment is much more pronounced in microcells than in macrocells. The significant effect of buildings on propagation is exemplified by the phenomenon called the *corner loss* effect, in which the received mean signal strength often decreases by as much as 30 dB when the mobile turns around a corner from a region where it had line-of-sight. With the signal subject to such large variability over very short distances, propagation prediction using macrocellular models could be in error by as much as 30 dB when employed for microcells.

11.6   GRADE OF SERVICE

If calls are to be handled without delay or loss, it is necessary to provide a very large number of full-duplex radio channels in each cell. For economic reasons this is unrealistic, and to limit the number of channels to a reasonable number it is necessary to tolerate a small amount of blocking in the system. Subscribers have to realise that a call attempt may fail when all channels are being used: they then have to wait and try calling the desired party at a later time. The grade of service (GOS) is used to quantify this situation, and GOS is defined as the ratio of unsuccessful calls to the total number of calls attempted.

Thus GOS is a measure of the inability of the network to cope with the demands placed upon it. In practice it is expressed as the percentage of calls that fail during the busy hour due to the limited availability of RF channels. In cellular radio the system design is usually based on a grade of service of 0.02 (2%) or better. A 0.02 GOS means that, on average, a subscriber will find an available channel 98% of the time during the busy hour. At other times of day the GOS will improve and in fact most systems will appear to be unblocked.

For other systems, such as wireless local loop (WLL), intended to compete against normal landline telephone systems, the required GOS is usually lower, about 0.5 to 1%. Any system that requires dedicated circuits effectively has a GOS of 0.0%, i.e. there is no blockage due to the unavailability of sufficient channels.

11.6.1 Milli-erlangs per subscriber

The number of erlangs per subscriber, E_m, is given by

$$E_m = \frac{\text{total number of calls arriving in 1 hour} \times \text{average call holding time in hours}}{\text{number of subscribers}}$$

Thus, if 100 subscribers use a total of 140 min of air time during the peak busy hour, the average call duration is 1.4 min or (0.023 h) per call; the number of erlangs per subscriber is then 0.023, or 23 milli-erlangs (mE).

Typical figures show considerable variation from one country to another. In Europe a busy-hour figure of 22–25 mE is usual, whereas 45–60 mE is more common in the Middle East. It is important to use appropriate figures so that the network can be dimensioned properly.

To model the concept of blocking, an appropriate traffic model is required. Erlang B and Erlang C are two widely used mathematical models that describe the relationship between the blocking probability (grade of service), traffic demand and the number of required channels. The Erlang B model assumes that blocked calls are cleared and that the caller tries again later. In other words, the caller whose call is blocked does not immediately reoriginate the call. This type of model is applicable to most cellular radio systems, including GSM, TACS, ETACS, NMT and WLL systems.

The Erlang C model assumes that a user whose call is blocked continues to reoriginate until the call is established. This is envisaged as a queuing system in which calls that are blocked are not lost, but are rather delayed until channels become available. This type of model is applicable to the Trans-European Trunked Radio (TETRA) system.

11.7   SUMMARY AND REVIEW

Cellular engineering encompasses different planning activities. Paramount among them is the dimensioning of the radio network, the configuration of radio sites, the radio frequency plans and the optimisation of the implemented networks. The main objective of these activities is to deliver a network which matches the operator's business and marketing plans in terms of service area, traffic handling capacity and quality of service, in a timely and cost-effective manner.

As a cellular radio network evolves, the challenge to the operator is to provide comprehensive coverage and to accommodate a high traffic density while ensuring the carrier-to-interference ratio remains acceptable within the finite amount of available spectrum. The strategy for any operator must be to remain flexible in order to react to rapid industry changes (regulatory, technological and competitive), and similar flexibility must therefore be inherent in the planning process.

We have briefly described some of the cellular planning processes and elaborated on a few aspects that are directly relevant to radio propagation. This chapter is meant as a guide not a definitive manual on radio planning methodology. In this final section we summarise and briefly review one or two additional aspects.

Before setting up a cellular network, potential operators need to establish the extent of coverage required for the various regions within the overall service area and the geographic regions in which service will be available to subscribers using different types of mobile terminals such as vehicle installations or hand-portables. These questions are of major strategic importance since approximately 70% of the total network infrastructure capital cost is expended in the delivery of the radio coverage, and system operating costs are dominated by the radio network infrastructure.

11.7.1   Cell site dimensioning

A key objective for the radio planning engineer is to provide coverage in any terrain environment by maximising the service area of each base station, hence minimising the number of required cell sites. While the base station can always transmit the ERP necessary to provide adequate coverage within the cell radius, the ERP of the subscriber unit is necessarily lower. It is the low power of the mobile unit, therefore, which is the limiting factor in determining the cell radius in any environment. Furthermore, in order to obtain viable balanced transmission between low-power mobile units and base stations, signal-enhancing techniques such as antenna diversity and increased receiver sensitivity are often used at the base station. Cell sizes, and therefore the number of cells required to cover a given area, depend on the intended coverage area and the associated traffic density.

By developing a radio link budget for a given base station configuration, it is possible to determine the maximum coverage of any cell and hence to determine the number of cells required. Alternatively the number of cells can be determined from the perspective of traffic demand. The systems designer has to calculate the expected number of cells and cell sizes using both approaches and then select the higher value.

The first step in dimensioning the cell radius for different environments is to create a link budget based on relevant parameters such as the standard deviation of the signal in various environments, the in-building penetration loss in urban and suburban areas, the in-car penetration loss, and the various characteristics of the base station antenna. These are important parameters which affect the received signal level at both the base station and mobile ends. The maximum cell coverage or service area of any base station transmitter is determined by the allowable system path loss figure calculated from an appropriate link budget, taking into account the effect of all relevant parameters.

The most important parameter in this context is the maximum allowable path loss. This can be used to dimension site coverage in different environments by considering an appropriate propagation prediction model, and generally the maximum allowable system path loss is about 143 dB. We have discussed the development of suitable models earlier in the book and showed in Section 11.5 how one such model can be optimised.

As an example, by taking typical base station antenna heights and substituting a maximum path loss of 143 dB into the equation, cell radii of approximately 2.4 km, 4.0 km and 11.1 km are obtained for GSM base stations at 900 MHz in urban, suburban and rural areas respectively. In practice, however, since the clutter is not homogeneous throughout the network coverage area and because a particular cell may include many clutter classes, individual cell radii may depart significantly from these figures.

The primary objective for subscriber traffic capacity planning is to deliver adequate capacity at the appropriate time. Reasonably accurate market projections are required for subscriber penetration figures on a regional basis, but this is a commercial rather than a technical matter.

The traffic handling capacity for the radio interface depends on the size of cells implemented (the geographical coverage), the quantity of equipment deployed at the base station sites and the transmission technique used (e.g. omnidirectional or sector antennas). Consequently, there are different options available to radio engineers when planning system capacity, and each will have associated costs, depth of coverage, timescales and risks. The goal is to develop an optimum growth pattern for the network.

Having obtained indications of the number of cells required to serve the traffic demand and the number of cells required to serve the coverage area, the higher of the two figures should be used. The designer will then be required to perform actual traffic planning; this involves determining the number of transceivers required at each base station site for a given spectrum allocation, grade of service and traffic demand. The traffic estimates obtained from a market survey, or from the network switch for an existing network, are normally used for this purpose.

Having established traffic figures for a given region, the radio planning tool will apportion the total traffic to various areas based upon clutter weightings within the area. This creates a non-uniform distribution with peaks in urban areas and dense urban areas – more realistic than assuming a uniform distribution. By comparing the coverage of each cell with the traffic estimates, it is possible to determine the number of erlangs that any cell needs to support and hence the number of transceivers required in that cell to meet the demand.

Note that the traffic planning process also involves some 'reassignment' of traffic from one cell to another in order to produce a more equally distributed traffic demand on the cells. This is achieved through a base station antenna reorientation (downtilt, azimuth and site location). To allow for anticipated expansion of a network, it is customary not to permit any of the cells to have the maximum allowed number of transceivers during the early stages. This would almost certainly lead to congestion and the immediate need to introduce an additional (unplanned) site.

11.7.2 Base station site planning

Both omnidirectional and sectored-cell site transmission techniques will typically be used in cellular networks to provide coverage in different terrain environments. Each of these techniques is associated with different characteristics from a planning perspective. Sector antennas can offer higher system gain than omnidirectional units, hence they can provide better system coverage. Typically, they will be used in areas of high traffic density, perhaps in locations where large numbers of subscribers require quality in-building service. Sector antennas provide a high degree of flexibility in defining coverage, with the directivity used to tailor the coverage and fill-in blackspots. The same characteristics enable interference problems to be minimised when using sector sites.

Sites at which omnidirectional antennas are used benefit from optimum trunking efficiency. Further advantages include minimised RF equipment and site-build costs (as fewer antennas will normally be required), plus a higher degree of environmental acceptance. Omnidirectional sites are typically less obtrusive than sector sites, with minimised planing regulations, civil engineering and ground rental costs.

Coverage in some isolated centres of population or in mountainous regions may be better achieved by an omnidirectional site rather than a sectored site, since the increase in range offered by sectored sites with directional antennas will only result in additional unpopulated areas being covered.

11.7.3 Frequency planning

It almost goes without saying that a major objective of cellular systems is to provide good quality communication to the highest possible number of mobile users in a given area. Using more radio channels can directly increase the system capacity but in practice the amount of authorised spectrum is limited. The challenge is therefore to design and plan the system to reuse frequencies as often as necessary in order to provide a specified minimum mean received signal level throughout the coverage area while keeping co-channel and adjacent channel interference within acceptable limits.

The classical approach to the frequency assignment problem is via the theory of regular hexagonal networks (Section 11.2), although several variants now exist. The regular grid basis, however, does not properly address real-world systems in which variations in radio propagation and radio traffic distribution produce a non-regular frequency demand. Propagation conditions, for example, are often such that interference levels do not depend on distance ratios alone; variations in traffic density in urban areas, near major road junctions or in shopping centres can lead to a demand for channels which varies from cell to cell, and environmental constraints often impose limits on the usability of certain frequencies.

There may also be other issues that have to be taken into account. In a start-up network, for example, there may be a desire to leave as much freedom as possible to adapt to future traffic changes; in a network that is expanding there may be a desire to preserve the existing channel allocations while accommodating increased demand. Whatever the design objectives, there will almost always be the need to trade off one requirement against another to obtain optimum frequency allocations.

A manual trial and error approach has often been used in the past but it requires skill and experience, and can be very time-consuming. Although it can cope with irregularities in frequency demand and even allow local exceptions to a regular grid layout, it still does not guarantee particularly good use of the available spectrum. Nowadays the use of high-speed computing systems allows the frequency planning problem to be addressed using automated techniques which provide a more adaptable approach to configuring the network. Automatic frequency planning tools make the process less labour-intensive and more reliable.

In mathematical terms, channel assignment is a combinatorial problem which is closely related to graph colouring [4]. Methods that have been used [5–8] in addition to the regular hexagonal scheme often provide acceptable results but have many drawbacks. Firstly, use of the graph theoretic approach needs, as an input, a 'hard' decision as to whether or not a particular channel can be used in two given cells. This is always difficult since in practice it really depends on factors such as the amount of traffic in the cells concerned. Potential for interference is only translated into real interference if the channel concerned is actually being used in both cells at the same time, and the probability of this happening needs to be estimated.

Secondly, in an allocation system which has strict rules or constraints, there is no basis for trade-offs. If an allocation which complies with the constraints cannot be found, there is no obvious way forward; the planners cannot decide which of the constraints they are prepared to violate. This precludes the planners examining some possible, but non-ideal, solutions and placing them in rank order. Thirdly, the aim of the graph theoretic approach is actually to minimise the used spectrum, which is not the real problem here. Planners are actually concerned with optimisation rather than with minimisation of spectrum use.

To address these points, it was suggested [4] that the problem was best formulated as a cost function optimisation problem and tackled by general discrete optimisation methods. Among such methods are genetic algorithms and simulated annealing; both are powerful techniques which have been successfully applied to a number of similar problems. In general terms the global optimisation problem is to find a solution, within a feasible set of solutions, which minimises a certain objective function.

In the case of frequency assignments, the feasible set is discrete but contains a number of possible solutions, so the problem is then termed a combinatorial problem. The theory of finding a minimum in the neighbourhood of an initial solution (a local minimum) is well developed [9] and includes classical hill-climbing methods. But what we really require is a solution that could be termed a global minimum, and this represents a formidable mathematical problem.

Genetic algorithms optimise a function using a process inspired by the mechanics of natural selection. The optimisation is accomplished by evolving a number of candidate solutions and incrementally improving the various possibilities. The convergence of the genetic algorithm approach to the global optimum is only guaranteed in a weak probabilistic sense, whereas it has been proved for the simulated annealing procedure.

Simulated annealing is based on the analogy between the physical annealing process in solids and the problem of finding the minimum of an objective function [10]; adaptive simulated annealing is also a possibility [11]. An algorithm proposed in the literature [4] provides a solution but does not deal with the problem of the strength of the interference.

Whatever the approach, ideally it should be able to take into account factors such as the number of available carrier frequencies, the neighbour list information, and the carrier separation requirements in neighbouring cells. Morphology or traffic weightings are also useful so that action can be taken to eliminate interference in important areas.

Generally a combinatorial optimisation problem consists of a set S of configurations of solutions and a cost function C, which determines for each configuration the cost $C(S)$. Furthermore, in order to perform a local search one needs to know the neighbours of each solution, i.e. one needs to define a neighbourhood structure $N(S)$; this determines, for each solution, a set of possible transitions which can be proposed by S.

Each optimisation problem has a number of degrees of freedom, and they determine the number of configurations which can be permitted as solutions. These restricted solution sets are in turn governed by a number of constraints. The objective of the optimisation process is to adopt a configuration which minimises the cost function. The procedure is basically as follows.

Beginning from an arbitrary starting solution S, in each iteration step a neighbour S' is proposed at random. An iterative improvement constitutes a situation where the cost function associated with the neighbour is less than that associated with the original solution, i.e. $C(S') < C(S)$. The mechanism through which a transition is made from one configuration to any one of its neighbours depends upon the type of algorithm used.

Some algorithms such as simulated annealing permit neighbourhood transitions which will yield a worse cost function so as to avoid being trapped in a local minimum, i.e. a configuration whose neighbours do not offer any improvement in the cost function but which does not itself constitute the global minimum. This is very important because the alternative to an intelligent search algorithm is a dumb search over all possible configurations in the solution space. For large data sets, this is an impossible task. The neighbourhood transition process is continually used to propose solutions and it is terminated after any transition to a neighbour which does not offer an improvement in the cost function.

Essential to this process are the usual computational desiderata: do it quickly, cheaply, in small memory and settle for the global minimum. These four issues are of course governed by the solution space, the neighbourhood transition methodology and the user's trade-off between the duration the program should run and the level of optimisation required.

The frequency planning problem is a multidimensional optimisation problem. Using the analogy described above, the set of configurations S represents different frequency assignments which can be applied to a network of n cells. In principle all permutations of frequencies can be assigned to the network. The cost function is generally expressed in terms of the cell-to-cell mutual interference between any two cells in the network; this is calculated using the area which may be affected if both cells share the same frequency or use adjacent frequencies, and the amount of traffic in erlangs which may be affected if both cells share the same frequency or use adjacent frequencies.

Both these parameters can easily be deduced through the use of a radio propagation model to predict the service area for any cell in the network. The neighbourhood structure refers to several other frequency assignments which can be made. The constraints on a cellular network require a minimum separation between the following items:

- Carriers within the same cell
- Carriers which belong to base stations which are co-sited
- Carriers which belong to cells which are neighbours with a high handover rate (e.g. 10% of all handovers occur between both cells)
- Carriers which belong to cells which are neighbours with a moderate handover rate (e.g. less than 10% of all handovers occur between both cells)
- Carriers which belong to cells which are not intentional neighbours but which may experience significant mutual interference

In addition there may be specific carriers that it is necessary to exclude on a cell-by-cell basis; this may be for border coordination, for regulatory purposes, or to prevent interference where a carrier is being unlawfully used by some other organisation. There may also be specific carriers to exclude on a global network basis, for example, those adjacent to the spectrum allocated to another network operator.

This information is available to all automatic frequency planning algorithms. What differentiates algorithms from each other is the neighbour transition mechanism, which essentially controls the speed of operation. Automatic frequency planning algorithms can take advantage of well-known cellular system constraints to reduce the number of assignments which must be examined before a solution is found. These include:

- Random assignments to a neighbouring solution should take advantage of the fact that, for any given cell in the network, there are a finite number of cells with which it either shares a boundary (e.g. co-sited cells or surrounding neighbours) or with which it has non-zero mutual interference. To reduce interference, these cells should not share a frequency with the serving cell.
- Although most optimisation algorithms in their original form would typically permit a random change of one frequency assignment in the transition from one neighbourhood structure to another, algorithm performance can be improved by permitting more than one frequency change to be made between one configuration and its neighbouring configuration. Extended neighbourhood structures are discussed in the literature [4].

With the plethora of parameters to be considered in the frequency planning process, such as cell coverage prediction, traffic analysis and mutual interference calculation, it has become increasingly necessary to use a radio planning tool to provide these inputs. This is even more important where the optimisation algorithm is being applied to a problem where a subset of the frequency assignment must be retained in the network and the rest are to be reallocated.

11.7.4 Outputs of planning

In the vast majority of cases, planners are attempting to deal with networks which are initially in a start-up phase and which then expand in a controlled way as demand increases. A number of phases are therefore involved and the outputs of the radio planning process include a summary roll-out coverage and subscriber distribution for each year. This will include the area covered in each phase, the percentage of total covered area and total covered population in each phase, the overall number of sites in each phase, the overall number of cells in each phase (a site may be used for more than one cell), the overall number of transceivers required for each phase and the overall number of subscribers supported in each phase.

11.7.5 Conclusion

As cellular and fixed radio access technologies evolve, newer radio planning and optimisation concepts will be introduced. It is likely that operators will have to continue to improve service quality by using adaptive radio planning techniques and more efficient utilisation of the system features provided by manufacturers. It is clear that automatic frequency planning and self-regulation concepts will dominate the optimisation and deployment methods in next-generation systems, and with the advent of computers running at speeds in excess of 800 MHz, it will not be long before problems are rectified by real-time frequency retuning.

11.8   A DESIGN EXAMPLE

The XYZ Mobilcom Company has been awarded a licence to operate a 900 MHz GSM cellular system in Western Ruritania, a typical European nation with several large cities and towns, and a good road and rail infrastructure. There are, however, some rural and mountainous areas where the population density is low. XYZ Mobilcom intends to provide a system which will give good service and will be accessible by over 95% of the population.

As a first step, the project engineers have commissioned a series of propagation trials in various parts of the country, using base stations which are representative of sites that might be used in the operational system. They have imported the results into a proprietary planning tool and produced a small number of optimised propagation models suitable for urban, rural and mountainous regions of the country.

In planning the coverage of the system on a national basis, XYZ Mobilcom takes as its starting point an ETSI document [12] which stipulates the margin to be used in calculations of cell-edge coverage in terms of two distinct and independent contributions:

- A lognormal fading margin which relates the location probability on the cell edge to area coverage within the cell.
- An interference margin (usually 3 dB) to allow for interference caused by intensive frequency reuse in urban areas.

It is decided to design the system by taking a graded approach to coverage using criteria which are relevant to the area under consideration. In the majority of areas it is decided to take 90% location probability as one of the major design criteria. We have seen in Section 11.3.2 that this requires a cell-edge coverage probability of about 72%, hence a lognormal fading margin of about 4 dB. For safety, a margin of 5 dB is used and a further 3 dB is added to allow for interference. XYZ Mobilcom assumes a mobile sensitivity of −104 dBm, so cells in these areas are designed for a median predicted signal on the cell edge of −96 dBm (−104 + 5 + 3).

In some small towns and villages where the offered radio traffic is likely to be lower and frequency reuse is not so intense, the 3 dB interference margin can be neglected. Cells can therefore be designed with a median edge coverage of −99 dBm. Because the lognormal margin remains the same, this can be done without reducing the percentage area coverage within the cell; it remains at 90%.

The planning engineers argue that steps can be taken to reduce the cell-edge median signal strength further in rural and mountainous areas, although this can only be done at the expense of area coverage. Their reasons are that these areas only contain a very small percentage of the national road and rail networks, and in any case the population density is low. Reducing the value by 2 dB to −101 dBm lowers the lognormal fading margin to 3 dB, and Figure 11.4 shows that the in-cell location probability is consequently reduced from 90% to about 76%. A further 2 dB reduction to −103 dBm lowers the fading margin to 1 dB and the in-cell location probability to 70%.

Whether either or both of these latter steps can be taken in practice, depends on the perception of the quality of service that is provided in the areas concerned. The rationale is that if the reduction to −101 dBm is considered reasonable in, say, 2% of the country, the consequent reduction of in-cell coverage from 90% to 76% represents a loss of 14% of 2%, i.e. 0.28%, of the total area. If the further reduction to −103 dBm takes place in another 1% of the country, coverage is lost in 20% of 1%, i.e. 0.2%, of the total area. These are very small figures in the context of common engineering practice in the provision of mobile services.

Although the level of the wanted signal, the amount of interference and the nature of the fading are very important factors in a mobile radio system, XYZ Mobilcom believes that in estimating the quality of service perceived by the users, it is also relevant to consider the percentage of the service area in which good quality is experienced.

It is well accepted that a good service will be available to subscribers in urban areas where the −96 dBm cell-edge signal level and a 5 dB lognormal margin lead to 90% in-cell location probability, with a further 3 dB margin added to account for interference due to intensive frequency reuse. The question to be addressed, therefore, is whether the level of service will be adequate in other areas such as those described above, where the network design parameters are different.

In dealing with quality of service, it is relevant to consider the number of failed call attempts. XYZ Mobilcom considers that it will have provided sufficient resource in both the radio and fixed network segments of the system if the number of failed call attempts is less than 5%. From this viewpoint it is almost certain that subscribers in the non-urban areas will experience an improved quality of service, and this can be seen as follows.

Failure of access to the network can occur due to a number of factors, all of which can be controlled. The lack of availability of radio (or network) resource, usually termed blocking, is one element. There are well-established procedures to measure the average blocking probability in a cell or in the complete network, knowing the number of available channels and the total offered traffic (Section 11.6). For a GSM system, calculations are made using the Erlang-B formula and refer to the busy hour. XYZ Mobilcom estimate the probability of blocking in a certain large group of urban cells within the proposed network to be such that the average blocking probability is 2.5%; this is considered satisfactory.

In the process of setting up a call, it is possible to come across a fault at one or more stages. One parameter which can be used as a measure of the actual user perception of blocking in a GSM system is the UTRNG (user TCH request not granted) or failed traffic channel assignment request rate. This is the actual blocking rate perceived by the subscriber. Its value is greater than the blocking rate discussed above and is estimated to be in the region of 6%.

Neither of the above factors depends on the in-cell location probability previously discussed. There is, however, one further parameter, the drop call rate, in which location probability does play a part. The drop call rate is a measure of the percentage of calls which are properly set up but then prematurely terminated for reasons such as failed handover or a network problem.

We can therefore summarise as follows. Location probability gives the percentage of the cell area where the median signal strength exceeds a preselected threshold related to the sensitivity of the mobile receiver. If this value is 90% then at 90% of locations within the cell there is a greater than 50% probability of initiating a connection and at 10% of locations the probability is less than 50%. Blocking and UTRNG are independent measures which depend only on the availability of radio channels in relation to the offered traffic. Drop call rate will be affected by a change in the location probability, and a reduction in coverage from 90% to 76% will cause an increase in the percentage of drop calls.

XYZ Mobilcom therefore decides to build a network in which the coverage, although not as intensive in some rural and mountainous regions as it is in the densely populated urban areas, will nevertheless be commensurate with the offered radio traffic and the extent of roads and railways that exist in these areas. The measure of coverage is related to the quality of service perceived by the subscriber; this is pragmatic and sensible because the subscriber does not think in terms of technical quantities such as signal strength or signal-to-interference ratios. Subscribers perceive coverage in terms of their ability to establish and maintain a connection with the network and the subjective quality of the conversation that ensues.

The vast majority of the network will be designed using a coverage border level at −96 dBm. This ensures that at 90% of locations within the cell the signal strength is greater than or equal to −96 dBm, which gives an 8 dB margin over the receiver sensitivity of −104 dBm. This coverage plan, however, is unduly stringent in some rural and mountainous regions and can be progressively relaxed. In certain parts of the country where the offered radio traffic is low and the frequency reuse is not intensive, it seems perfectly reasonable to dispense with the 3 dB interference margin and design for a cell-edge signal strength of −99 dBm. This leaves the lognormal margin untouched, so the in-cell location probability remains at 90%. The quality of service is at least as good as in urban areas and may even be better.

Although further reductions in border coverage level will eat into the lognormal margin and reduce the in-cell location probability, a reduction to −101 dBm is proposed over an additional small percentage of the country, reducing the in-cell location probability to 76% in these areas. A further reduction to −103 dBm is proposed in a further even smaller percentage, reducing the location probability to 70% in these areas. Although the reduction in location probability from 90% to 76% and 70% many seem large, the actual area involved is very small since only very small percentages of the country are being considered and these areas contain very small percentages of the overall road and rail networks. This represents a consistent strategy of tailoring the network design to be appropriate to the region concerned and the likely subscriber density.

It remains to provide a measure of service quality, and hence to demonstrate that the needs of the subscriber are met equally well in all areas. The drop call rate will increase marginally where the location probability is reduced, but the absolute number of affected calls in these areas will be small and the value is expected to remain comfortably within the normal limits of engineering practice (5%). Blocking and UTRNG do not depend on location probability, but because of the low amount of offered radio traffic in the areas of relevance, the blocking and UTRNG probabilities will fall virtually to zero, representing an enormous improvement over highly urbanised areas.

It is highly likely therefore, that no reduction in the quality of service will be perceived by users in the marginal areas where the location probability is reduced; to the subscriber, the 'coverage' in these areas is at least as good as in the urban areas. Indeed, it may even be better because there is a much lower (practically zero) blocking probability, even though the radio propagation characteristics marginally increase the possibility of a failed access attempt, particularly at or near the cell, edge and the drop call rate also worsens slightly.

11.9 THE FUTURE

Several factors have emerged in recent years with regard to the future of mobile communications systems. Indeed mobile radio in general, and cellular radio in particular, has been one of the outstanding success stories of the past two decades. The number of GSM subscribers worldwide is now greater than 100 million and in some countries the number of mobile phones exceeds the number of fixed-line telephones. The mobile phone may well become the principal personal communications device within the next ten years if technological advances can overcome the inherent capacity and quality issues.

Although the GSM standard is constantly being enhanced, the future lies with a new third generation of systems, briefly mentioned in the earlier parts of the book. The European vision of such systems is embodied in the Universal Mobile Telephone System (UMTS), and standards for this have been set by ETSI committees, which are also responsible for GSM. This standard and others worldwide will come within the purview of the International Telecommunications Union (ITU), which will decide on the family of standards that will apply to the next generation of mobile telephones – perhaps mobile communication terminals would be a better description.

It is not our purpose here to discuss UMTS in any detail, but it is obvious that current planning tools (which already support the planning and development of GSM and other second-generation systems) need to include an enhanced wideband code division multiple access (W-CDMA) tool which facilitates studies of UMTS and other third-generation networks. This will enable operators to continue to develop existing networks while preparing for new technology and also to investigate the evolutionary path from second generation to third generation.

11.9.1   A UMTS planning tool

A useful approach in evaluating the performance of W-CDMA networks is to employ a Monte Carlo simulation technique. The ensemble of Monte Carlo trials can then be used to generate information on predicted capacity, coverage and system quality, and to generate reports and performance displays that allow the user to assess the effectiveness of the system design. Features can be provided that allow system designers to obtain a very large amount of information about an existing or proposed system, including the different services and the different subscriber types; this will allow the system resources to be allocated with great flexibility. The output is a wide range of statistical results that describe system coverage and capacity, call quality, and detailed blocking statistics.

Services and subscriber types

Third-generation systems will support a wide range of services such as the provision of high-speed data to vehicular users at 144 kbit/s, to pedestrians at 384 kbit/s and to stationary users (who might be indoors) at 2.048 Mbit/s. Each of these services is associated with one or more subscriber types for which different design settings need to be used. The planning tool therefore has to be flexible enough to simulate a wide range of subscriber types, for example the three or more types that will use basic vehicular and portable services. Relevant factors include the following:

- Portables use smaller maximum transmit power because of battery life and safety issues.
- The receiver sensitivity of a vehicular installation is different to that of a portable because of the potential for antenna diversity in a vehicular system.
- Vehicular installations will be served by a macrocellular system whereas portables will be served by macro-, micro- and picocellular (in-building) systems.
- Portables are expected to operate within a macrocellular system from the interior of a moving vehicle, and this implies some penetration loss.

A system designer might therefore create user types such as pedestrians with portables, high-speed in-vehicle portable users and high-speed vehicular system users. A further type might be added if the service provider is planning to provide in-building coverage from macrocells. The designer needs to identify the traffic channel rate for the different types and specify, for each type, the appropriate maximum transmit power, antenna gain and cable loss. Different values for downlink and uplink CNR must also be specified, and building and vehicle penetration loss have to be included in all relevant calculations.

Different radio traffic maps have to be created according to the expected traffic distribution. High-speed vehicular system types might be associated with traffic distributions on highways and major roads whereas pedestrians with portables might be associated with traffic distributions measured near hotels, train and bus stations, shopping malls, etc. The tool should be able to report performance statistics for an individual user type and for combinations of user types.

Served-user limits

A useful measure of performance for multiple-access communication systems is the peak load that can be supported with a given voice quality and with a given availability of service as measured by the blocking probability. As we have seen earlier, for GSM systems, blocking occurs when all frequencies or time slots are occupied. For W-CDMA systems, new users can still be added as long as channel elements at the cell site are available to support new users, power amplifier (PA) demand is within the maximum PA capacity, and as long as adding a new user does not significantly degrade the overall performance of the system. The overall performance of the system is related to cell loading or noise rise, which affects voice quality for all users in the system. The planning tool must allow the designer to specify limits on the number of captured users based on available hardware and other call quality parameters.

Multiple-carriers

As traffic demand increases, operators will wish to utilise more carrier frequencies in order to increase capacity and, in principle, additional W-CDMA carriers can be added and reused at each cell site. In multiple-carrier systems the distribution of the call arrival process at a specific cell, using a specific carrier, depends on the algorithm used for assigning calls to carriers. The planner must specify how traffic is divided between carriers and define rules for multiple-carrier interaction. Key elements are as follows:

- Any number of carriers can be created within the planning tool and carrier assignment to cells can be many-to-one or one-to-many. The number of carriers per cell can be different for different cells. If more than one carrier exists in a cell, it is important to allow for the cell parameters to be set for each carrier individually.
- Carrier preference weightings to describe how offered traffic is assigned to the different carriers can be specified. This will result in a smoothing out of the traffic load on each carrier.
- Carrier loading thresholds can be used as soft limits before rejecting calls. Specifically, power amplifier (PA) threshold and noise rise threshold, for all cells on a particular carrier, can be used to determine when mobile stations attempting to obtain service on that carrier should be moved to other carriers with a smaller load. If other carriers are loaded then the soft limits can be ignored and a call accepted, as long as other rules and limits such as the maximum PA power or the maximum noise rise are not violated.
- Multiple carriers can be analysed simultaneously using the above rules. Analysis results, statistics and displays are made available for each carrier individually, and a system report combines statistics for all carriers.

Save, load and export

A very useful feature in planning tools is the capability to store or load multiple analyses for comparison, since each analysis may represent a totally different system scenario. Key elements include:

- The ability to save an existing analysis by storing some calculated *operating points*. These operating points can then be loaded into the W-CDMA tool, at a later time, to view the analysis results.
- After creating the display layers for a given analysis, these layers can be saved into an analysis folder. They can be loaded back into the W-CDMA tool without the need to recreate them, or they can be exported to an external tool if needed.

Users can choose to append more runs to an existing analysis if they are required to accurately characterise the performance of the system under consideration.

Parameter settings and analysis

Analysis of W-CDMA involves a large number of user-definable, W-CDMA-specific parameters that describe many aspects of the system infrastructure and service types; they include individual settings specific to each site, cell and carrier throughout the network. By changing the values of some of these parameters and running the analysis, the planner can evaluate the effects of the different parameters on the system design. This flexibility allows investigation of the concerns that face operators both at the initial system design stage and when the system matures. Analysis usually proceeds along the following lines.

The traffic density in each bin or pixel, obtained from a traffic map associated with a service type, is passed into a random number generator which uses the Poisson distribution to generate a random number of users randomly distributed throughput the analysis area (this is known as a snapshot of the system). To simulate a W-CDMA system, taking into account variations of traffic, many snapshots must be analysed. The number of snapshots used in a given analysis is stipulated by the planner.

For each subscriber-based Monte Carlo run, there will be N snapshots, where N is the number of service types. A system with two traffic maps associated with two service types will therefore have two snapshots in each run. Subscriber-based runs adopt the user-supplied parameter settings to compute information such as transmit powers and noise rise on a per cell basis. Combined uplink and downlink analyses are then used to compute the optimum required system parameters and to evaluate system performance. At the end of each subscriber-based analysis, a number of parameters can be produced for examination.

After completing the subscriber-based runs for all the snapshots, the calculated operating points can be averaged and fed into an area analysis process to provide a realistic representation of the system being modelled. This analysis uses the average values of the operating points, along with the path loss information and the user-supplied parameter settings, to provide, on a per pixel basis, results such as the received signal strength, the received downlink interference power and the required subscriber transmit power for all the cells. These results are then used to produce coverage plots and statistics relating to areas rather than individual users.

After all Monte Carlo runs are complete, the calculated operating points are averaged and used to generate displays of various aspects of system performance. Displays are generated using the averaged operating points and a hypothetical probe mobile that does not disturb the system. To compute the value of a display at a given pixel, the probe mobile is placed at this pixel, path loss predictions from all cells to this pixel are examined, and the required transmit and receive powers associated with a hypothetical link are computed. Loading and PA power at each cell assume values obtained from the operating points that are based on the Monte Carlo runs. The display module can draw different layers of information for any carrier and service type pair, some examples being as follows:

- *Uplink best server*: uses a distinct colour to show the uplink coverage area for each cell within the analysis area.
- *Uplink coverage probability*: gives the probability of uplink coverage at each bin and includes the effects of shadowing.
- *Uplink required mobile ERP*: shows the required mobile ERP at each bin.
- *Uplink required ERP margin*: displays the difference (dB) between the maximum possible mobile ERP and the actual required ERP for each bin.
- *Uplink load*: displays the cell loading for each bin.

For the downlink, displays can be generated to show the voice best server, voice channel coverage, voice channel CNR and the total downlink received power including signal, interference and thermal noise. A display can also be generated to show the handover status of each bin within the analysis area. The path balance displays the balance between the downlink and uplinks. A good system design should have balanced downlink and uplinks such that cell boundaries on the uplink and downlink coincide.

Finally the tool also produces statistical reports giving information on overall system access, the number of users served or dropped, cell blocking, and downlink and uplink performance for the selected analysis. For example, the *system report* combines users from all carriers. It shows the total number of users that attempted to obtain service and the number of those users that were served by the system. It also shows the various handover categories indicated by the number of users that fall into each category as well as the number of users dropped and the reasons why. Other reports can be generated for served-user statistics, dropped-user statistics (blocking and failed-handover statistics) and performance figures for both the uplinks and downlinks.

11.9.2 Ray tracing models

In Chapter 7, we referred to deterministic methods of propagation prediction using ray tracing methods. These methods have the advantage of being able to cope with three-dimensional scenarios and are already finding application as planning tools. However, the accuracy of ray tracing methods for outdoor scenarios depends crucially on database that are detailed and up to date. In the first instance they are being developed for microcell propagation prediction, but they should also provide good accuracy in urban macrocells. Key features, which will lead to improved prediction accuracy, are the ability to take rooftop diffraction into account, to estimate diffraction loss around more than one corner and to make predictions at different heights, e.g. at various levels within a building.

Provided that a detailed building and terrain database is available and the basic propagation theory is correctly modelled, replacing the empirical components of a propagation model with deterministic components is likely to increase accuracy. Furthermore, the need for model tuning will be reduced. On the other hand, deterministic models tend to be computationally slower than empirical models, so it is very important to optimise them for speed. To make this possible, the building data needs to be stored in a spatially partitioned, hierarchical data structure which can be accessed very quickly.

At present, most outdoor ray tracing models still include empirical components. There are several reasons for this, including the fact that the two-component path loss encountered in line-of-sight situations in microcells is costly to model deterministically. The direct and ground-reflected rays have to be added vectorially, and when predicting the median signal strength in a small area, the large signal fluctuations have to be averaged to obtain a local mean of the predicted signal strength. An empirical expression based on the dual-slope model (Chapter 7) is therefore favoured.

Without the empirical approach it is likely that the signal strength would be overestimated in line-of-sight areas and underestimated in non-line-of-sight areas. The global distance dependence is therefore determined by the two-component reference path loss, and added to this are the diffraction loss and the antenna loss. If the prediction point is indoors, the building penetration loss is also added. The signal strength is equal to the effective isotropic radiated power (EIRP) minus the total loss.

Diffraction loss is usually estimated using the uniform theory of diffraction (UTD). This models building edges as perfectly absorbing knife-edges and it considers only 'convex' diffraction edges, i.e. edges that would touch an imaginary rubber band stretched between the base station and the mobile. Very often the diffraction loss as predicted by UTD is excessive when compared to measurements, so it is not taken at full value in the model but multiplied by an empirical weighting factor.

Rooftop-diffracted rays propagate via roof edges that lie in the vertical plane defined by the base station and the mobile station, or the image of the mobile station if there is a reflection. If the ground obstructs the path, diffraction over the ground has to be included. Calculations of losses due to vertical diffraction are usually performed first.

Diffraction losses in rays which propagate around buildings are found using ray tracing techniques and, as indicated in Chapter 7, it is usual to introduce a maximum path loss parameter in order to limit the computation time; if a ray has not reached the pixel to be predicted before the limit is reached, then it is discarded. There is no need to place a limit on vertical diffraction, since the computation time is much shorter. The street-corner loss predicted using this model is often excessive when compared with measurements, so an empirical correction factor is introduced to control the steepness of the corner loss.

Building walls are used as reflectors if the reflection point is within the mobile's field of view. To limit the computation time, an upper limit is introduced on the

Figure 11.6 An example of ray tracing. It shows the dominant rays for the pixels along a line just in front of a row of buildings. Notice the vertical, transversal and reflected components (Courtesy Mobile Systems International plc).

distance between the mobile station and the reflection point. Figure 11.6 gives an illustration of the mechanisms involved.

A simple, two-parameter model is used for calculating building penetration loss. When a ray penetrates a building, it first suffers a loss due to the external wall, but between the external wall and the prediction point there is an additional distance-dependent loss which has to be added. The building penetration loss can be applied in both directions, so the base stations can be placed inside buildings (picocells) and the resulting signal outside can be predicted.

Antenna polar diagrams are normally measured under highly controlled conditions in open areas in two planes: the horizontal (azimuth) plane and the vertical (elevation) plane. However, radiation patterns are affected by nearby obstacles in a real environment, so the signal strength should be calculated with care. There is normally no problem when the base station is well above local rooftop height (as in macrocells), but when low antenna heights are used (as in a microcell situation) the radiation pattern may be considerably modified.

In reality, outdoor propagation models for macrocells using deterministic techniques are still in their infancy. Although progress is being made rapidly and advances are being reported [13], there is a very long way to go before empiricism becomes a thing of the past.

REFERENCES

1. MacDonald V.H. (1979) The cellular concept. *Bell Syst. Tech. J.*, **58**(1), 15–41.

2. Jakes W.C. (ed.) (1974) *Microwave Mobile Communications*. John Wiley, New York.
3. Ng E.W. and Geller M. (1969) A table of integrals of the error functions. *J. Res. NBSB*, **73**.
4. Duque-Anton M., Kunz D. and Ruber B. (1993) Channel assignment for cellular radio using simulated annealing. *IEEE Trans.*, **VT42**(1), 14–21.
5. Gamst A., Zinn E.-G., Beck R. and Simon R. (1986) Cellular radio network planning. *IEEE AES Mag.*, **1**, 8–11.
6. Gamst A. (1986) Some lower bounds for a class of frequency assignment problems. *IEEE Trans.*, **VT35**, 8–14.
7. Sivarajan K.N., McEliece R.J. and Ketchum J.W. (1989) Channel assignment in mobile radio. *Proc. IEEE VT Conference*, San Francisco CA, pp. 846–50.
8. Hale W.K. (1980) Frequency assignment: theory and applications. *Proc IEEE.*, **68**(12), 1497–514.
9. Fletcher R. (1987) *Practical Methods of Optimisation*. John Wiley, New York.
10. Aarts E. and Korst J. (1989) *Simulated Annealing and Boltzmann Machines*. John Wiley, New York.
11. Ingber L. (1989) Very fast simulated annealing. *Math. Comput. Modeling*, **12**(8), 967–73.
12. ETSI (1996) Digital cellular telecommunications system: radio network planning aspects. *GSM Specification 03.30 Version 5.0.0*, ETR 364.
13. Tameh E.K. (1999) The development and evaluation of a deterministic mixed cell propagation model based on radar cross-section theory. PhD thesis, University of Bristol.

Appendix A

Rayleigh Graph Paper and Receiver Noise Figure

The noise at the IF output of a linear narrowband receiver is Gaussian in nature and has a probability density function

$$p(n) = \frac{1}{\sqrt{2\pi N}} \exp\left(-\frac{n^2}{2N}\right) \tag{A.1}$$

The standard deviation of this noise is $\sqrt{N}$ and the mean noise power that would be developed across a $1\,\Omega$ resistor is N. The envelope r of such noise has a Rayleigh probability density function

$$p_r(r) = \frac{r}{N} \exp\left(-\frac{r^2}{2N}\right) \tag{A.2}$$

The cumulative distribution function is

$$\text{prob}[r > R] = P(R) = \exp(-R^2/2N) \tag{A.3}$$

The mean square value of r is $2N$, so if the receiver IF amplifier is followed by an ideal envelope detector then the RMS output voltage will be

$$(r^2)^{1/2} = \sqrt{2N}$$

We can write equation (A.3) in the form

$$R = (N[2 \ln(1/P(R))])^{1/2}$$

or

$$20 \log R = 10 \log N + 10 \log 2 + 10 \log[\ln(1/P(R))]$$

If R (dB) is plotted against $\log[\ln(1/P(R))]$ the result will be a straight line having a slope that is independent of N, but a position that varies with $10 \log N$.
  For $\ln(1/P(R)) = 1$, i.e. for $P(R) = \exp(-1) = 0.368$ then

$$20 \log R = 10 \log N + 3.01 \tag{A.4}$$

The value of R corresponding to 36.8% (cumulative) probability is therefore 3 dB greater than the standard deviation of IF noise and this particular value of R corresponds to the RMS value of r. A plot of R (dB) as ordinate against a scaled version of $\log[\ln(1/P(R))]$ as abscissa is often called Rayleigh graph paper [1].

The absolute position of the straight line representing receiver noise depends on the value of N and should be chosen such that the available noise power, expressed in decibels referred to the input, corresponds to 36.8% cumulative probability on the scaled abscissa. The envelope of receiver noise has a short-term power (averaged over one cycle) of $R^2/2$, so equation (A.4) can be written as

$$10 \log(R^2/2) = 10 \log N$$

Since the short-term power level exceeded for 36.8% of the time corresponds to the variance of the IF noise, the ordinate scale can be calibrated in $dBkT_0B$ by identifying the 36.8% point on the abscissa with the receiver noise figure; it is assumed that the receiver input is correctly terminated.

REFERENCE

1. Beckmann P. (1964) Amplitude probability distribution of atmospheric radio noise. *J. Res. Nat. Bur. Stand.*, **68D**(6), 723–36.

Appendix B

Rayleigh Distribution (dB) and CNR in a Rayleigh Fading Environment

Practical measurements of signal strength are often made using a receiver having a logarithmic characteristic. This can be calibrated directly in decibels with respect to a given reference level (one milliwatt is commonly used and the abbreviation dBm denotes dB with respect to 1 mW; the power dissipated in a 50 Ω resistor by a signal of 1 μV is -107 dBm). The output of such a receiver can be written as

$$r_{dB} = 20 \log_{10} r = \alpha \ln r^2$$

where $\alpha = 10/\ln 10 = 4.34$.

The mean value of the output, i.e. the mean value of the dB-record is

$$E\{r_{dB}\} = \int_0^\infty \alpha \ln r^2 p_r(r) \, dr$$

which, for a Rayleigh-distributed envelope, is

$$E\{r_{dB}\} = \alpha\{\ln(2\sigma^2) - C\}$$

where C is Euler's constant $= 0.5772$, thus

$$E\{r_{dB}\} = 10 \log_{10}(2\sigma^2) - 2.51 \tag{B.1}$$

The mean of the dB-record, $E\{r_{dB}\}$, is not the same as the mean value of r, *expressed in decibels*, which should be written $[E\{r\}]_{dB}$. However, equation (B.1) shows the relationship between $E\{r_{dB}\}$ and the mean power, expressed in decibels. Since the mean square value (mean power) is $2\sigma^2$, this can be expressed in decibels as $10 \log(2\sigma^2)$ and hence equation (B.1) can be written as

$$E\{r_{dB}\} = (\text{mean power})_{dB} - 2.51$$

The mean power of a Rayleigh fading signal can therefore be estimated directly from $E\{r_{dB}\}$ by adding 2.51 dB.

The mean square value of the dB-record is given by

$$E\{r_{dB}^2\} = \int_0^\infty \alpha^2 (\ln r^2)^2 p_r(r) \, dr$$

which for a Rayleigh variable is

$$\int_0^\infty \alpha^2 (\ln r^2)^2 \frac{r}{\sigma^2} \exp\left(-\frac{r^2}{2\sigma^2}\right) dr$$

Use of the substitution $r^2 = y$ puts this in the form

$$\int_0^\infty \frac{\alpha^2}{2\sigma^2} (\ln y)^2 \exp\left(-\frac{y}{2\sigma^2}\right) dy$$

which is a standard integral having the solution

$$E\{r_{dB}^2\} = \alpha^2 \left(\frac{\pi^2}{6} + (C - \ln 2\sigma^2)^2\right) \tag{B.2}$$

The variance σ_{dB}^2 is given by $E\{r_{dB}^2\} - E\{r_{dB}\}^2$, hence

$$\sigma_{dB}^2 = \frac{\alpha^2 \pi^2}{6} \tag{B.3}$$

Thus the standard deviation of a Rayleigh variable is

$$\sigma_{dB} = \frac{\alpha \pi}{\sqrt{6}} = 5.57\, dB \tag{B.4}$$

The median value of the Rayleigh-distributed variable, expressed in decibels, is obtained from equation (5.24) as

$$20 \log_{10} r_M = 10 \log(2\sigma^2) - 1.59\, dB \tag{B.5}$$

and is the same as the median of the dB-record.

The Rayleigh distribution, as expressed by equation (5.19) describes the envelope of the received signal when that envelope is described in *volts*. We are, however, often concerned with signal power because of the need to work in terms of the ratio between signal power and noise power (SNR) or between carrier power and noise power (CNR). To make the transformation, we note that the mean power of a signal with an envelope r, averaged over one RF cycle, is $r^2/2$. If such a signal is received in the presence of additive Gaussian noise of mean power N, then the short-term carrier-to-noise ratio (CNR) is

$$\gamma = \frac{r^2}{2N} \tag{B.6}$$

We now define a mean CNR as

$$\gamma_0 = \bar{\gamma} = \frac{\text{mean carrier power}}{\text{mean noise power}} = \frac{\sigma^2}{N}$$

The PDF of γ can be found through the transformation

$$p_r(r)|dr| = p_\gamma(\gamma)|d\gamma|$$

and hence

$$p_\gamma(\gamma) = \frac{1}{\gamma_0} \exp(-\gamma/\gamma_0) \tag{B.7}$$

also, the probability that a specific value Γ is not exceeded is

$$P_\gamma(\Gamma) = 1 - \exp(-\Gamma/\gamma_0)$$

Note that $p_r(r)$ and $p_\gamma(\gamma)$ are not identical distributions although they describe the same phenomenon (i.e. Rayleigh fading). Equation (B.7) is an exponential distribution with a mean and standard deviation quite different from those of equation (5.19). Once again however the median values are the same.

Appendix C

Deriving PDFs for Variables in Logarithmic Units

This appendix deals with calculations of probability density functions for a variable expressed in logarithmic units (dB). Such functions are useful because the signal strength is often expressed in dBm or dBμV.

The mixture of Rayleigh plus lognormal has now become known as the Suzuki distribution. In order to deal with this it is first necessary to express the Rayleigh signal PDF as a function of its own mean value. In the 'fading and shadowing' situation this mean signal voltage is lognormally distributed. However, as the Rayleigh distribution is derived for the signal measured in decibels, the mean value, expressed in decibels, should be normally distributed. Hence the general PDF can be found by multiplying the distribution for a Rayleigh variable, expressed in decibels, by a normal distribution and integrating over the range of the normal distribution. In a similar way, the proposed lognormal–Rician distribution assumes that the mean voltage of the Rician PDF is lognormally distributed. This compound distribution should be more flexible than the Suzuki distribution. Indeed, for specific values of the parameters, the lognormal–Rician distribution will reduce to the Suzuki PDF. The mixture of a lognormal–Rician distribution has not been discussed in any literature.

In the following derivations $y = 20 \log r$ where r is the signal strength in linear units, $\bar{y} = 20 \log \bar{r}$ where $\bar{r}$ is the mean signal strength in volts, and $M = 20/\ln 10$.

PDF FOR A RAYLEIGH VARIABLE EXPRESSED IN DB TERMS

The Rayleigh PDF is given by equation (5.19) as

$$p(r) = \frac{r}{\sigma^2} \exp\left(-\frac{r^2}{2\sigma^2}\right) \tag{C.1}$$

If $y = 20 \log r$ then

$$\frac{dy}{dr} = \frac{M}{r} \tag{C.2}$$

The PDF of y, which has logarithmic units, will be given by

$$p(y) = p(r) \left| \frac{dr}{dy} \right|_{r=\exp(y/M)} \tag{C.3}$$

hence

$$p(y) = \frac{r^2}{M\sigma^2} \exp\left(-\frac{r^2}{2\sigma^2}\right) \tag{C.4}$$

and since

$$r = \exp\left(\frac{y}{M}\right) \tag{C.5}$$

the distribution of the logarithm of a Rayleigh variable can be expressed as

$$p(y) = \frac{1}{M\sigma^2} \exp\left[\frac{2y}{M} - \frac{1}{2\sigma^2}\exp\left(\frac{2y}{M}\right)\right] \tag{C.6}$$

PDF FOR A RICIAN VARIABLE EXPRESSED IN DECIBELS

The Rician PDF is given by equation (5.59) as

$$p(r) = \frac{r}{\sigma^2} \exp\left[-\left(\frac{r^2 + r_s^2}{2\sigma^2}\right)\right] I_0\left(\frac{rr_s}{\sigma^2}\right) \tag{C.7}$$

where r is the signal envelope, r_s is the magnitude of a steady line-of-sight or specularly reflected component, σ^2 is the variance of the random vector component which represents multipath or noise, and $I_0(\cdot)$ is the modified Bessel function of the first kind. If $y = 20 \log r$ then

$$\frac{dy}{dr} = \frac{M}{r} \tag{C.8}$$

The PDF of y will be given by

$$p(y) = p(r) \left| \frac{dr}{dy} \right|_{r=\exp(y/M)} \tag{C.9}$$

hence

$$p(y) = \frac{r^2}{M\sigma^2} \exp\left[-\left(\frac{r^2 + r_s^2}{2\sigma^2}\right)\right] I_0\left(\frac{rr_s}{\sigma^2}\right) \tag{C.10}$$

and because $r = \exp(y/M)$ the PDF of the logarithm of a Rician variable can be expressed as

$$p(y) = \frac{1}{M\sigma^2} \exp\left[\frac{2y}{M} - \frac{1}{2\sigma^2}\left\{r_s^2 + \exp\left(\frac{2y}{M}\right)\right\}\right] I_0\left[\frac{r_s}{\sigma^2}\exp\left(\frac{y}{M}\right)\right] \tag{C.11}$$

PDF FOR A SUZUKI VARIABLE EXPRESSED IN DECIBELS

To determine the Suzuki PDF it is necessary to express the Rayleigh signal PDF as a function of its own mean value. Thus, equation (C.1) can be rewritten as equation (5.26):

$$p(r) = \frac{\pi r}{2\bar{r}^2} \exp\left(-\frac{\pi r^2}{4\bar{r}^2}\right) \tag{C.12}$$

The PDF for the signal strength y (dB) can be obtained by using

$$p(y) = p(r) \left.\left|\frac{dr}{dy}\right|\right|_{r=\exp(y/M)} \tag{C.13}$$

As $dr/dy = r/M$ and $r/\bar{r} = \exp[(y - \bar{y})/M]$, the distribution for a Rayleigh variable, expressed in decibels, is given by

$$p(y) = \frac{\pi}{2M} \exp\left[\frac{2}{M}(y - \bar{y}) - \frac{\pi}{4}\exp\left(\frac{2}{M}(y - \bar{y})\right)\right] \tag{C.14}$$

Now, if we assume that $\bar{y}$ is normally distributed, the resultant distribution for a Suzuki variable, expressed in decibels, is given by

$$p(y) = \sqrt{\frac{\pi}{8}} \frac{1}{M\sigma_{\bar{y}}} \int_{-\infty}^{\infty} \exp\left[\frac{2}{M}(y - \bar{y}) - \frac{(\bar{y} - m)^2}{2\sigma_{\bar{y}}^2}\right] \exp\left[-\frac{\pi}{4}\exp\left(\frac{2}{M}(y - \bar{y})\right)\right] d\bar{y} \tag{C.15}$$

where m is the mean of the Rayleigh distribution, and $\sigma_{\bar{y}}$ is the standard deviation of this mean signal. Both parameters are measured in decibels.

PDF FOR A LOGNORMAL RICIAN VARIABLE EXPRESSED IN DECIBELS

The mean value of the Rician distribution is given as follows [1]:

$$\bar{r} = \sqrt{\frac{\pi}{2}} \sigma_1 F_1\left[-\frac{1}{2}; 1; -\frac{r_s}{2\sigma^2}\right] \tag{C.16}$$

where $_1F_1(\cdot)$ is the confluent hypergeometric function. If we write $_1F_1[-1/2; 1; -r_s/2\sigma^2] = F$ then

$$2\sigma^2 = \frac{\bar{r}^2}{F^2} \frac{4}{\pi} \tag{C.17}$$

Substituting this result into equation (C.7) we get

$$p(r) = \frac{\pi r F^2}{2\bar{r}^2} \exp\left[-\frac{\pi F^2}{4}\left(\frac{r^2 + r_s^2}{\bar{r}^2}\right)\right] I_0\left[\frac{\pi r_s F^2 r}{2\bar{r}^2}\right] \tag{C.18}$$

If there is no dominant component, the value of r_s in equation (C.18) is zero, so $F^2 = 1, I_0 = 1$ and equation (C.18) reduces to equation (C.12). If we express this distribution in terms of a logarithmic variable, we obtain

$$p(y) = \frac{\pi F^2}{2M} \exp\left[\frac{2}{M}(y - \bar{y}) - \frac{\pi F^2}{4} \exp\left(\frac{2}{M}(y - \bar{y})\right) - \frac{\pi F^2 r_s^2}{4} \exp\left(-\frac{2\bar{y}}{M}\right)\right]$$

$$\times I_0\left[\frac{\pi F^2 r_s}{2} \exp\left(\frac{y - 2\bar{y}}{M}\right)\right] \tag{C.19}$$

Finally, if we assume that the mean of the Rician PDF is normally distributed, we obtain the following compound distribution for a lognormal–Rician variable expressed in decibels:

$$p(y) = \sqrt{\frac{\pi}{8}} \frac{F^2}{Ms_L} \int_{-\infty}^{\infty} \exp\left[\frac{2}{M}(y - \bar{y}) - \frac{(\bar{y} - m_L)^2}{2s_L^2} - \frac{\pi F^2}{4} \exp\left(\frac{2}{M}(y - \bar{y})\right)\right]$$

$$\times \exp\left[r_s^2 \exp\left(-\frac{2\bar{y}}{M}\right)\right] I_0\left[\frac{\pi F^2 r_s}{2} \exp\left(\frac{y - 2\bar{y}}{M}\right)\right] d\bar{y} \tag{C.20}$$

where m_L and s_L are the mean and standard deviation of the Rician distribution, both in decibels.

If $r_s = 0$ then equation (C.20) reduces to equation (C.15), i.e. the PDF for a lognormal–Rician variable in decibels reduces to the PDF for a Suzuki variable in decibels.

REFERENCE

1. Urkowitz H. (1983) *Signal Theory and Random Processes*. Artech House, London, pp. 328–34.

Appendix D

Effective Signal Envelope

To compare different diversity combiners we introduce the concept of an effective signal envelope, which is determined from the resultant CNR after combining. Assume that the envelopes of the signals on the various branches are $r_1(t), r_2(t), r_3(t), \ldots, r_M(t)$ in an M-branch system.

Maximal ratio combiner

In this case the output CNR is

$$\gamma_0 = \sum_{k=1}^{M} \gamma_k,$$

so we can write

$$\gamma_0 = \frac{r_1^2}{2N} + \frac{r_2^2}{2N} + \cdots + \frac{r_M^2}{2N} = \frac{1}{2N} \sum_{k=1}^{M} r_k^2$$

We can define an equivalent envelope r_0 such that

$$\gamma_0 = \frac{r_0^2}{2N} = \frac{1}{2N} \sum_{k=1}^{M} r_k^2$$

so that

$$r_0 = \sqrt{2N\gamma_0} = \sqrt{\sum_{k=1}^{M} r_k^2} = \sqrt{r_1^2 + r_2^2 + r_3^2 + \cdots + r_M^2} \qquad \text{(D.1)}$$

Equal-gain combiner

In this case the output CNR is given by

$$\gamma_0 = \frac{1}{M} \left(\sum_{k=1}^{M} \gamma_k^{1/2} \right)^2,$$

so

$$\gamma_0 = \frac{1}{M}\left(\frac{r_1}{\sqrt{2N}} + \frac{r_2}{\sqrt{2N}} + \ldots + \frac{r_M}{\sqrt{2N}}\right)^2 = \frac{1}{2MN}\left(\sum_{k=1}^{M} r_k\right)^2$$

Again we define an equivalent output envelope r_0 such that

$$\gamma_0 = \frac{r_0^2}{2N} = \frac{1}{2MN}\left(\sum_{k=1}^{M} r_k\right)^2$$

or

$$r_0 = \sqrt{2N\gamma_0} = \frac{1}{\sqrt{M}}\sum_{k=1}^{M} r_k = \frac{r_1 + r_2 + r_3 + \ldots + r_M}{\sqrt{M}} \tag{D.2}$$

Selection diversity

In this case $\gamma_0 = \max\{\gamma_1, \gamma_2, \gamma_3, \ldots\}$, so

$$r_0 = \max\{r_1, r_2, r_3, \ldots\} \tag{D.3}$$

Index